Advances in Healthcare and Nanoparticle Toxicology

Edited by

Jorddy Neves Cruz

Institute of Biological Sciences, Federal University of Pará, Belém, Pará, Brazil

Copyright © 2024 by the authors

Published by **Materials Research Forum LLC**
Millersville, PA 17551, USA

Published as part of the book series
Materials Research Foundations
Volume 171 (2024)
ISSN 2471-8890 (Print)
ISSN 2471-8904 (Online)

Print ISBN 978-1-64490-332-2
eBook ISBN 978-1-64490-333-9

This book contains information obtained from authentic and highly regarded sources. Reasonable efforts have been made to publish reliable data and information, but the author and publisher cannot assume responsibility for the validity of all materials or the consequences of their use. The authors and publishers have attempted to trace the copyright holders of all material reproduced in this publication and apologize to copyright holders if permission to publish in this form has not been obtained. If any copyright material has not been acknowledged please write and let us know so we may rectify this in any future reprints.

Distributed worldwide by

Materials Research Forum LLC
105 Springdale Lane
Millersville, PA 17551
USA
https://www.mrforum.com

Manufactured in the United States of America
10 9 8 7 6 5 4 3 2 1

Table of Contents

Preface

Nanoparticles, with their remarkable versatility and unique physicochemical properties, have emerged as transformative tools across diverse scientific disciplines. Their capacity to interact with biological systems at the molecular and cellular levels has opened doors to revolutionary advancements in medicine, environmental science, and beyond. This book, provides a cohesive exploration of these innovations, focusing on the multifaceted roles of nanoparticles in health and disease.

The chapters in this volume are thoughtfully curated to guide readers through the foundational principles, therapeutic applications, and potential risks associated with nanoparticles. Chapter 1, Photoresponsive Nanoparticles for One Health, explores how light-responsive nanoparticles can address global health challenges by integrating human, animal, and environmental health strategies. Chapter 2, Nanoparticles Based Therapeutics Approaches for Cancer, examines cutting-edge applications of nanoparticles in cancer treatment, from drug delivery systems to precision oncology.

The interaction of nanoparticles with living organisms is a recurring theme throughout this book. Chapter 3, Nanoparticles Interaction with Bacteria and Viruses, delves into their role in combating microbial infections, while Chapter 4, Whispers of Healing: Navigation of Nanomaterials for Wound Restoration and Inflammation Control, highlights the therapeutic potential of nanomaterials in regenerative medicine and inflammation management.

At the molecular level, Chapter 5, Interaction of Nanoparticles with Nucleic Acids, focuses on their ability to engage with genetic material, paving the way for advances in gene therapy and diagnostics. Chapters 6 and 7, Molecular Interactions between Nanoparticles and Biomolecules and Interactions of Nanoparticles with Lipid and Cell Membranes, respectively, provide detailed insights into the mechanisms by which nanoparticles influence biomolecular structures and cellular interfaces.

From a physiological perspective, Chapter 8, Nanoparticle Interactions with Endothelial Cells, sheds light on their impact on vascular health and their potential as therapeutic agents. Finally, Chapter 9, Nanoparticles in Focus: Understanding Genotoxicity and Carcinogenicity, addresses critical concerns regarding nanoparticle safety, offering a balanced view of their risks and benefits.

Advances in Healthcare and Nanoparticle Toxicology
Materials Research Foundations 171 (2024) 1-21

Materials Research Forum LLC
https://doi.org/10.21741/9781644903339-1

Chapter 1

Photoresponsive Nanoparticles for One Health

Estelle Leonard[1,*] and Suresh Sagadevan[2]

[1] Laboratoire TIMR UTC-ESCOM, Centre de recherche de Royallieu, rue du docteur Schweitzer, 60203 Compiègne Cedex France

[2] Nanotechnology & Catalysis Research Centre, University of Malaya, 50603, Kuala Lumpur, Malaysia

*e.leonard@escom.fr

Abstract

Photoresponsive nanoparticles (NPs) have emerged as innovative tools within the "One Health" framework, addressing interconnected health issues spanning humans, animals, and the environment. "One Health" represents an integrative approach that underscores the interdependence of human, animal, and environmental health, tackling challenges such as zoonoses, food safety, and antimicrobial resistance. This study explores the role of photoresponsive NPs, focusing on photochromic and photothermal NPs, in advancing applications relevant to One Health. We describe the synthesis and functional characteristics of photochromic and photothermal NPs, emphasizing their responsiveness to light as a means of controlled action. Photoresponsive NPs such as zinc oxide, magnesium oxide, and titanium dioxide exhibit significant activity against phytopathogens, providing sustainable solutions to agricultural health threats and reducing reliance on chemical pesticides. Furthermore, the study details advancements in photoresponsive NPs for human health applications, including their utility in targeted drug delivery and release, cancer phototherapy, diagnostic imaging, biosensing, and wound healing. These applications showcase the potential of photoresponsive NPs to enhance treatment precision and efficacy while minimizing side effects.

Keywords

Photoresponsive Nanoparticles, Photochromic Nanoparticles, Photothermal Nanoparticles, Drug Delivery, Drug Release

Contents

1.   Introduction

In chemistry, reactive oxygen species (ROS) are highly reactive chemicals formed from diatomic oxygen (O_2). Examples of ROS include peroxides, superoxide, hydroxyl radical and singlet oxygen [1]. The reduction of molecular oxygen (O_2) produces superoxide ($^{\cdot}O_2$), which is the precursor to most other reactive oxygen species. ROS can be generated in several manners (Fenton or Fenton-like reaction, Haber-Weiss reaction, electro-or photoinduced species) and especially on the surface of metal oxide nanoparticles (NPs) when exposed to light in aqueous or physiological solutions (Figure 1). Indeed, in this case, as the NPs are photoexcited, electron (e^-) transfers from valence band to conduction band, leaving behind a hole (h^+). At the conduction band electrons can lead to reduction of oxygen molecules whereas at the valence band oxidation occurs leading to hydroxyle radical or hydrogen peroxide formation. Alternatively, radiative recombination of electron-hole pair results in the emission of photon that transforms the oxygen molecule in the ground state to an excited singlet oxygen [2].

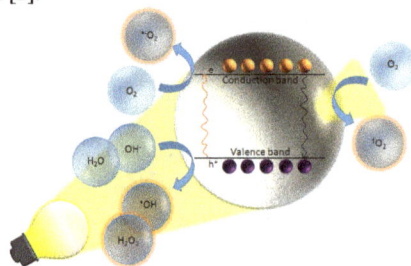

Figure 1 Simplified mechanism of photo-generated ROS formation

Some of photo-generating ROS producers are shown in Table 1.

Table 1 Some photo-ROS-generating metal-based NPs (in alphabetical order) and band gap

NPs	Computed Average Total photo-ROS production (μM)	NPs band gap (eV)
Ag	NA	3.4 [3]
Al_2O_3	158.5 ± 8.0 [b]	8.5 [b]
Au	NA	2.45-2.7 [4]
CeO_2	8.4 ± 0.2 [b]	3.3 [b]
CuO^a	0 [b]	1.7 [b]
Fe_2O_3	20.4 ± 1.2 [b]	2.2 [b]
MgO	NA	5.15 [5]
Si	NA	1.23 [6]
SiO_2	56.5 ± 2.5 [b]	9 [b]
TiO_2	442.9 ± 20.0 [b]	3.2 [b]
ZnO	277.3 ± 15.6 [b]	3.2 [b]
[a]The presence of surface defects on metal NPs can, via catalytic reactions, result in ROS formation, even under dark conditions [7]. [b] From [8] NA: Not described in literature		

This generation of ROS imply that numerous NPs can be used as antimicrobial agents, which kill bacteria by inducing ROS bursts [9]. However, interactions between metallic NPs and cells can indirectly induce ROS formation via different biological responses. H_2O_2 can also be generated by a cell due to inflammation, induced by interactions with metallic NPs or released metal species that can initiate Fenton(-like) and Haber–Weiss reactions forming various radicals [7]. The mechanism of nanoparticles for One Health activity has maybe to be investigated further on.

2.   What is "One Health"?

2.1   History

One Health is an emerging concept that aims to bring together human, animal, and environmental health [10]. Coming from the "One Medicine" concept [11] that plead for a combination of human and veterinary medicine [12] in response to zoonoses, the "One Health" concept [13] therefore constitutes a global strategy (Figure 2). It highlights the need for an approach that is holistic and transdisciplinary, and incorporates multisector expertise in dealing with the human health, animals, and ecosystems [14].

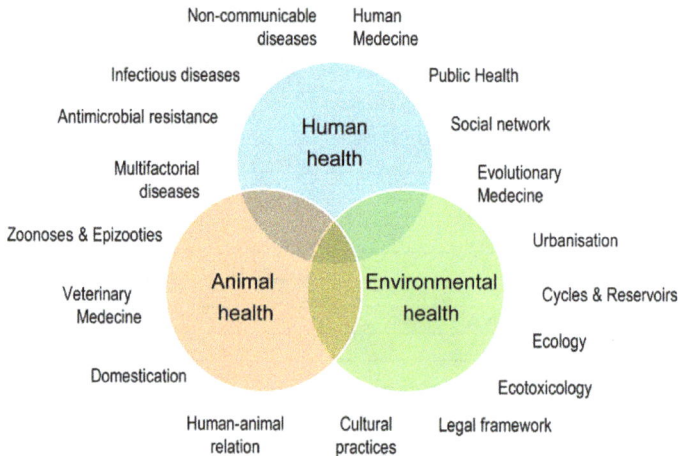

Figure 2 The global approach of One Health concept. Under Creative Commons Attribution licence (CC-BY). Destoumieux-Garzón, D., Mavingui, P., Boetsch, G., Boissier, J., Darriet, F., Duboz, P., Fritsch, C., Giraudoux, P., Le Roux, F., Morand, S., Paillard, C., Pontier, D., Sueur, C., Voituron, Y., 2018. The One Health Concept: 10 Years Old and a Long Road Ahead. Frontiers in Veterinary Science 5, online.

2.2   Zoonoses

Zoonotic diseases are infections that can be transmitted between animals and humans [15]. There are approximately 1500 pathogens, which are known to infect humans. If wild animals can infect humans, pets can also transmit bacteria for example by cat or dog bite attack or scratch. For example, in dogs and cat oral cavity can be found pathogenic bacteria including *Pasteurella sp.* and alpha-hemolytic streptococci which are the most frequent isolates from human bite wounds [16,17]. Wild animals can also transmit pathogens to farm animals, and then to humans. For example, in the Malaysian epidemic "nipah" the virus seems to be transmitted from pigs to humans through direct contact with body fluids of pigs, and severe cases can result in coma and death [18,19]. But restrict a zoonose to a single cause is a mistake as the lifecycle of the living is really complex [20].

In this example (Figure 3) the cycling of microorganisms from the soil to plants, animals and humans, and back into the environment seems to be fundamental. Thus, the health of all organisms in an ecosystem is interconnected in time and space.

Materials Research Forum LLC
https://doi.org/10.21741/9781644903339-1

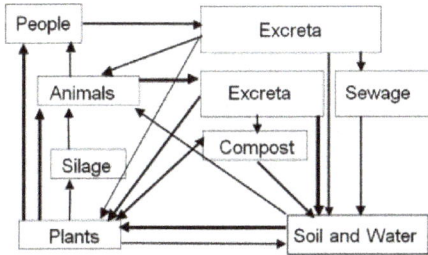

Figure 3 Complex microbial lifecycle. Reproduced with permission of van Bruggen, A.H.C., Goss, E.M., Havelaar, A., van Diepeningen, A.D., Finckh, M.R., Morris, J.G., 2019. One Health - Cycling of diverse microbial communities as a connecting force for soil, plant, animal, human and ecosystem health. Science of The Total Environment 664, 927–937. https://doi.org/10.1016/j.scitotenv.2019.02.091, copyright Elsevier (2023).

2.3   Food

Food is also at the heart of the problem. It was found that foods associated with negative environmental impact are associated with the largest increases in human disease risk (Figure 4). Thus, healthier foods would generally improve environmental sustainability [21].

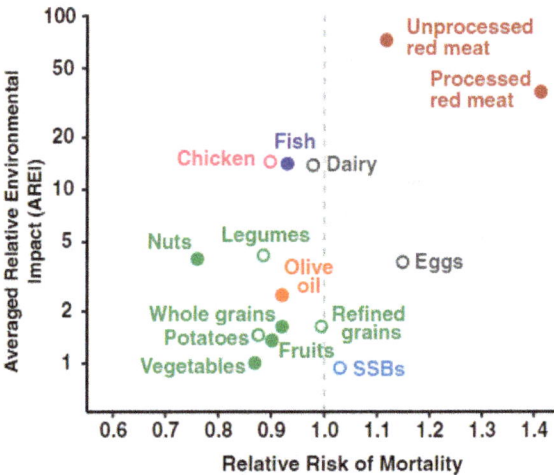

Figure 4 Association between a food group's impact on mortality and its AREI. Under Creative Commons Attribution licence (CC-BY). Clark, M.A., Springmann, M., Hill, J., Tilman, D., 2019. Multiple health and environmental impacts of foods. Proceedings of the National Academy of Sciences 116, 23357–23362. https://doi.org/10.1073/pnas.1906908116

2.4   Antimicrobial resistance

One Health is closely associated to AMR (Antimicrobial Resistance). If AMR current trends continue to increase at this rate, there could be 10 million estimated annual AMR-associated deaths from a wide array of infections by 2050 [22]. The literature describing the impact of the misuse and the release of antibiotics in the environment on the antimicrobial resistance [23] is important. For example, the overuse of antibiotic drugs in multiple sectors (human, animal, agriculture) is one of the main selection pressure problem and one that must be addressed [24,25]. Another way for releasing antibiotics in the environment is the open defecation. Indeed manure from humans and animals contaminate water, crops, and food animal products, threatening water quality and food safety. Lowering or stopping antimicrobial resistance must include addressing global sanitation and hygiene [26]. But how the resistance can be transmitted from microbial strains to others, and induce a global resistance to antibiotics? The resistome [27] is described as genes of resistance to several antibiotics in an environmental reservoir. For example, while some microbes are naturally resistant to antibiotics, others acquire resistance from fellow microbes. ARGs (Antibiotic Resistant Genes) are spread through several forms of horizontal gene transfer [28] including:

- conjugation (the transfer of DNA between cells, often via mobile genetic elements like plasmids and transposons),
- transduction (bacteriophage-mediated transfer of genes between bacterial hosts)
- transformation (a process through which bacteria absorb DNA from the environment and incorporate it into their genome).

Based on the recent achievements of resistome studies, future directions for research could be taken to improve the understanding and control of ARG transmission [29]:

- ranking the critical ARGs and their hosts;
- understanding ARG transmission at the interfaces of One-Health sectors;
- identifying selective pressures affecting the emergence, transmission, and evolution of ARGs;
- elucidating the mechanisms that allow an organism to overcome taxonomic barriers in ARG transmission.

The other way to avoid the antibioresistance is to treat the pathologies with non resistance inducing drugs. Thermo- or photoresponsive molecules can be the solution as they can act at will [30]. Here photoresponsive nanoparticles will be explored.

3.   Photoresponsive nanoparticles, definition

Photoresponsive nanomaterials are a class of nanomaterials that can react to exposure to light by undergoing reversible alterations in their chemical compositions and/or their physical characteristics. This ability to respond to light has made light-responsive nanomaterials an attractive area of research in many fields, including chemistry, materials science, and biology. Figure 5 shows the overview of the types of photoresponsive nanoparticles and their potential applications in human health. To date, two main types of photoresponsive nanomaterials that have been extensively researched for human health related applications are categorized below:

4.   Photochromic nanoparticles

Photochromic nanoparticles refer to nanoparticles that exhibit photochromism, a phenomenon where a substance can change its color or optical properties in response to exposure to light [31]. Some common types of photochromic nanoparticles include organic nanoparticles, such as spiropyran-based nanoparticles, diarylethene nanoparticles, and azobenzene nanoparticles. In addition, inorganic materials like metal oxides, semiconductors, and quantum dots can also exhibit photochromism. These photochromic nanoparticles find applications in various fields, such as the development of smart materials, sensors, and imaging techniques [32,33]. They can be engineered to respond to specific wavelengths of light, enabling precise control and manipulation of their optical properties [34,35]. This versatility makes them valuable in areas like drug delivery, where they can release drugs in response to light, or in the creation of displays and sensors that can be controlled with light signals. For instance, Zheng et al. developed a target-responsive, label-free aptamer 'signal-on' fluorescent biosensor for the detection of carcinoembryonic antigen based on the synthesis of DNA-CdTe QDs through a one-pot route [36,37].

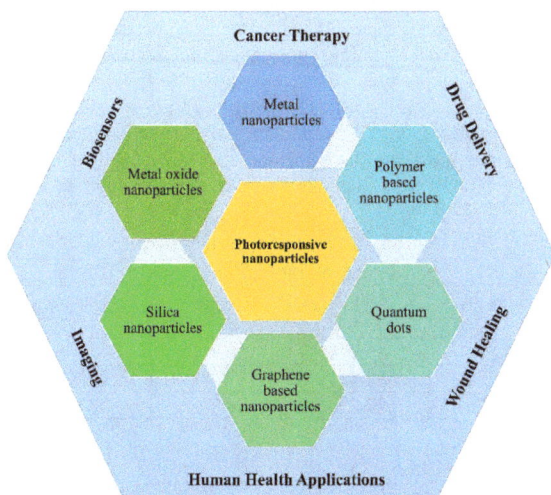

Figure 5 Types of photoresponsive nanoparticles and their human health related applications human health applications of photoresponsive nanoparticles

5.   Photothermal nanoparticles

Certain metal nanoparticles, such as gold, iron, and silica, as well as natural and synthetic polymer nanoparticles, and carbon nanomaterials like graphene and carbon nanoparticles, can efficiently convert absorbed light energy into heat [38,39]. These materials are often used in photothermal therapy (PTT) for various applications, including cancer treatment and imaging. Some key properties of photothermal nanoparticles, such as strong light absorption, high photothermal conversion efficiency, biocompatibility, and easy functionalization, make it easier to integrate them into medical practice.

6.   Photoresponsive NPs against phytopathogens

6.1   Zinc oxide

ZnO NPs possess photoresponsive toxicity, especially against phytopathogens [40]. Indeed, photoresponsive activities of zinc oxide can imply a photocatalytic degradation of environmental pollutants as well as antimicrobial activity [41]. Zinc oxide (ZnO) has a wide band gap (3.37 eV) and thus gained substantial interest for a wide range of applications [42] including those needing the generation of ROS (Reactive Oxygen Species) under UV_A (around 370 nm) illumination [43,44]. Due to the large surface to volume ratio, ZnO nanoparticles facilitate better interaction with microbes thus possess an excellent antifungal and antibacterial properties [45]. The antimicrobial action of ZnO can be related to disruption of the cell membrane, production of strong oxidizing agents such as hydrogen peroxide (H_2O_2) which are lethal and can penetrate bacterial cells, or the generation of reactive oxygen species [46]. For example, ZnO NPs were found to have an activity [47] against *Fusarium solani* [48] affecting peas, beans, potatoes, and many types of cucurbits, *Fusarium oxysporum* affecting banana, cotton, melon, tomato and *Colletotrichum gloeosporioides* pathogen of papayas, citrus, and coffee [49]. These nanoparticles were also effective [50] against *Botrytis cinerea* affecting tomato, raspberry, strawberry, bean, cucumber, salad, roses, gerberas, peonies, carnations, lilies and chrysanthemums [51] and *Sclerotinia sclerotiorum* affecting rapeseed, sunflower, beans and carrots [52]. Concerning coffee fungi *Mycena citricolor* the ZnO NPs were found to reach 92.3% inhibition rate at 9 mmol.L^{-1} [53].

Table 2 *Extracted data for the ZnO NPs inhibition rate against Mycena citricolor [53]*

ZnO NPs concentration	Inhibition rate (%) compared to control.				
	Day 1	Day 3	Day 6	Day 9	Day 12
12 mmol.L^{-1}	76.8%	85.8%	86.9%	90.1%	91.1%
9 mmol.L^{-1}	81.2%	86.8%	87.9%	90.9%	92.3%
6 mmol.L^{-1}	80.6%	87.7%	84.7%	87.5%	87.1%

To merge the concepts of One Health and Green Chemistry, biosynthetic approach involve the use of zinc-tolerant microbes. Indeed, to tolerate the high toxicity of zinc ions, microbes synthetize zinc oxide nanoparticles from zinc ions under stress [54]. For example, the biosynthesis of ZnO nanoparticles from the zinc-tolerant bacterial strain *Serratia nematodiphila* was performed and their antifungal potential against phytopathogens were examined. Nanoparticles were uniform in size ranging from 10 to 30 nm and their zeta potential value of −33.4 mV indicated that the nanoparticles synthesized were highly stable. It was found that the green-synthesized nanoparticles gave 85.93% inhibition of mycelial growth of *Alternaria sp* (being known worldwide both as a common phytopathogenic organism and as an airborne allergen) and 92.22% inhibition of its spore germination [55].

But one can ask question about the role of light in the formation of ROS. It was previously shown that photo-ROS can be produced by ZnO illumination, but is the antimicrobial activity better under illumination? That was effectively proven against *Botrytis cinerea*. *B. cinerea* is a species of necrotrophic fungus and is responsible for gray mold, a fungal disease that affects several crops of major agronomic interest, such as vines, sunflowers, tomatoes and strawberries. This fungus is also responsible for noble rot, which produces certain sweet wines such as Sauternes or Tokay. The mycelial growth of this fungus was measured without application of ZnO NPs, with ZnO NPs application in the dark and under light illumination at (34 J/cm²) at 405 nm, 6 cm from the sample. It was found that compared to Control, zinc oxide in the dark inhibited only by 12% the radial growth of *B. cinerea*, whereas under light illumination the inhibition rate reached 80% [56].

6.2 Magnesium oxide

Ralstonia solanacearum is a soil-borne and nonsporing bacterium that can infect several hundred host plant species around the world, including potatoes, tomatoes, eggplants, groundnuts, olives, bananas, and ginger [57]. While in contact with magnesium oxide nanoparticles, MgO NPs directly adhered to the cell wall of *R. solanacearum*, inducing a significant increase in the Mg element content, intracellular and extracellular ROS were generated, causing cytoplasm leakage, such as of protein and DNA, and led to cellular inactivation [58]. *Acidovorax oryzae* is a rice sezrious pathogen. The maximum antibacterial effect of MgO NPs was observed at 20 $\mu g\ mL^{-1}$ (35.2 $\pm$ 0.61 mm inhibition zone diameter, 65.38% swarming motility and 80.83% MIC at OD600) with a clear damage in bacterial cells observed by SEM and TEM (Figure 5) at this concentration [59].

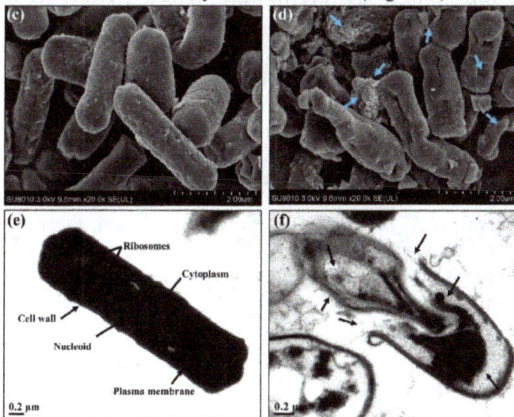

Figure 6 SEM and TEM images of A. oryzae cells treated with water (c, e) and damaged bacterial cells of A. oryzae treated with 20 $\mu g\ mL^{-1}$ MgO NPs (d, f). Reproduced with permission of Ahmed, T., Noman, M., Shahid, M., Shahid, M.S., Li, B., 2021. Antibacterial potential of green magnesium oxide nanoparticles against rice pathogen Acidovorax oryzae. Materials Letters 282, 128839. https://doi.org/10.1016/j.matlet.2020.128839. Copyright Elsevier (2023).

Band gaps can also be linked to the capability of MgO NPs to form antimicrobial photo-generated ROS compounds. For instance, it was determined that against *Ralstonia solanacearum*, the smallest size and band gap values led to the best antimicrobial activity [60]. Several samples of MgO NPs were bio-synthesized and their band gap were calculated leading to the conclusion that S3 and S4 gave the best results (Table 3), according to the relationship between size/band gap, and antimicrobial properties.

Table 3 Calculated band gaps and experimental ZOI against Ralstonia solanacearum.

	S1	S2	S3	S4	S5	S6	S7	S8	S9
band gap (eV)	4.2081	4.1758	3.7616	3.5660	5.0678	4.7680	4.9780	4.9780	5.2468
ZOI*	7 mm	6 mm	9 mm	9 mm	5 mm	6 mm	6 mm	6 mm	5 mm
* Zone of Inhibition, MgO NPs concentration of 3 mg/mL									

6.3 Titanium dioxide

With a band gap of ~3.2 eV, research on titanium oxide photocatalysts are still ongoing and of great interest. For example, under UV_A irradiation, TiO_2 NPs A310C had a better photodegradation capability than the commercial Degussa P25, known as a high photocatalyst, even if it was not possible to determine which parameter is primarily responsible for the enhanced photocatalytic activity, part from the largest crystallite size. However, in this case, it was shown that against *Erwinia amylovora, Xanthomonas arboricola, Pseudomonas syringae* and *Allorhizobium vitis,* A310C was inactive in the dark for at least 15 minutes [61], that only UV_A treatment did not inhibit the growth of these plant pathogens, but A310C had a better antimicrobial effect than P25. So the photo-ROS activation was mainly responsible for this effect (Figure 6 Antibacterial activity of A310C and P25 TiO2 photocatalysts at 0.5 mg/mL).

The doping, duration and intensity of light can also influence results. Indeed, time dependence of photocatalytic activity of TiO_2, TiO_2/Zn, and TiO_2/Ag NPs coated glass on *Xanthomonas perforans* (responsible of tomato plant disease) gave the result that TiO_2/Ag and TiO_2/Zn reduced significantly the survival of *X. perforans* (Table 4). At 15 min, the photocatalytic effect of TiO_2/Ag, and TiO_2/Zn NPs was 0 and <1 log CFU/ml respectively. At 20 min, there was no bacterial survival with TiO_2/Ag and TiO_2/Zn. At this duration of photocatalysis, a bacterial population of 1 log CFU/ml was recovered from TiO_2. There was no significant reduction in the number of bacterial cells on controls in illuminated condition compared with non-illuminated conditions [62].

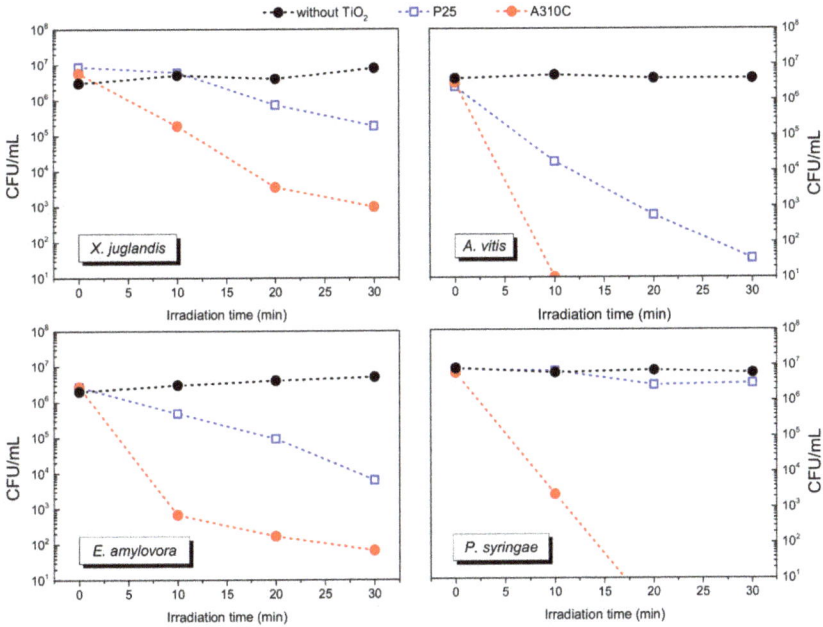

Figure 7 Antibacterial activity of A310C and P25 TiO2 photocatalysts at 0.5 mg/mL. Under Creative Commons Attribution licence (CC-BY). Kőrösi, L., Pertics, B., Schneider, G., Bognár, B., Kovács, J., Meynen, V., Scarpellini, A., Pasquale, L., Prato, M., 2020. Photocatalytic Inactivation of Plant Pathogenic Bacteria Using TiO2 Nanoparticles Prepared Hydrothermally. Nanomaterials 10, 1730. https://doi.org/10.3390/nano10091730

Table 4 Decrease of the Xanthomonas perforans population by photo-ROS production. Data from [62]

Light duration*	Population of *Xanthomonas perforans* (log CFU/ml)			
	TiO_2	TiO_2/Zn	TiO_2/Ag	control
10 min	6.5	3	<1	7
15 min	3.5	<1	0	7
20 min	1	0	0	6.5
dark (20 min)	7	6	7	7
* 60 Watt incandescent bulb, intensity of $3x10^4$ lux at 4°C				

7.   Photoresponsive Nanomaterials for human health applications

7.1   Drug delivery and release

Photoresponsive nanoparticles have been used to design smart drug delivery systems. These nanoparticles can be engineered to release therapeutic agents in response to light, allowing for precise and controlled drug administration [63–65]. This approach minimizes side effects and enhances the effectiveness of treatments, particularly in cancer therapy. Li et al. fabricated a multifunctional nanocomplex for the delivery of a classic chemotherapy drug (Doxorubicin, DOX) and a near-infrared (NIR) dye (indocyanine green, ICG) based on mesoporous silica-coated Fe3O4 nanoparticles [66]. Hyaluronic acid (HA) was conjugated onto the surface of the nanocomplex to respond to hyaluronidase (HAase). The final complex, M-MSN/HA/DI showed preeminent T2 MR and fluorescence imaging ability, and the release of DOX was accelerated in the presence of HAase.

7.2   Cancer therapy

Photothermal therapy (PTT) is a prominent application of photoresponsive nanoparticles in oncology. Nanoparticles like gold or carbon-based materials can selectively accumulate in tumor tissues and, upon exposure to near-infrared (NIR) light, generate heat, leading to localized hyperthermia and the destruction of cancer cells (Figure 8) [39,67]. This approach minimizes damage to healthy tissues and offers a promising alternative to conventional cancer treatments. For instance, Liang and colleagues developed a novel system involving folic acid-TiO2-Al (III) phthalocyanine chloride tetra-sulfonic acid (FA-TiO2-Pc) nanoparticles for targeting folate receptor-positive cancer cells [68]. In this system, FA-TiO2-Pc nanoparticles displayed a remarkable capability for photodynamic therapy (PDT) and demonstrated biocompatibility in laboratory-based studies. In vitro investigations revealed that the growth of tumors was effectively inhibited with minimal adverse effects when mice bearing HeLa xenograft tumors were subjected to treatment using low doses of FA-TiO2-Pc in combination with low-intensity light irradiation.

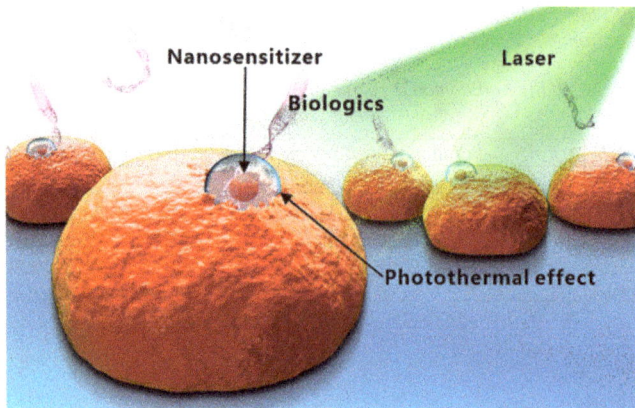

Figure 8 Schematic representation of photothermal nanomaterial-mediated photoporation (Xiong et al. 2023).

7.3   Imaging and diagnostic

Photoresponsive nanoparticles can serve as contrast agents in various imaging techniques, including fluorescence microscopy, photoacoustic imaging, and magnetic resonance imaging (MRI) [69,70]. They enable real-time visualization of tissues and cells, aiding in the early detection and diagnosis of diseases. A nanoparticle consisting of a bimetallic composition of Pt and Pd was developed by Jia et al., possessing exceptional photothermal capabilities and a mesoporous structure, ideal for loading the drug doxorubicin (DOX) [71]. They went a step further by designing a PtPd-ethylene glycol (PEG)-folic acid (FA)-doxorubicin (DOX) nanoparticle, incorporating folic acid engineering for active targeting, to enable chemo-photothermal therapy specifically for MCF-7 tumors as shown in Figure 9. Furthermore, the PtPd-PEG-FA-DOX nanoparticles exhibited excellent photoacoustic (PA) imaging, which allowed precise *in vivo* tracking and evaluation of their targeting efficiency. Both in vitro and in vivo experiments demonstrated that PtPd-PEG-FA-DOX nanoparticles represent a secure and promising approach for effectively treating MCF-7 tumors.

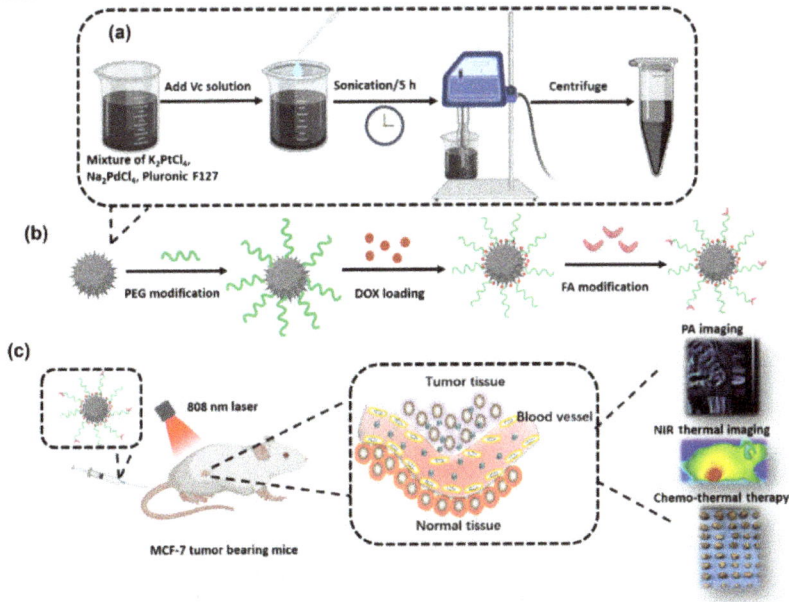

Figure 9 Schematic illustration of the preparation process of multifunctional PtPd nanoparticles and their application in chemo-photothermal therapy for MCF-7 breast cancer. This figure is adapted from (Jia et al. 2020).

7.4   Biosensors

The ability of photoresponsive nanoparticles to respond to light provides a versatile platform for sensitive and selective detection in areas such as medical diagnostics [72]. For example, a rapid and colorimetric nano-biosensor was constructed employing gold nanoparticles (AuNPs) to target platelet-derived growth factor (PDGF), a circulating biomarker that is upregulated in plasma in prevalent ovarian cancer [73]. In this proposed principle, AuNPs are mixed with PDGF specific aptamer and employed to identify PDGF by screening changes in the color as well as absorbance of the Aptamer and AuNPs caused by aggregation. The AuNPs color changes from pinkish to light purple at a higher level of concentration as demonstrated in Figure 10. Signal-output exhibited for PDGF was in the linear range of 0.01–10 µg/ml under the optimum conditions.

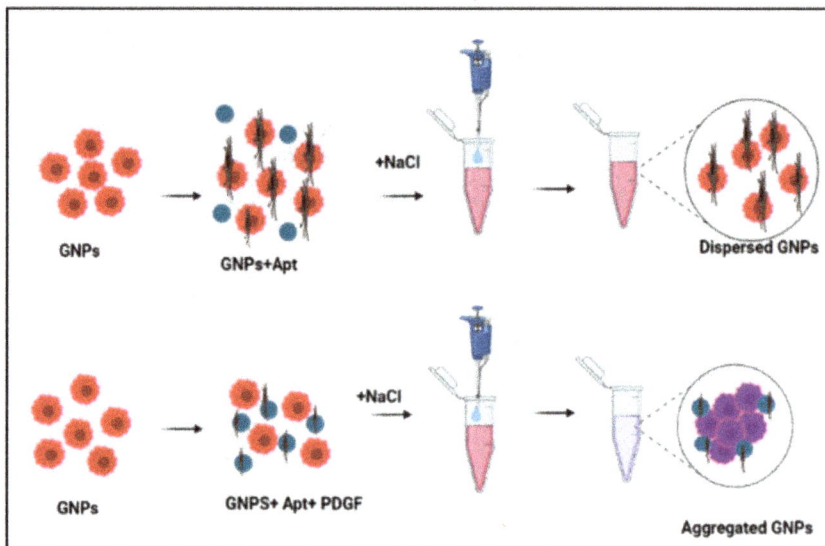

Figure 10 Schematic Representation of the GNPs-based colorimetric assay to diagnose PDGF. This figure is adapted from [73].

7.5   Wound healing

Injuries to the skin can lead to wounds that are susceptible to bacterial infections, which can hinder the natural process of skin regeneration and wound healing [38]. This is particularly problematic for patients with severe burns or chronic conditions like diabetes, where skin infections caused by drug-resistant bacteria can be life-threatening. Consequently, there is a significant demand for a broad-spectrum therapy that can effectively eliminate bacterial infections through a mechanism different from traditional antibiotics. Qiu et al. have successfully developed antibacterial photodynamic gold nanoparticles (AP-AuNPs), which are self-assembled nanocomposites composed of an antibacterial photodynamic peptide and polyethylene glycol [74,75]. These AP-AuNPs exhibit stability in aqueous environments and under light exposure. They also demonstrated a satisfactory ability to generate reactive oxygen species and exhibited a remarkable antibacterial effect against both Gram-positive bacteria (Staphylococcus aureus) and Gram-negative bacteria (Escherichia coli) when exposed to light. Through a combination of the peptide's bactericidal properties, the photosensitizing effects of a photosensitizer, and the clustering of multiple antibacterial agents on AuNPs to maximize their antibacterial effects under light irradiation, AP-AuNPs have developed as a nanomaterial for wound dressing in cases of skin infections.

Conclusion

In chemistry, reactive oxygen species (ROS) are highly reactive chemicals formed from diatomic oxygen (O_2) and examples of ROS can be found including peroxides, superoxide, hydroxyl radical and singlet oxygen. As reduction of molecular oxygen (O_2) produces superoxide ($\cdot O_2$), which is the precursor to most other reactive oxygen species, ROS can be generated in several manner and especially on the surface of metal oxide nanoparticles (NPs) when exposed to light in aqueous or physiological solutions. In this context, nanoparticles capable of forming ROS species are of choice as their applications in plant or human health are wide.

As the "One Health" concept is of great interest to help people heal their wounds, cancers or detect diseases without interfering with environment, it can also allow treating plants with a benign ecological impact. As seen in this chapter, the use of photoresponsive nanoparticles can do both application, either used for Human health and against phytopathogens, but with a light switching which can allow people to use them at will.

References

[1] M. Hayyan, M.A. Hashim, I.M. AlNashef, Superoxide Ion: Generation and Chemical Implications, Chem. Rev. 116 (2016) 3029–3085. https://doi.org/10.1021/acs.chemrev.5b00407

[2] J. Bogdan, J. Pławińska-Czarnak, J. Zarzyńska, Nanoparticles of Titanium and Zinc Oxides as Novel Agents in Tumor Treatment: a Review, Nanoscale Res Lett 12 (2017) 225. https://doi.org/10.1186/s11671-017-2007-y

[3] A.J. Das, R. Kumar, S.P. Goutam, Sunlight Irradiation Induced Synthesis of Silver Nanoparticles using Glycolipid Bio-surfactant and Exploring the Antibacterial Activity, J Bioengineer & Biomedical Sci 06 (2016). https://doi.org/10.4172/2155-9538.1000208

[4] L.L.G. Al-mahamad, Analytical study to determine the optical properties of gold nanoparticles in the visible solar spectrum, Heliyon 8 (2022) e09966. https://doi.org/10.1016/j.heliyon.2022.e09966

[5] I. Apostolova, A. Apostolov, J. Wesselinowa, Magnetic, Optical and Phonon Properties of Ion-Doped MgO Nanoparticles. Application for Magnetic Hyperthermia, Materials 16 (2023) 2353. https://doi.org/10.3390/ma16062353

[6] J. Low, M. Kreider, D. Pulsifer, A. Jones, T. Gilani, Band Gap Energy in Silicon, AJUR 7 (2008). https://doi.org/10.33697/ajur.2008.010

[7] A. Kessler, J. Hedberg, E. Blomberg, I. Odnevall, Reactive Oxygen Species Formed by Metal and Metal Oxide Nanoparticles in Physiological Media—A Review of Reactions of Importance to Nanotoxicity and Proposal for Categorization, Nanomaterials 12 (2022) 1922. https://doi.org/10.3390/nano12111922

[8] Y. Li, W. Zhang, J. Niu, Y. Chen, Mechanism of Photogenerated Reactive Oxygen Species and Correlation with the Antibacterial Properties of Engineered Metal-Oxide Nanoparticles, ACS Nano 6 (2012) 5164–5173. https://doi.org/10.1021/nn300934k

[9] Z. Yu, Q. Li, J. Wang, Y. Yu, Y. Wang, Q. Zhou, P. Li, Reactive Oxygen Species-Related Nanoparticle Toxicity in the Biomedical Field, Nanoscale Res Lett 15 (2020) 115. https://doi.org/10.1186/s11671-020-03344-7

[10] R.M. Atlas, One Health: Its Origins and Future, in: J.S. Mackenzie, M. Jeggo, P. Daszak, J.A. Richt (Eds.), One Health: The Human-Animal-Environment Interfaces in Emerging Infectious Diseases: The Concept and Examples of a One Health Approach, Springer, Berlin, Heidelberg, 2013: pp. 1–13. https://doi.org/10.1007/82_2012_223

[11] J. Zinsstag, E. Schelling, D. Waltner-Toews, M. Tanner, From "one medicine" to "one health" and systemic approaches to health and well-being, Preventive Veterinary Medicine 101 (2011) 148–156. https://doi.org/10.1016/j.prevetmed.2010.07.003

[12] C.W. Schwabe, Veterinary medicine and human health., Baltimore: Williams & Wilkins, 1984.

[13] D. Destoumieux-Garzón, P. Mavingui, G. Boetsch, J. Boissier, F. Darriet, P. Duboz, C. Fritsch, P. Giraudoux, F. Le Roux, S. Morand, C. Paillard, D. Pontier, C. Sueur, Y. Voituron, The One Health Concept: 10 Years Old and a Long Road Ahead, Frontiers in Veterinary Science 5 (2018) online.

[14] M. Hristovski, A. Cvetkovik, I. Cvetkovik, V. Dukoska, Concept of one health-a new professional imperative, Maced J Med Sci 3 (2010) 229–232.

[15] L. Cantas, K. Suer, Review: The Important Bacterial Zoonoses in "One Health" Concept, Frontiers in Public Health 2 (2014) online.

[16] R.D. Griego, T. Rosen, I.F. Orengo, J.E. Wolf, Dog, cat, and human bites: A review, Journal of the American Academy of Dermatology 33 (1995) 1019–1029. https://doi.org/10.1016/0190-9622(95)90296-1

[17] M. Assis, M.O. Gonçalves, C.C. de Foggi, M. Burck, S. dos Passos Ramos, L.O. Libero, A.R.C. Braga, E. Longo, C.P. de Sousa, Applications of (nano)encapsulated natural products by physical and chemical methods, in: J.N. Cruz (Ed.), Drug Discovery and Design Using Natural Products, Springer Nature Switzerland, Cham, 2023: pp. 323–374. https://doi.org/10.1007/978-3-031-35205-8_11

[18] A. Easton, New virus is identified in Malaysia epidemic, BMJ 318 (1999) 1232–1232. https://doi.org/10.1136/bmj.318.7193.1232b

[19] S. Muzammil, J. Neves Cruz, R. Mumtaz, I. Rasul, S. Hayat, M.A. Khan, A.M. Khan, M.U. Ijaz, R.R. Lima, M. Zubair, Effects of Drying Temperature and Solvents on In Vitro Diabetic Wound Healing Potential of Moringa oleifera Leaf Extracts, Molecules 28 (2023) 710. https://doi.org/10.3390/molecules28020710

[20] A.H.C. van Bruggen, E.M. Goss, A. Havelaar, A.D. van Diepeningen, M.R. Finckh, J.G. Morris, One Health - Cycling of diverse microbial communities as a connecting force for soil, plant, animal, human and ecosystem health, Science of The Total Environment 664 (2019) 927–937. https://doi.org/10.1016/j.scitotenv.2019.02.091

[21]   M.A. Clark, M. Springmann, J. Hill, D. Tilman, Multiple health and environmental impacts of foods, Proceedings of the National Academy of Sciences 116 (2019) 23357–23362. https://doi.org/10.1073/pnas.1906908116

[22]   A. White, J.M. Hughes, Critical Importance of a One Health Approach to Antimicrobial Resistance, EcoHealth 16 (2019) 404–409. https://doi.org/10.1007/s10393-019-01415-5

[23]   S. Hernando-Amado, T.M. Coque, F. Baquero, J.L. Martínez, Defining and combating antibiotic resistance from One Health and Global Health perspectives, Nat Microbiol 4 (2019) 1432–1442. https://doi.org/10.1038/s41564-019-0503-9

[24]   S.A. McEwen, P.J. Collignon, Antimicrobial Resistance: a One Health Perspective, in: Antimicrobial Resistance in Bacteria from Livestock and Companion Animals, John Wiley & Sons, Ltd, 2018: pp. 521–547. https://doi.org/10.1128/9781555819804.ch25

[25]   J.N. Cruz, S. Muzammil, A. Ashraf, M.U. Ijaz, M.H. Siddique, R. Abbas, M. Sadia, Saba, S. Hayat, R.R. Lima, A review on mycogenic metallic nanoparticles and their potential role as antioxidant, antibiofilm and quorum quenching agents, Heliyon 10 (2024). https://doi.org/10.1016/j.heliyon.2024.e29500

[26]   L.H. Kahn, Antimicrobial resistance: a One Health perspective, Transactions of The Royal Society of Tropical Medicine and Hygiene 111 (2017) 255–260. https://doi.org/10.1093/trstmh/trx050

[27]   M.O.A. Sommer, G. Dantas, G.M. Church, Functional Characterization of the Antibiotic Resistance Reservoir in the Human Microflora, Science 325 (2009) 1128–1131. https://doi.org/10.1126/science.1176950

[28]   M. Barron, The Gut Resistome and the Spread of Antimicrobial Resistance, ASM.Org (2022). https://asm.org:443/Articles/2022/June/The-Gut-Resistome-and-the-Spread-of-Antimicrobial (accessed July 31, 2023).

[29]   D.-W. Kim, C.-J. Cha, Antibiotic resistome from the One-Health perspective: understanding and controlling antimicrobial resistance transmission, Exp Mol Med 53 (2021) 301–309. https://doi.org/10.1038/s12276-021-00569-z

[30]   W.A. Velema, J.P. van der Berg, M.J. Hansen, W. Szymanski, A.J.M. Driessen, B.L. Feringa, Optical control of antibacterial activity., Nat. Chem. 5 (2013) 924–928. https://doi.org/10.1038/nchem.1750

[31]   V.A. Barachevsky, Photochromic Nanoparticles and Their Properties, Crystallography Reports 63 (2018) 271–275. https://doi.org/10.1134/S1063774518020025/METRICS

[32]   B. Hatamluyi, M. Rezayi, S. Amel Jamehdar, K.S. Rizi, M. Mojarrad, Z. Meshkat, H. Choobin, S. Soleimanpour, M.T. Boroushaki, Sensitive and specific clinically diagnosis of SARS-CoV-2 employing a novel biosensor based on boron nitride quantum dots/flower-like gold nanostructures signal amplification, Biosensors and Bioelectronics 207 (2022) 114209. https://doi.org/10.1016/J.BIOS.2022.114209

[33]   N.G. Naga, M.I. Shaaban, Quorum sensing and quorum sensing inhibitors of natural origin, in: J.N. Cruz (Ed.), Drug Discovery and Design Using Natural Products, Springer Nature Switzerland, Cham, 2023: pp. 395–416. https://doi.org/10.1007/978-3-031-35205-8_13

[34]   M. Babazadeh-Mamaqani, H. Roghani-Mamaqani, A. Abdollahi, M. Salami-Kalajahi, Optical Chemosensors based on Spiropyran-Doped Polymer Nanoparticles for Sensing pH of Aqueous Media, Langmuir 38 (2022) 9410–9420. https://doi.org/10.1021/acs.langmuir.2c01389

[35]   D. Lu, S. Liu, H. Zhang, X. Zhang, W. Li, Synthesis and characterization of photochromic polyurethane nanoparticles by miniemulsion polymerization, Dyes and Pigments 219 (2023) 111647. https://doi.org/10.1016/j.dyepig.2023.111647

[36]   J. Zheng, Y. Di, L. Gao, X. Kong, Y. Zheng, J. Han, J. Wang, Construction of fluorescent biosensing system based on DNA templated quantum dots- graphene oxide interactions for detecting carcinoembryonic antigen, Materials Technology 37 (2022) 2116–2122. https://doi.org/10.1080/10667857.2021.1943120

[37]   V.S. Nagtode, C. Cardoza, H.K.A. Yasin, S.N. Mali, S.M. Tambe, P. Roy, K. Singh, A. Goel, P.D. Amin, B.R. Thorat, J.N. Cruz, A.P. Pratap, Green Surfactants (Biosurfactants): A Petroleum-Free Substitute for Sustainability—Comparison, Applications, Market, and Future Prospects, ACS Omega 8 (2023) 11674–11699. https://doi.org/10.1021/acsomega.3c00591

[38]   S.G. Alamdari, M. Amini, N. Jalilzadeh, B. Baradaran, R. Mohammadzadeh, A. Mokhtarzadeh, F. Oroojalian, Recent advances in nanoparticle-based photothermal therapy for breast cancer, Journal of Controlled Release 349 (2022) 269–303. https://doi.org/10.1016/j.jconrel.2022.06.050

[39]   R. Xiong, F. Sauvage, J.C. Fraire, C. Huang, S.C. De Smedt, K. Braeckmans, Photothermal Nanomaterial-Mediated Photoporation, Accounts of Chemical Research 56 (2023) 631–643. https://doi.org/10.1021/acs.accounts.2c00770

[40]   W. Zhao, Y. Liu, P. Zhang, P. Zhou, Z. Wu, B. Lou, Y. Jiang, N. Shakoor, M. Li, Y. Li, I. Lynch, Y. Rui, Z. Tan, Engineered Zn-based nano-pesticides as an opportunity for treatment of phytopathogens in agriculture, NanoImpact 28 (2022) 100420. https://doi.org/10.1016/j.impact.2022.100420

[41]   K. Qi, B. Cheng, J. Yu, W. Ho, Review on the improvement of the photocatalytic and antibacterial activities of ZnO, Journal of Alloys and Compounds 727 (2017) 792–820. https://doi.org/10.1016/j.jallcom.2017.08.142

[42]   R. Raji, K.G. Gopchandran, ZnO nanostructures with tunable visible luminescence: Effects of kinetics of chemical reduction and annealing, Journal of Science: Advanced Materials and Devices 2 (2017) 51–58. https://doi.org/10.1016/j.jsamd.2017.02.002

[43]   E. Kanakari, C. Dendrinou-Samara, Fighting Phytopathogens with Engineered Inorganic-Based Nanoparticles, Materials 16 (2023) 2388. https://doi.org/10.3390/ma16062388

[44]   C. Wang, X. Hu, Y. Gao, Y. Ji, ZnO Nanoparticles Treatment Induces Apoptosis by Increasing Intracellular ROS Levels in LTEP-a-2 Cells, BioMed Research International 2015 (2015) 1–9. https://doi.org/10.1155/2015/423287

[45]   A. Raghunath, E. Perumal, Metal oxide nanoparticles as antimicrobial agents: a promise for the future, International Journal of Antimicrobial Agents 49 (2017) 137–152. https://doi.org/10.1016/j.ijantimicag.2016.11.011

[46]   Y. Xie, Y. He, P.L. Irwin, T. Jin, X. Shi, Antibacterial Activity and Mechanism of Action of Zinc Oxide Nanoparticles against *Campylobacter jejuni*, Appl Environ Microbiol 77 (2011) 2325–2331. https://doi.org/10.1128/AEM.02149-10

[47]   N. Pariona, F. Paraguay-Delgado, S. Basurto-Cereceda, J.E. Morales-Mendoza, L.A. Hermida-Montero, A.I. Mtz-Enriquez, Shape-dependent antifungal activity of ZnO particles against phytopathogenic fungi, Appl Nanosci 10 (2020) 435–443. https://doi.org/10.1007/s13204-019-01127-w

[48]   J.A. Wrather, S.R. Koenning, Estimates of Disease Effects on Soybean Yields in the United States 2003 to 2005, J Nematol 38 (2006) 173–180.

[49]   R. Dean, J. a. L. Van Kan, Z.A. Pretorius, K.E. Hammond-Kosack, A. Di Pietro, P.D. Spanu, J.J. Rudd, M. Dickman, R. Kahmann, J. Ellis, G.D. Foster, The Top 10 fungal pathogens in molecular plant pathology, Molecular Plant Pathology 13 (2012) 414–430. https://doi.org/10.1111/j.1364-3703.2011.00783.x

[50]   P. Tryfon, N.N. Kamou, S. Mourdikoudis, K. Karamanoli, U. Menkissoglu-Spiroudi, C. Dendrinou-Samara, CuZn and ZnO Nanoflowers as Nano-Fungicides against Botrytis cinerea and Sclerotinia sclerotiorum: Phytoprotection, Translocation, and Impact after Foliar Application, Materials 14 (2021) 7600. https://doi.org/10.3390/ma14247600

[51]   J.W. Kronstad, ed., Fungal pathology, Kluwer Academic Publishers, Dordrecht, 2000.

[52]   G.S. Abawi, R.G. Grogan, Epidemiology of diseases caused by Sclerotinia spp., Phytopathology 69 (1979) 899–904.

[53]   P.A. Arciniegas-Grijalba, M.C. Patiño-Portela, L.P. Mosquera-Sánchez, B.E. Guerra Sierra, J.E. Muñoz-Florez, L.A. Erazo-Castillo, J.E. Rodríguez-Páez, ZnO-based nanofungicides: Synthesis, characterization and their effect on the coffee fungi Mycena citricolor and Colletotrichum sp., Materials Science and Engineering: C 98 (2019) 808–825. https://doi.org/10.1016/j.msec.2019.01.031

[54]   H. Mohd Yusof, R. Mohamad, U.H. Zaidan, N.A. Abdul Rahman, Microbial synthesis of zinc oxide nanoparticles and their potential application as an antimicrobial agent and a feed supplement in animal industry: a review, J Animal Sci Biotechnol 10 (2019) 57. https://doi.org/10.1186/s40104-019-0368-z

[55]   D. Jain, Shivani, A.A. Bhojiya, H. Singh, H.K. Daima, M. Singh, S.R. Mohanty, B.J. Stephen, A. Singh, Microbial Fabrication of Zinc Oxide Nanoparticles and Evaluation of Their Antimicrobial and Photocatalytic Properties, Front. Chem. 8 (2020) 778. https://doi.org/10.3389/fchem.2020.00778

[56]   Z. Luksiene, N. Rasiukeviciute, B. Zudyte, N. Uselis, Innovative approach to sunlight activated biofungicides for strawberry crop protection: ZnO nanoparticles, Journal of Photochemistry and Photobiology B: Biology 203 (2020) 111656. https://doi.org/10.1016/j.jphotobiol.2019.111656

[57]   M.A. Schell, Control of Virulence and Pathogenicity Genes of *Ralstonia Solanacearum* by an Elaborate Sensory Network, Annu. Rev. Phytopathol. 38 (2000) 263–292. https://doi.org/10.1146/annurev.phyto.38.1.263

[58]   L. Cai, J. Chen, Z. Liu, H. Wang, H. Yang, W. Ding, Magnesium Oxide Nanoparticles: Effective Agricultural Antibacterial Agent Against Ralstonia solanacearum, Front. Microbiol. 9 (2018) 790. https://doi.org/10.3389/fmicb.2018.00790

[59]   T. Ahmed, M. Noman, M. Shahid, M.S. Shahid, B. Li, Antibacterial potential of green magnesium oxide nanoparticles against rice pathogen Acidovorax oryzae, Materials Letters 282 (2021) 128839. https://doi.org/10.1016/j.matlet.2020.128839

[60]   M.I. Khan, M.N. Akhtar, N. Ashraf, J. Najeeb, H. Munir, T.I. Awan, M.B. Tahir, M.R. Kabli, Green synthesis of magnesium oxide nanoparticles using Dalbergia sissoo extract for photocatalytic activity and antibacterial efficacy, Appl Nanosci 10 (2020) 2351–2364. https://doi.org/10.1007/s13204-020-01414-x

[61]   L. Kőrösi, B. Pertics, G. Schneider, B. Bognár, J. Kovács, V. Meynen, A. Scarpellini, L. Pasquale, M. Prato, Photocatalytic Inactivation of Plant Pathogenic Bacteria Using TiO2 Nanoparticles Prepared Hydrothermally, Nanomaterials 10 (2020) 1730. https://doi.org/10.3390/nano10091730

[62]   M.L. Paret, G.E. Vallad, D.R. Averett, J.B. Jones, S.M. Olson, Photocatalysis: Effect of Light-Activated Nanoscale Formulations of TiO2 on Xanthomonas perforans and Control of Bacterial Spot of Tomato, Phytopathology® 103 (2013) 228–236. https://doi.org/10.1094/PHYTO-08-12-0183-R

[63]   P. Pan, D. Svirskis, S.W.P. Rees, D. Barker, G.I.N. Waterhouse, Z. Wu, Photosensitive drug delivery systems for cancer therapy: Mechanisms and applications, Journal of Controlled Release 338 (2021) 446–461. https://doi.org/10.1016/j.jconrel.2021.08.053

[64]   G. Shim, S. Jeong, J.L. Oh, Y. Kang, Lipid-based nanoparticles for photosensitive drug delivery systems, Journal of Pharmaceutical Investigation 52 (2022) 151–160. https://doi.org/10.1007/s40005-021-00553-9

[65]   F.S. Alves, J.N. Cruz, I.N. de Farias Ramos, D.L. do Nascimento Brandão, R.N. Queiroz, G.V. da Silva, G.V. da Silva, M.F. Dolabela, M.L. da Costa, A.S. Khayat, J. de Arimatéia Rodrigues do Rego, D. do Socorro Barros Brasil, Evaluation of Antimicrobial Activity and Cytotoxicity Effects of Extracts of Piper nigrum L. and Piperine, Separations 10 (2023) 21. https://doi.org/10.3390/separations10010021

[66]   T. Li, Y. Geng, H. Zhang, J. Wang, Y. Feng, Z. Chen, X. Xie, X. Qin, S. Li, C. Wu, Y. Liu, H. Yang, A versatile nanoplatform for synergistic chemo-photothermal therapy and multimodal imaging against breast cancer, Taylor & Francis, 2020. https://doi.org/10.1080/17425247.2020.1736033

[67]   E.G.R. Dias, K. Davis, M.S. do Nascimento Remígio, T.S. Rabelo, M.S.M. da Silva, J.K.L. Vale, Essential oil as a source of bioactive compounds for the pharmaceutical industry, in: J.N. Cruz (Ed.), Drug Discovery and Design Using Natural Products, Springer Nature Switzerland, Cham, 2023: pp. 501–524. https://doi.org/10.1007/978-3-031-35205-8_18

[68]   X. Liang, Y. Xie, J. Wu, J. Wang, M. Petković, M. Stepić, J. Zhao, J. Ma, L. Mi, Functional titanium dioxide nanoparticle conjugated with phthalocyanine and folic acid as a promising photosensitizer for targeted photodynamic therapy in vitro and in vivo, Journal of Photochemistry and Photobiology. B, Biology 215 (2021). https://doi.org/10.1016/J.JPHOTOBIOL.2020.112122

[69]   D.R. Sánchez-Ramírez, R. Domínguez-Ríos, J. Juárez, M. Valdés, N. Hassan, A. Quintero-Ramos, A. del Toro-Arreola, S. Barbosa, P. Taboada, A. Topete, A. Daneri-Navarro, Biodegradable photoresponsive nanoparticles for chemo-, photothermal- and photodynamic therapy of ovarian cancer, Materials Science and Engineering: C 116 (2020) 111196. https://doi.org/10.1016/J.MSEC.2020.111196

[70]   M.H. Sarfraz, M. Zubair, B. Aslam, A. Ashraf, M.H. Siddique, S. Hayat, J.N. Cruz, S. Muzammil, M. Khurshid, M.F. Sarfraz, A. Hashem, T.M. Dawoud, G.D. Avila-Quezada, E.F. Abd_Allah, Comparative analysis of phyto-fabricated chitosan, copper oxide, and chitosan-based CuO nanoparticles: antibacterial potential against Acinetobacter baumannii isolates and anticancer activity against HepG2 cell lines, Frontiers in Microbiology 14 (2023) 1188743. https://doi.org/10.3389/fmicb.2023.1188743

[71]   Y. Jia, Y. Song, Y. Qu, J. Peng, K. Shi, D. Du, H. Li, Y. Lin, Z. Qian, Mesoporous PtPd nanoparticles for ligand-mediated and imaging-guided chemo-photothermal therapy of breast cancer, Nano Research 13 (2020) 1739–1748. https://doi.org/10.1007/s12274-020-2800-2

[72]   B. Purohit, P.R. Vernekar, N.P. Shetti, P. Chandra, Biosensor nanoengineering: Design, operation, and implementation for biomolecular analysis, Sensors International 1 (2020) 100040. https://doi.org/10.1016/J.SINTL.2020.100040

[73]   M.R. Hasan, P. Sharma, R. Pilloton, M. Khanuja, J. Narang, Colorimetric biosensor for the naked-eye detection of ovarian cancer biomarker PDGF using citrate modified gold nanoparticles, Biosensors and Bioelectronics: X 11 (2022) 100142. https://doi.org/10.1016/j.biosx.2022.100142

[74]   L. Qiu, C. Wang, M. Lan, Q. Guo, X. Du, S. Zhou, P. Cui, T. Hong, P. Jiang, J. Wang, J. Xia, Antibacterial Photodynamic Gold Nanoparticles for Skin Infection, ACS Applied Bio Materials 4 (2021) 3124–3132. https://doi.org/10.1021/acsabm.0c01505

[75]   I.N. de F. Ramos, M.F. da Silva, J.M.S. Lopes, J.N. Cruz, F.S. Alves, J. de A.R. do Rego, M.L. da Costa, P.P. de Assumpção, D. do S. Barros Brasil, A.S. Khayat, Extraction, Characterization, and Evaluation of the Cytotoxic Activity of Piperine in Its Isolated form and in Combination with Chemotherapeutics against Gastric Cancer, Molecules 28 (2023) 5587. https://doi.org/10.3390/molecules28145587

Advances in Healthcare and Nanoparticle Toxicology
Materials Research Foundations 171 (2024) 22-75

Materials Research Forum LLC
https://doi.org/10.21741/9781644903339-2

Chapter 2

Nanoparticles based Therapeutics Approaches for Cancer

Somya Ranjan Dash[1] and Chanakya Nath Kundu[1*]

[1]Cancer Biology Division, School of Biotechnology, Kalinga Institute of Industrial Technology (KIIT), Deemed to be University, Campus-11, Patia, Bhubaneswar, Odisha 751024, India

* cnkundu@kiitbiotech.ac.in

Abstract

Surgery, chemotherapy, radiotherapy, targeted therapy and immunotherapy are all examples of standard cancer treatments. However, achieving successful cancer treatment is complicated by issues including insufficient selectivity, cytotoxicity, and chemo resistance. Nanotechnology's introduction has brought about seismic shifts in the ways in which cancer is detected and treated. Biocompatibility, lower toxicity, better stability, increased permeability and retention effect, precision targeting, and other benefits of nanoparticles (NPs) make them a promising option for cancer therapy. This book chapter provides a comprehensive overview of both organic and inorganic NPs and how they might be used in the treatment of cancer. In addition, we have also emphasized various aspects of NPs targeting cancer stem cells (CSCs), reprogramming the tumor microenvironment (TME), cancer angiogenesis and how they can be utilized in overcoming drug resistance. Finally, the future of NPs in non-invasive cancer treatment has been discussed along with its potential risk to the environment.

Keywords

Nanoparticles in Cancer Therapy, Cancer Stem Cells Targeting, Tumor Microenvironment Reprogramming, Drug Resistance in Cancer, Non-Invasive Cancer Treatment

Contents

1.   Introduction

Although chemotherapy can destroy rapidly dividing cells but it also has some substantial negative effects, such as suppressing the bone marrow normal activity, causing hair loss, and triggering gastrointestinal related disorders [1]. A greater number of research in cancer over the past few years has been to produce treatments that more specifically kills malignant cells, rather than healthy cells. The development of personalized treatments has greatly advanced precision medicine, yet there remain many unintended consequences and the perennial problem of drug resistance. As a result, there has been a rise in the number of research aimed at developing targeted cancer therapies and overcoming drug resistance. Nanotechnology uses in medicine have expanded over the past few decades, with recent examples including safer, more precise cancer cell targeting potentiality, faster diagnostics and more successful therapy. Better pharmacokinetics, site specific targeting of malignant cells, decreased toxic effects, and lowered the drug resistance property are just a few of the benefits of nanoparticles (NPs) based nano systems that have been seen in cancer treatment [2]. NPs are excellent nano-carrier that when conjugated with a chemical group specifically bind to cancer cells. The medications are then released to kill off cancer cells [3]. Traditional chemotherapeutic drugs are encapsulated within the nano-carriers, demonstrating their potential utility in cancer treatment. In addition, NPs generate a platform for encapsulating and delivering into the blood circulation for some of the poorly soluble drugs. NPs increases the half-life of medications and inducing their target specific accumulation in tumor tissues due to their enhanced permeability and retention effect. Various studies have reported the use of NPs in cancer immunotherapy, in addition to chemotherapy and gene therapy. When it comes to treating cancer, NPs are preferable than microparticles since they degrade more quickly. Normal blood arteries contain a strong extracellular matrix that prevents NPs from passing through. Tumor-induced angiogenesis causes the creation of new blood vessels; however, these vessels are immature and hinder lymphatic outflow as the tumor expands. NPs are able to enter their intended cells because lymphatic drainage is inhibited. This is known as the "enhanced permeability and retention effect" (EPR), and it is crucial to the passive targeting of NPs [3].

Previous studies reported that many types of cancer have been successfully treated with light-activated, NPs based photosensitizer therapy. There are primarily two options: photodynamic treatment, which causes chemical induced damage in the targeted site and photothermal therapy, which causes thermal damage at the targeted lesions [4]. NPs based phototherapeutic medications and devices have come a long way as cancer therapies over the past several decades, but there are still some major obstacles preventing them from being widely used outside of dermatological applications. To overcome from the problem multifunctional hybrid NPs has been used for the enhancement in the tumor selectivity of photosensitizers compounds by site specific activation, may produce good results with lesser unpleasant side effects. Multiple forms of cancer have been reported to have promise in overcoming multi drug resistance (MDR) using nanoparticle-based treatment in recent years [4]. Nanotechnology in medicine has ushered in a new era in the fight against cancer, and the intersection of these two disciplines is worthy of further investigation. Here, in this chapter, we will talk about the several kinds of NPs that may be used to treat cancer, each with its own set of advantages, as well as the mechanistic approach to targeting specific stages of the disease. We have also emphasized the current developments in photo treatment and immuno therapy with NPs. At long last, our work has thrown light on the potential of NPs-mediated precision medicine for cancer therapy.

2.   Cancer treatment using various NPs

Therapeutic efficacy of a given NPs based therapy is typically controlled by its size, shape, and surface property all of which are carefully tailored for optimal nano-drug delivery to the target site. Cancer therapy often employs NPs with a diameter between $10 \sim 100$ nm due to their capacity to efficiently distribute medications and attain an EPR effect. NPs whose size is lower than 1-2 nm can easily pass from normal blood vessel and can also cause harm to healthy cells, while NPs of size greater than 100 nm can be easily removed from blood circulation by phagocytes [5,6]. In addition, NPs bioavailability and half-life might be affected by their surface properties. Coating NPs with hydrophilic materials like polyethylene glycol (PEG) reduces opsonization by other immune cells, allowing them to evade immune system clearance. Thus, NPs are often altered to become hydrophilic, which extends the half-life of drugs and enhances drug incorporation and accumulation in malignant cells [7]. The therapeutic potentiality of NPs in cancer therapy is determined by a combination of their distinct features. Majorly the NPs are classified into organic, inorganic and hybrid NPs.

2.1   Organic NPs in cancer treatment

Organic NPs are 2 or 3 dimensional compounds that range in size from 1 to 100 nm. The optical, magnetic, catalytic, thermodynamic, and electrochemical characteristics of NPs, for example, vary with their size. Organic NPs have unique features because of their chemical make-up and their geometrical structure. Liposomes, micelles, protein/peptide based, and dendrimers are the primary categories of organic NPs [8]. Dendrimers are synthetic polymers with a high branching degree and a multilayer architecture consisting of a core, an internal area, and multiple terminal groups. Many different materials may be found in organic NPs, which have been studied extensively over many years for cancer treatment and prevention as well. The lipid bilayer around the core entraps the medicine, which can be hydrophobic or hydrophilic, and makes liposomes the first ever nanotechnology based drug licensed to be used in clinics [8]. By altering the structure of the lipid layer, liposomes can mimic the biophysical characteristics of living cells, thus leading to more efficient therapeutic drug delivery to the target site [8].

Liposomes

Over the course of several decades of study, various generations of liposomes have been developed. In the field of oncology, liposomes have proven to be an effective vehicle for the *in vivo* administration of a wide variety of chemotherapeutic medicines and nucleic acids. An growing number of studies have shown that liposomes are useful in treating various cancer types. Recent progress in liposomal vesicle research has allowed for both time-released dosing and

Advances in Healthcare and Nanoparticle Toxicology | Materials Research Forum LLC
Materials Research Foundations 171 (2024) 22-75 | https://doi.org/10.21741/9781644903339-2

disease-specific localization of medication delivery. As cancer's first-line treatment is usually surgery, radiation therapy, or chemotherapy, this quality is incredibly useful. Chemotherapy must be given systemically in some cases of cancer. The majority of chemotherapeutic used to date are extremely cytotoxic to both cancerous and healthy cells. Since the free medicine is injected straight into a patient's veins, it has several unwanted side effects and restrictions. Cancerous and healthy tissues alike may absorb the chemotherapeutic chemical, which can have devastating effects on vital organs including the heart, kidneys, and liver. Liposomal encapsulation of chemotherapeutic drugs has been shown to improve the therapeutic potentiality of the anti-cancer drugs by reducing its absorption by healthy tissues. Liposomes can concentrate preferentially on the tumor by passive targeting due to the increased EPR effect, which occurs when leaky tumor arteries combine with missing lymph drainage. Liposomes are passively targeted when they migrate through fluids and enter the tumor interstitium via the leaky tumor vasculature [9]. Using an antibody-based strategy, liposomes may be directed specifically toward the tumor tissues. To do this, immunoliposomes (ILP) can be created by attaching antibodies to the surface of liposomes that bind to either the malignant cells themselves or the endothelial cells that line the blood vessels that supply the tumor. These ILPs were more effective than regular liposomes in reaching their intended target (lung endothelial cells and tumor tissue) [10]. However, not all cancer cells express the antigen required for the targeted antibody to attach to the tumor. Therefore, this method can only work with antigen-antibody pairs. Local hyperthermia could make liposomal drug delivery more effective in several ways by releasing the drug at a particular temperature which is very much close to the lipid transition phase of the liposomes, by increasing the blood flow by making it easier for the liposomes to accumulate at specific sites by making endothelial cells more permeable to liposomes, and by making it easier for the drug that is released from liposomes to get into the target cells [11,12]. One of the thermosensitive liposomes were engineered by Yatvin et al. in 1978, in which two lipids DPPC:DSPC was encapsulated with neomycin. A localized hyperthermia promotes the sustain release of the drug from the liposomes [13]. Weinstein et al. conducted the first ever animal experiment of a thermosensitive liposome-based anticancer formulation in 1979. Thermosensitive liposomes encapsulating methotrexate were used to treat lung cancer in a mouse model. It was observed that the anti-cancer drugs accumulated four times as much in tumor tissue. However, the liposomes were degraded in the body after an hour, which was a major drawback of this formulation. Enzyme-responsive liposomes are another way to get anticancer drugs into the body. For instance, some of the extracellular enzymes, such as released sPLA2, MMP-2 and MMP-9, which increases in varieties of cancers like breast, pancreatic, lung and colorectal cancer.

Immunotherapy for cancer through vaccine is not yet a major way to treat cancer. Immunotherapy could be used to treat or avoid cancer, but many experts are looking for new ways to use it and new ways to make it work. Allison and Gregoriadis were the first people to say that antigens trapped in liposomes could cause immune reactions [14]. There are several types of cancer vaccines that are being looked into right now. These include vaccines based on tumor cells, antigens, dendritic cells, and vectors [15,16]. For example, a mix of Rituximab® and non-PEGylated liposomal encapsulated DOX is suggested as preliminary treatment for aggressive non-Hodgkin lymphoma in people who are in the age group of 80 [17]. Chemotherapy is the first line of defense when it comes to treating cancer. Most of them are restricted, though, because they are too toxic, don't target the right cells well enough, have a narrow therapeutic index, or have a high chance of making the body resistant to the drug. These things can make cancer treatment fail in a big way. It has been shown that making tiny liposomal formulations helps the drug get to the tumor

Advances in Healthcare and Nanoparticle Toxicology   Materials Research Forum LLC
Materials Research Foundations 171 (2024) 22-75   https://doi.org/10.21741/9781644903339-2

cells in a targeted way [14]. This means that the EPR effect won't cause harm in places other than the target. Some drug delivery systems made with liposomes were allowed by the FDA because they worked well against cancer. **Table 1** is a summary of some liposomal products that are used to treat cancer.

Table 1 *Other liposomal products are summarized in the below table for their usage in cancer therapy.*

Liposome type	Drug encapsulated	Size (nm) of liposome	The Type of Cancer Being Targeted	Reference
Egg phosphatidylcholine: cholesterol: TPGS1000-TPP	Paclitaxel	80	Lung cancer	[18]
HSPC: DSPE-PEG2000: cholesterol: anacardic acid	Mitoxantrone	112	Melanoma cancer	[19]
HSPC/DSPE/cholesterol	DOX	130	Colorectal cancer	[20]
Cholesterol, DSPC, DSPE and DSPE-PEG2000 (10 µmol total phospholipid).	DOX	100	Prostate cancer	[14]
DPPC:cholesterol:1,2-distearoyl-sn-glycero-3-phosphoethanolamine - Methoxy PEG2000	all-trans retinoic acid	200	Human Thyroid carcinoma	[21]
DOTAP, cholesterol and ATRA	all-trans retinoic acid	263	Lung cancer	[22]

In conclusion, liposomes are an appealing vehicle for the selective distribution of anticancer drugs. These nanocarriers not only transport chemotherapeutic agents to the site of action, but also increase their bioavailability and reduces secondary side effects. To accomplish the necessary receptor-specific targeting, liposomes can be surface-functionalized with a variety of biomolecules.

Polymeric micelles (PMs)

Interest in polymeric micelles (PMs) as innovative drug delivery systems for the diagnostic and treatment of cancer has increased for a number of reasons. This includes increased specificity for tumor tissue, lower toxicity, longer circulation duration and higher accumulation in the tumor tissue. Poly(ethylene glycol) (PEG), along with newer polymers like poly(vinylpyrrolidone) and poly(trimethylene carbonate), is the most often used hydrophilic component. Polyesters, poly(propylene oxide) and co-polymers of glycolic and lactic acids are common examples of hydrophobic components. In addition, the co-polymer blocks may be adjusted in order to meet the ultimate goal by tinkering with various hydrophobic/hydrophilic component combinations and ratios. Drug adverse effects can be mitigated thanks to the size of PMs, which aids in bioavailability, target specificity, controlled release, and longer circulation duration [23]. Tumors abnormal blood vessel network allows for a certain size range of NPs (30-100 nm) to passively accumulate at the tumor site. For instance, micelles with a positively charged surface are more easily absorbed by the body when taken orally, and their penetration and mucoadhesion capabilities are enhanced during the process. Negatively charged surfaces, on the other hand, tend to aggregate in vivo due to nonspecific protein binding [24,25]. As a result, developing stable micelles requires a careful consideration of the surface charge balance. Micelles composed of multiple drugs for simultaneous administration are shown in **Table 2**.

Table 2 *Displays micelles that are encapsulated with various drugs for cancer treatment.*

Polymeric micelles type	Drugs encapsulated	Model type	Reference
PEG-b-p(Asp-Hyd)[a]	DOX/wortmannin	MCF-7	[26]
PEG-b-PLA	Paclitaxel/17-AAG/rapamycin	A549 NSCLC and MDA-MB-231	[27]
PEG-b-p (γ-benzyl L-glutamate) + PEG-b-p(L-lactide)	DOX/etoposide, DOX/paclitaxel	CT-26 murine colorectal cancer	[28]
PEG-b-PLGA	DOX/paclitaxel	A549 NSCLC, B16 mouse melanoma, and HepG2	[29]
Poly(2-methyl-2-oxazoline) -b-poly(2-butyl-2-oxazoline) -b-poly(2-methyl-2-oxazoline)	Paclitaxel/17-AAG/etoposide, Paclitaxel/17-AAG/bortezomib	MCF-7 and MDA-MB-231, PC3, and HepG2	[30]
PEG-DSPE/TPGS	Paclitaxel/17-AAG	SKOV-3	[31]
PEG-b-poly-(glutamic acid) -b-poly(phenylalanine)	Paclitaxel/cisplatin	A2780	[32]
PLGA-b-PEG-b-PLGA	Paclitaxel/17-AAG/rapamycin	ES-2-luc	[33]

Drug combinations have shown to be very beneficial in cancer therapy, which has increased the importance of targeting aberrant signaling pathways in solid tumors. Synergy, selectivity, pharmacokinetic (PK) profile and safety are all important factors to consider when justifying a anti-cancer drug combination. By targeting solid tumors simultaneously, anticancer drugs with similar pharmacokinetic properties may achieve synergy. The PK and pharmacodynamics (PD) of anticancer drugs may be altered by concomitant multi-drug administration, which may have a negative impact on antitumor effectiveness and safety [34]. A PEG-b-p (Asp-Hyd) micelles can be employed for simultaneous multi-drug administration and contains dangling anticancer drugs. The EPR effect can be used to target tumors specifically using chemically conjugated anti-cancer drugs. In both circumstances, polymeric micelles degrade into monomers that are eliminated by the kidneys. However, crosslinking ways exist to keep PMs together in the bloodstream, where they can be used to target tumors.

Dendrimers

Likewise, the well-defined size and form of dendrimers make them a promising nanomaterial for use in biological applications. Dendrimers are spherical molecules with a clearly defined chemical structure, and they are extremely branched. They are characterized as a highly branching polymer that is monodispersed in three dimensions. Dendrimers show promise as a means to increase the drug's solubility and decrease its toxicity [35]. Dendrimers have been investigated for the encapsulation and controlled administration of numerous anti-cancer drugs due to their high drug loading capacity, simple synthesis and high stability. The success of dendrimer research has led to their application in the delivery of various anti-cancer medications and in theranostic approach. Bypassing the efflux transporter, delivering the medication intracellularly and increasing the bioavailability of loaded molecular payload are all the possible capabilities demonstrated by dendrimeric conjugates of anticancer drugs. Cisplatin dendrimer complexes have demonstrated potent anti-proliferative action with decreased cytotoxicity in other healthy cells. Dendrimers are not only being studied for the targeted administration of anti-cancer medicines, but also for the targeted imaging of cancer cells [36] [37].

Covalent conjugation to dendrimer end groups or entrapment of drug molecules within the dendrimer core via hydrophobic linkage, hydrogen bonding or electro-static interactions are all possible. The drug loading capacity is affected by the number of generations, with a higher generation number allowing for more space for the guest medications and more surface functional groups for drug conjugation. Most of the research into anticancer drug delivery has focused on peptide-based and amine-based polymers (PPI, PAMAM, PLL), polyesters, polyetherdendrimers and dendrimers based on polyethylene glycol (PEG) or carbohydrates [38]. The benefits of dendrimers and how anticancer drugs interact with them are summarized in **Table 3**.

Table 3 *Provides a concise overview of the benefits of anti-cancer drugs based on dendrimers for the treatment of cancer.*

Drug type	Classification of drug-dendrimer interactions	Applications of Drug-dendrimer NPs	Helpful in cancer types	Reference
Cisplatin	Physical encapsulation	Reduced toxicity	Pancreatic, Bladder, ovarian and cervical cancer	[39]
5-Fluorouracil	Chemical conjugation	Increase the solubility of 5-Fu	Colorectal and breast cancer	[40]
Doxorubicin	Physical encapsulation	Doxorubicin solubility improvement	Breast, Gastric, Ovarian and Thyroid cancer	[41]
Methotrexate	Physical encapsulation	Improve bioavailability and reduces toxicity.	Head & neck and Breast cancer	[42]

In order to enable controlled drug release at the lysosomal compartment at pH 5, Wrobel *et al.* [43] created varieties of PAMAM G3 conjugates with nucleotides and anti-nucleosides attached through phosphamide pH-sensitive linker. The anti-cancer efficacy of these above stated conjugates was then compared to that of free medicines utilizing *in vitro* cellular models of leukemia. Dendrimer-based DDS, especially when dexamethasone is present, enables the decrease of anticancer medication effective concentrations by roughly a 10-fold range in specific experimental circumstances without compromising the medicines anticancer potential. **Table 4** displays the available dendrimers for use in combating different cancers type.

Table 4 list of synthesized dendrimers for cancer therapy.

Dendrimer type	Drug type	Cancer type	Benefits	Reference
Polyethylenimine dendrimer-grafted tungsten disulfide	Letrozole	Breast cancer	pH, thermosensitive polymer and drug release is enhanced under the NIR irradiation at the target specific site	[44]
Core-shell tecto dendrimer (CSTD)	Cu (II) and disulfiram	Breast cancer	Improved MR imaging and faster tumor eradication are achieved by increasing the accumulation of medicines, inducing apoptosis in cancer cells, and amplifying synergistic chemodynamic therapeutic impact.	[45]
Carbosilane metallodendrimer G_1-[[NCPh(o-N)Ru(η6-p-$cymene$)Cl]Cl]$_4$ (CRD13)	DOX, 5-Fluorouracil (5-Fu), and Methotrexate (MTX)	Triple negative breast cancer	Forms stable nanocomplexes with all those anti-cancer drugs, enhancing their effectiveness	[46]
G5-PBA@CuS/cGAMP/antigen	-	Melanoma cancer	Induce antitumor immune response to inhibit the distal tumors	[47]
peptide dendrimers	DOX	Pancreatic ductal adenocarcinoma (PDAC)	Facilitating the accumulation of free drugs within tumors and their uptake by cancer cells.	[48]
PAMAM dendrimer G3	Lapatinib and fulvestrant	Triple negative and HER2 positive breast cancer	Reduced cytotoxicity	[49]

PAMAM dendrimers	Vismodegib	Basal cell carcinoma.	Penetrated into the skin reaching the viable epidermis and reducing toxicity	[50]
MPA-based dendrimers	DOX	Various models of cancer cell lines and human mesenchymal stem cells	Deliver the drug in a more sustained manner	[51]
aptamer-modified fluorinated dendrimer	Gefitinib and hematoporphyrin	Non-small cell lung cancer	Platform to overcome hypoxia-related resistance	[52]
PAMAM G4.5 dendrimers	Paclitaxel and sorafenib	Liver cancer	Provides better stability and a rapid drug release under acidic condition	[53]

Dendrimer is an innovative drug attachment platform because of its many binding and release processes. Dendrimer has the potential to facilitate the transport of medicines across biological membranes, making it an ideal delivery vehicle for water-insoluble anticancer treatments. Dendrimer can form a strong covalent and cleavable binding with drugs [54]. Due to the creation of a stable bond, there are issues with medications that have been covalently attached to dendrimers. The dendrimer can prevent the breakdown of medicines and slow down their clearance by the kidneys. Dendrimers, like other NPs, must demonstrate unambiguous safety for regulatory approval of any potential biological or therapeutic uses [54].

2.2   Inorganic NPs in cancer treatment

Metals and metal oxides are examples of inorganic NPs. Gold (Au), silver (Ag), iron oxide (IO), titanium oxide (TiO_2), copper oxide (CuO), and zinc oxide (ZnO) are some of the commonly used metals and metal oxide for cancer treatment. Liposomes are frequently utilized as carriers, especially for passive targeting and are a kind of organic NP. Although liposomes and lipid NPs have made significant contributions to cancer, no liposome-based medicinal treatments have yet been marketed. Liposomes have a number of problems that make them a poor choice for transporting actively targeted nanocarriers, such as their susceptibility to oxidation and hydrolysis, their structural instability, and the large amount of drug that leaks out of the carriers [55]. When heated, organic NPs tend to destabilize. For these reasons, inorganic NPs may be the better option for treating cancer. Gold nanoshell, IO, and hafnium oxide NPs are just a few of the inorganic nanomaterials being investigated on cancer patients. In Europe, IO NPs are already on the market for use against glioblastoma. Nanodiamond and other new inorganic NPs have recently been investigated as potential cancer treatments.

Advances in Healthcare and Nanoparticle Toxicology   Materials Research Forum LLC
Materials Research Foundations 171 (2024) 22-75   https://doi.org/10.21741/9781644903339-2

Silver (Ag) NPs

Fundamental and translational research on the use of nano-sized materials in cancer treatment has received considerable attention over the past decade. So far, various therapy approaches based on NPs have been implemented into clinical trials with promising results against cancer. Recent years have seen the emergence of various inorganic based NPs, particularly metal-based nanostructures, as therapeutically effective agents. Silver NPs (AgNPs) are among the most commonly employed in therapeutic applications because of their well-recognized anti-microbial effects [56]. AgNPs show both anti-microbial and unusual cytotoxic effects against mammalian cells, making them potentially suitable in tumor treatment. Endocytosis is the primary route for AgNPs absorption and once inside the cell, the endosomes collect the particles before sending them on to fuse with lysosomes [57]. When AgNPs enter the acidic lysosomal environment, they release more silver ions, which disrupt the normal cellular function of the cell which cause the cell to induce apoptosis [58].

Despite the fact that several research teams have shown the extraordinary potential of AgNPs as anticancer agents, it has been underlined that the harmful effects of these nanomaterials makes them relatively comparable to traditional chemotherapy medications. However, the "nano" form of AgNPs distinguishes them from ordinary tiny molecular medications and eventually aids in lessening the severity of unwanted side effects. In nanomedicine, this effect has been used for medication design and is known as EPR effect. Shape, size, and surface capping compounds applied to NPs have all been proposed as means of maximizing their passive accumulation. For instance, accumulating nanomedicine in tumors has been proven to be more effective in small animal xenograft models than in individuals with their own naturally existing cancers. However, there is evidence that the EPR effect also causes drug accumulation in patient-derived tumors, however the efficiency of this strategy varies greatly depending on the unique anatomical characteristics of each tumor. Since the EPR effect has lately been a point of contention, it is important to approach it with caution even if it has the potential to be a game-changing idea in nanoparticle tumor-targeting.

Rajivgandhi et al. [59] synthesized AgNPs and tested the efficacy of the NPs in MCF-7 cancer cell line. The result of cytotoxicity assay showed that the synthesized AgNPs induces apoptosis via ROS generation, mitochondrial membrane damage and nucleus cleavages in MCF-7 cancer cell line. Hence, the present result suggested that AgNPs were effective against MCF-7 breast cancer cell line. Le et al. [60] synthesized AgNPs using *Ardisia gigantiflia* extract (Arg-AgNPs) having an average diameter of 6 nm with added functional groups conjugated on the surface of the NPs. The IC50 values for Arg-AgNPs were found to be 1.37 and 0.65 µg/mL for AGS cells and 1.03 and 0.96 µg/mL for MKN45 cells, respectively, from the results of the viability experiments. Cell migration was inhibited and the cell cycle was arrested in the G0/G1 phase after exposure to Arg-AgNPs. Furthermore, compared to the control, Arg-AgNPs considerably incresaes the proportion of dead cells and promotes the generation of ROS. This research supports the idea that Arg-AgNPs are a viable therapeutic option for treating human gastric cancer.

To further understand the mechanisms of AgNPs cytotoxicity, early studies looked at whether the AgNPs themselves or the Ag ions produced from AgNPs influenced the apoptotic processes with AgNPs exposure. Study showed that Ag ions are the primary triggers of AgNPs-induced effects, as ionization of the AgNPs is aided by an acidic pH. It is also known that Ag ions cause ROS to be formed, which in turn causes severe oxidative stress and activates the other cellular pathways that ultimately result in cell death. However, it appears that generating oxidative stress in the cells is insufficient to fully imitate AgNP toxicity. Literature suggested that both cisplatin and AgNPs treatments lead to equal amounts of reactive oxygen species (ROS) formation meanwhile, cisplatin treatments trigger necrosis in addition of apoptosis, whereas AgNP-treatments produce apoptosis exclusively [61]. This finding may have therapeutic implications because, although cisplatin treatment promotes fast cellular death, AgNPs have a longer delayed effect on cell survival.

Recent research has identified dysregulated autophagy as a unique characteristic of cancer, suggesting that blocking autophagic flux using AgNPs might be an effective technique in contemporary cancer therapy. Several studies have shown that AgNPs can influence autophagy in a variety of cell types, including malignant ones. Furthermore, it has been reported that AgNPs of varying diameters may induce autophagy process in breast cancer cells. Indeed, inducing autophagy process appears to be an important mechanism by which AgNPs exert their lethal effects in both normal and malignant cells. To counteract the cytotoxic effects of AgNPs, transformation-related deregulation of the autophagy machinery can shift the autophagic response toward a pro-survival stance. Several malignancies have been shown to use this strategy to evade from various treatments. As a pro-survival strategy, biosynthesized protein-capped AgNPs reinstated autophagy in osteosarcoma and primary liver cancer cells [62,63].

Combinations of chemotherapeutics are commonly utilized in clinical practice to increase efficacy. The neoplastic mass can be eradicated more thoroughly when many medications with synergistic effects are used to attack the cancer cells from diverse angles. Despite remarkable in vitro and in vivo anti-cancer activity, AgNPs cytotoxicity potentiality was enhanced when used in combination with other anti-cancer drugs. As a result, many investigations into AgNPs biological effects on cancer cells have been conducted, both alone and in tandem with chemotherapeutic drugs (**Table 5**).

Table 5 The list of AgNPs used in the pre-clinical cancer model system is summarized in the table.

NPs type	Model type	Benefit	Reference
Quinacrine-based hybrid AgNPs	SCC-9 head & neck cancer cells and tumorigenic mice	Reduced tumor size and restore body weight	[64]
Ag/alisertib@polymeric NPs conjugated with chlorotoxin	U87MG and xenograft mice	Reduced the size of the tumor	[65]
Au shell on AgNPs	PC-3 prostate Cell line and tumorigenic mice	Inhibited tumor size by photothermal therapy	[66]
AgNP	C6-glioma bearing Rat tumor model	Enhances life span, And its anti-tumor efficacy	[67]
AgNP-TAT	B16 melanoma xenograft	Reduced tumor growth	[68]

Tat-FeAgNP-Dox	MCF-7 based tumorigenic mice model	Decreased tumor growth	[69]
rTL/ABZ@BSA/Ag NP	A549/T based xenograft mice	Inhibited metastasis and reduced the size of the tumor	[70]
Aptamer As1411-functionalized AgNP	C6-glioma based xenograft mice	Improves the success of radiation therapy	[71]
Ag@TiO₂NP	B16-F10 mleanoma tumorigenic mice model	Reduce tumor growth	[72]
pGAgNPs	HCT116 colorectal cancer tumorigenic mice model	Reduced tumour growth and increased life span of tumor mice	[73]

Gold (Au) NPs

Recent research has shown a growing interest in the medicinal applications of noble metal NPs, notably gold NPs (AuNPs), with a focus on cancer. To begin, AuNPs come in a wide variety of geometric forms that may be adjusted using relatively elementary synthetic methods. Organic NPs are notoriously difficult to tailor for optimal performance and application. In addition to their size and shape-dependent optical character, AuNPs are distinguished by the surface plasma resonance (SPR) phenomena [74]. Furthermore, AuNPs capacity for functionalization and high drug loading capacity are a result of their higher surface area. Through either direct or indirect conjugation, drugs, nucleic acids, proteins and peptides can interact with AuNPs [74]. In addition, the inert nature of gold makes AuNPs biocompatible, and studies have shown that they are relatively durable in physiological media after amphiphilic material modification. Because of these characteristics, AuNPs have become widely used as nano-vectors in cancer treatment.

Colloidal gold was originally reported to be used as a delivery vector in 2004 by Paciotti and colleagues. Mice tumors were targeted by conjugating tumor necrosis factor (TNF) to the surface of AuNPs. When compared to natural TNF, the AuNP-TNF conjugate was found to have greater tumor accumulation and less damage in healthy organs. Since then, research into AuNPs potential as delivery vehicles has been extensive [75]. Synthetic compounds, phytochemicals and therapeutic peptides are only some of the anticancer substances that have been found to be delivered by AuNPs in recent years. Poor solubility, a short half-life, drug resistance and poor tumor selectivity limit the use of these anticancer agents despite their enhanced cytotoxic effects on cancer cells. Conjugating the anticancer compounds to NPs, especially those with a "hard" core like AuNPs, is one of the successful ways [75].

As drug delivery vehicles, AuNPs have been shown to possess a number of essential features. The EPR effect predicts that the NPs will concentrate preferentially in tumors. Because of their compatibility with a broad variety of molecules including drugs, nucleic acids and other targeting ligands, AuNPs may be useful for active targeted delivery [76]. By virtue of these characteristics, AuNPs have attractive possibilities as drug carriers (**Table 6**).

Table 6 *Displays AuNPs used to deliver different anti-cancer drugs in pre-clinical cancer model.*

NPs type	Model type	Benefit	Reference
DOX-PECAuNPs	HepG2 cell line	More potent cytotoxicity than free DOX	42
DTX@HA-cl-AuNPs	HeLa and MCF-7 Cell line	Greater anti-cancer potentiality when combined with NIR	43
K-AuNPs	MCF-7 cell line	Promotes greater cell apoptosis and inhibition of angiogenesis	26
AuNPs-PEG-5-FU-FA	M139 and M213 cell line	Higher cytotoxic effects as compared to free 5-FU and FA.	44
P1-AuNPs	HT-29 and MDA- MB-231 cells	Greater DNA disintegration in both cells and promotes subsequent cell apoptosis	29
QAuNP	SCC-9 cancer stem cells and patient-derived primary breast cancer cells	Decrease metastatic and anti-angiogenic potentiality of cancer stem cells by inducing apoptosis and decreased TGF-β mediated angiogenic response.	
AuNPs-MTX	SJNKP (a non-amplified NB cell line) and IMR5 (a n-Myc amplified NB cell line)	Enhancement of the drugs ability to enter cancer cells	[77]
AuNR@PDA	HeLa cells	The uniform polydopamine shell improves the biocompatibility of gold nanorods upon NIR exposure	[78]
DOX-Apts-CS-AuNPs complex	A549 and 4T1 cells	Provides a more potent anti-tumor action than free DOX while being distributed into fewer tissues	[79]

PE@AuNFs	HeLa cells and tumor bearing mice model	Offers excellent inhibition efficacy of tumor with laser irradiation	[80]
Dtxl-GHANPs	HepG2 cells	Dtxl-GHANPs were shown to have increased cytotoxicity in assays against HepG2 human liver cancer cells	[81]
MET-H-AuNPs	HepG2 cells	Showed improved targeting and enhanced regressive action in HepG2 cells compared to free MET	[82]
EGCG–pNG	MBT-2 murine bladder tumor cells and mice model	Induce apoptosis in cancer cells and downregulates the expression of VEGF	[83]

Using exogenous DNA and RNA in the form of gene therapy to treat or prevent cancer is an excellent strategy. Nucleic acid medications tend to be more unstable than small-molecule therapies. During gene modification and transfection, nucleic acid therapeutics are exposed to a number of environmental hazards, including enzymatic, chemical, and physical degradation. However, as biologic agents, these medications are more likely to provoke an innate immunological response. Thus, can be used as optimal delivery system to transport such anti-cancer drugs into cells, where they can limit the degradation of nucleic acid drugs and improve transfection efficiency. However, the host's immune response can be triggered by viral vector systems, reducing the efficacy of future gene therapy. The use of non-viral vectors systems, such AuNPs, for the transport of nucleic acids provides a solution to this problem. AuNPs have a more adaptable surface design than viral vectors, making them more suitable for functionalization and biocompatibility in the body. Despite their surface negative charge, cells absorb more than 99% of AuNPs, and they can shield nucleic acid from destruction by nucleases and physical damage. AuNPs coupled with nucleic acid have the aforementioned characteristics, making them suitable for silencing of the gene for the treatment in tumor models. For the purpose of treating breast cancer, for example, Tunc et al. encapsulated morpholino antisense oligonucleotides within a DNA-tile-AuNPs framework. The DNA-tile-AuNPs structure was shown to be superior to the liposome-based approach for delivering morpholinos and silencing HER2 and ER genes in breast cancer model [84]. Since AuNPs have a photothermal impact, the conjugate can serve as a nanoplatform for delivery of both gene silencing and photothermal treatment [85]. After nucleic acid functionalization, the compound retains its potent photothermal action. After being exposed to a laser, the composite greatly slowed the development of tumors with no obvious harm to vital organs.60 In addition, AuNPs can be loaded with both genes and chemotherapeutic medicines at once for a more potent therapeutic impact. To combine photothermal treatment, chemotherapy, and gene therapy, Huang et al. created an AuNP-based multifunctional nanoplatform that co-delivered microRNA-122 and DOX. This delivery system demonstrated superior anticancer impact than any single medication, and it was able to preferentially target hepatoma carcinoma cells by using PEG and HA [86].

Titanium and Iron Oxide NPs

Titanium dioxide (TiO_2) is a semiconductor type material having useful physicochemical properties that has been used in paint pigments and sunscreen creams. These properties include a greater refractive index, the ability to absorb UV radiation while reflecting or scattering visible light, and a high chemical stability. However, photoexcitation has been intensively researched to accelerate photodegradation processes, for instance of pollutants in water, by inducing charge separation through its semiconductor characteristics, resulting in the generation of high-energy.

Thevenot et al. found that the toxicity effects of NPs on malignant cells varied depending on the cell line and the chemicals conjugated to the NPs. Although uncoated TiO_2 NPs (at concentrations as high as 10 mg/ml) significantly decreased the vitality of lung carcinomas in mice. Polymer conjugation to the NPs surface lowered their cytotoxicity, which was seen in LLC cells [87]. When given intravenously to rats, uncoated TiO2 NPs aggregated in 0.9% of saline solution and accumulated mostly in the lungs, liver and spleen although at doses of 7.7-9.4 mg/kg, they were nontoxic. Mesoporous TiO2 NPs were coated with RBC-derived membranes by Li et al. to increase their stability and decrease their clearance and accumulation in the spleen, liver and kidney of tumorigenic mice. Using this approach, NPs were generated that accumulated more favourably in an in vivo model of MCF-7 cancer cells [88].

Therapeutic methods that use the semiconducting properties of TiO2 NPs after being conjugated with appropriate molecules include photodynamic therapy (PDT), which is thought to be less expensive, more effective, and less invasive than more traditional methods [89]. Although PDT is most commonly used to treat oral, head and neck cancer, it is being researched as a potential treatment for other forms of cancer as well. The active species in photodynamic therapy (PDT) is accelerated molecular oxygen to a highly reactive singlet state by light in the presence of a photosensitizer (PS) [89]. Most PS glow on their own, making it simpler to locate them precisely where they need to be for PDT, but they are only active when exposed to light of the right wavelength. Even though PS is administered systemically, ROS generation is limited to the site where the NPs were photoexcited, significantly reducing side effects in healthy tissues [89]. TiO_2 NPs are being investigated as potential theranostic agents against a wide range of cancer subtypes [90]. These NPs are typically found in conjunction with photoreactive organic based dyes including phthalocyanines and porphyrins.

TiO_2 NPs based nano-biomaterials have also been studied for cancer treatment, but to a lesser extent than IO NPs based materials. One example is the use of TiO_2 NPs in conjunction with anti-cancer drugs as a chemotherapeutic drug delivery platform. There has been research on a variety of drug release methods, including pH-triggered and irradiation-triggered delivery, among others. When compared to the transfer of bioactive molecules into healthy cells, the secretion of anti-cancer drugs induced by a stimulus is more efficient and regulated when targeting tumor cells [91]. **Table 7** provides illustrative examples of NPs having potential applications in oncological therapy, all of which point to promising futures for the development of effective TiO2-based NPs for cancer treatment.

Table 7 *Listed of different types of TiO2-based NPs for cancer treatment.*

NPs type	Model type	Benefit	Reference
Ti-Ce NPs	MCF-7 cell lines	Showed lower IC50 value in MCF-7 cells	[92]
(Ir–B–TiO2@CCM + NIR‑II	HepG2, MCF7, and A549 cell lines	NPs selectively accumulate in the mitochondria	[93]
TiO2-NPs	MDA-MB-231 and Hs578T human breast cancer cell lines	Inhibited EGFR autophosphorylation	[94]
PAA-Starch-TiO2-Cur NPs	MCF-7 cancer cells	Decreases the potential side effects	[95]
TiO2 NPs	human neuroblastoma (SH-SY5Y) cells	Induces ROS generation and ER stress and promotes apoptosis	[96]

IO NPs elicited a wide range of biological responses that appeared to be size and surface chemistry specific. Smaller PEG-coated NPs were shown by Feng et al. to have much reduced cellular uptake in vitro, but delayed the removal and increased its accumulation in SKOV-3 ovarian tumor model. The biocompatibility of NPs may be investigated in a number of different ways, including in vitro, in vivo, and ex vivo [97]. Incubation with human blood samples in a hemolysis test verifies their hemocompatibility.

It is of interest to note that Magnetic hyperthermia (MHT) is a therapeutic modality made possible by IO NPs magnetic characteristics. Here, an alternating magnetic field (AMF) is used to provide enough heat to raise the site-specific temperature, which in turn may cause cell death or immune system activation. Using this technique to selectively trigger malignant cell death while sparing the nearby normal cells is a major benefit for solid tumor treatment [98]. NanoTherm® magnetic fluid, the first product based on IO NPs, was licensed by the European Medicines Agency in 2010 for use in magnetic hyperthermia therapy (MHT) through direct injection into the tumor, followed by sessions of alternating magnetic fields (AMF) [99]. Combining MHT with radiation to treat patients with glioblastoma has been the subject of clinical trials. **Table 8** enumerates the additional IO used to treat various forms of cancer. Cancer therapy with both organic and inorganic NPs has been graphically represented in the **figure 1**.

Table 8 *Represents lists of additional IO used to treat diverse cancer types.*

NPs type	Model type	Benefit	Reference
PEI-SPION-siRNA	Pancreatic cancer cell lines	Showed a good silencing effect on MUC4	[100]
Pani/γ-Fe2O3 NPs	MCF-7 human breast cancer cells	To convert macrophages into an anti-tumor M1 phenotype, NPs are used as an immunomodulatory treatment.	[101]
Fu-IO NPs	P-selectin-overexpressing lung cancer cells	Increases greater NPs uptake by cancer cells	[102]
SPIONs	C6 and U87 glioma cancer cell lines	Disrupts the mitochondrial function	[103]
IONPs	EGFR-positive FaDu and 93-Vu head and neck cancer cell lines	Enhance internalization of NPs	[104]
AIOoxp	orthotopic pancreatic cancer	promoted ROS production, inducing ferroptosis and enhancing apoptosis	[105]
Green synthesized IONPs	MDA-MB-231 breast cancer cells	Induce ROS production and deregulates mitochondrial membrane potentiality	[106]
β-CD-SH (IONPs)	MCF-7 cancer cell line	Improves the delivery of hydrophobic drugs with greater theranostic potentiality	[107]

Advances in Healthcare and Nanoparticle Toxicology   Materials Research Forum LLC
Materials Research Foundations 171 (2024) 22-75   https://doi.org/10.21741/9781644903339-2

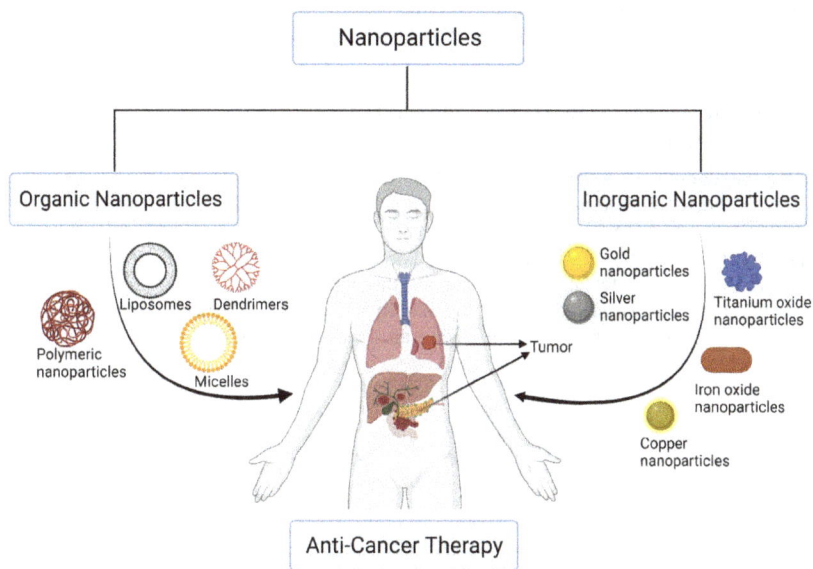

Figure 1: Therapeutic use of organic and inorganic NPs for cancer treatment.

3. Mechanistic action of NPs in targeting cancer stem cells (CSCs)

Cancers are time-related illnesses that pose a significant risk to human health and survival. Cancer may be treated in a variety of ways, with techniques including surgery, radiation therapy, chemotherapy, and targeted therapy among the most recent innovations. The cancer death rate has been holding steady because to the availability of effective therapies. However, conventional cancer treatments are only successful against a subset of aggressive tumors. Metastasis, recurrence, heterogeneity, drug/radiation resistance and supressing immunological memory are the main causes of therapeutic failure in the treatment of malignant cells. All of these setbacks may have a common cause: cancer stem cells (CSCs) [108,109]. By being stuck in the G0 phase and giving birth to new tumors, CSCs are majorly responsible for cancer recurrence. Therefore, CSCs may be the best candidates for future targeted cancer therapies [108].

Since CSCs are majorly drug-resistant and higher invasivness, novel delivery strategies for cancer therapies are required. In an effort to direct chemotherapeutic drugs toward surface biomarkers, biomolecules in CSCs signaling pathways several NPs have been created in recent years. NPs have several advantages over conventional drug delivery vehicles, including increased biodistribution of hydrophobic anti-cancer medicines in physiological fluids, enhanced drug stability and controlled release at the target site [110] . Using NPs, chemotherapeutic drugs can be directed to tumor cells in two basic ways **(Figure 2)**:

1. Passive targeting: Pathophysiological conditions such as impaired angiogenesis and a high demand for nutrients and oxygen by proliferating tumor cells lead to the development of abnormal tumor vessels characterized by relatively large gaps between the endothelial cells lining the lumen. Because of its wide intercellular clefts and insufficient lymphatic outflow, tumor tissue collects and maintains NPs in the size range of 100-200 nm. This EPR phenomenon makes it possible for drug-loaded NPs to be delivered to the tumor vasculature without invasive surgery.

2. Active targeting: Cancer stem cells (CSCs) can be targeted in tumor tissue using ligands that bind to one or more surface markers on CSCs. By taking this measure, medication toxicity and unwanted absorption by normal cells are drastically reduced [111]. Despite their usefulness in transporting drugs to CSCs, NPs have a number of drawbacks, including a propensity to aggregate, rapid clearance from the circulation, passive uptake through pinocytosis, pulmonary inflammation, translocation to other tissues, and so on. Drug toxicity and adverse effects can be greatly reduced with the use of NPs loaded with drugs and surface-attached with multifunctional ligands that target many CSC biomarkers simultaneously.

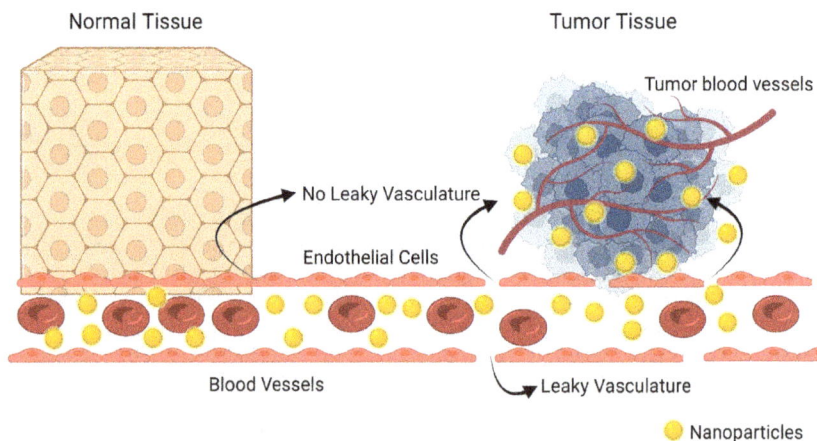

Figure 2: *Graphical representation of enhanced permeability and retention effect (EPR) mediated through NPs.*

Yang et al. [112] synthesized PLGA@NPs, surface modified with lipids, to successfully co-deliver paclitaxel (PTX) and curcumin (CUR) to nude mice with breast cancers. By targeting the CSCs in particular, CUR-PLGA@NPs slowed the progression of breast cancer, and PTX got rid of the tumor in its entirety. To specifically target HER2-positive CSCs, Li et al. used PLGA@NPs loaded with Salinomycin (SLM) and an antibody targeting HER2 in a tumorigenic breast cancer mice

Materials Research Forum LLC
https://doi.org/10.21741/9781644903339-2

model [113]. This strategy decreased the proportion of tumor cells that were CSCs in culture and animal models. PTX-loaded PLGA@NPs coupled with folic acid (FA) suppressed ABCG2 and MDR1 expression and upregulated cell death markers in tumor cells in ovarian tumorigenic mice model [114]. Antimicrobial drug anthothecol encapsulated in PLGA NPs suppressed pancreatic CSC migration and proliferation and triggered death in a research using Kras-mutant mice CSCs through regulating the sonic hedgehog pathway [115]. Co-delivery of DOX and SN38 to malignant cells was investigated in one research, and pH-sensitive PEG NPs were produced for this purpose [116]. Using a passive EPR effect, the systemic examination of metabolites demonstrated increased drug accumulation in tumor tissue, leading to the death. Both the liver tumor cells and CSCs were shown to be sensitive to SLM-loaded PEGceramide nano-micelles (SCM) [117].

A similar approach was taken in a research [118] to target CD133+ ovarian CSCs by loading SLM-loaded PLGA-PEG NPs with an surface conjugated antibody that target CD-133+ markers. The encapsulated drug in the NPs increased the bioavailability and decreased the proportion of CD133+ CSCs in a mice with ovarian tumor xenograft. Docetaxel (DTXL) and BMI-1-targeting tiny interfering RNA (miRNA)-loaded PLGA-b-PEG NPs were investigated in a MDA-MB-231 induced orthotopic tumor mice model. Polycomb repressor complex-1, of which BMI-1 is a component, has been linked to CSC self-renewal because of its role in mediating gene silencing and controlling chromatin structure. The release of DTXL killed off the majority of tumor cells, whereas the miRNA made CSCs more sensitive to DTXL by reducing BMI-1 oncogene expression, hence decreasing stemness marker expression [119]. To specifically kill CD-133+ CSCs in vitro and in a xenograft mouse system of lung cancer, Zhang et al. developed a mixture of SLM and gefitinib-loaded NPs [120,121]. The combination of SLM and gefitinib-NPs was more effective in killing CD133+ CSCs and decreases the volume of tumor.

It has been shown that inorganic NPs with diameters as small as a few nanometres and a homogeneous dispersion are very attractive as passive or active carriers in targeting tumor [122]. However, they are not very useful since they are quickly recognized and eliminated by Kupffer cells in the liver and the reticuloendothelial system (RES) in the blood. These restrictions can be circumvented by optimizing their physicochemical features for a certain cancer therapeutic application such as surface chemistry and size. It's important to keep in mind that different inorganic nanomaterials have different characteristics that influence how they interact with cells, especially in terms of how well they're taken up by cells [122].

Gold (Au) NPs have attracted a lot of interest as passive or active anti-cancer drug carriers in cancer treatment because of their biocompatibility, non-toxicity, and limited size distribution. As a carrier for tumor-specific drug delivery, the in vivo efficacy of AuNPs was enhanced by surface modification using ligand immobilization techniques. The biocompatibility of AuNPs in physiological fluids has also been enhanced by surface functionalization with biocompatible coatings [123]. One of the research showed synthesis of 20 nm AuNPs by using sodium citrate reduction and then covering them in thiol-terminated PEG [124]. PEGylation not only improved the stability and biocompatibility of AuNPs, but it also decreased their aggregation. Drug accumulation by breast CSCs expressing CD24/CD44+ markers and drug-induced malignant cell death were both improved by conjugating SLM to PEGylated AuNPs. Because cancer cells need more glucose (Glu) than healthy cells do, Glu has been considered as a reagent for targeting the tumor [125]. Earlier study reported that Glu-AuNPs with an average size of 50 nm were synthesized. Glu-AuNPs were formed by immobilizing the synthesized uPICs on the surface of AuNPs via Au-S coordination between the Au surface and the thiol groups of Glu. Overexpressed

GLUT1 on the surface of CSCs allowed the Glu moieties on the AuNPs to be recognized. Antitumor efficacy against GLUT1-overexpressing MDA-MB-231 breast cancer spheroids and MDA-MB-231 induced orthotropic breast tumors was enhanced by the Glu functionalized AuNPs. Using uPIC-AuNPs as a carrier for systemic delivery of siRNA to solid tumors was previously described by the same group [126] [127]. **Table 9** represents the list of other NPs that target the cancer stem cells. The many kinds of NPs that particularly target cancer stem cells (also known as CSCs) in the majority of tumor cells are represented schematically in **figure 3**.

Table 9 Contains a list of additional NPs that target cancer stem cells.

NPs type	Model type	Benefit	Reference
Sal-AuNP	breast cancer stem cells	induction of oxidative stress, mitochondrial dysfunction, and lipid oxidation promotes ferroptosis mediated cell death	[124]
GSH-bioimprinted NPs	leukemic stem cells	induce ferroptosis by disrupting intracellular redox status	[128]
CD133Ab-NPs-SN-38	Colorectal cancer stem cells	Specifically target CSCs and inhibit colony formation	[129]
CFH-DCLK1 NPs	colon cancer stem cells	Promotes greater cytotoxicity, inhibited the colonosphere growth and inhibited the migration and invasion	[130]
S-HA-NPs	breast cancer stem cells	Altered GSK-3β-related COX-independent pathway and promotes apoptosis	[131]
QAuNP + NIR-I	Oral cancer stem cells	Induces apoptosis by activating DISC complex and inhibiting the expression of HSP-70	[132]
QAuNP + NIR-I	Patient-derived breast cancer stem cells	Reduced TGF-β production in the TME, thereby inhibiting the angiogenesis response of CSCs.	[133]

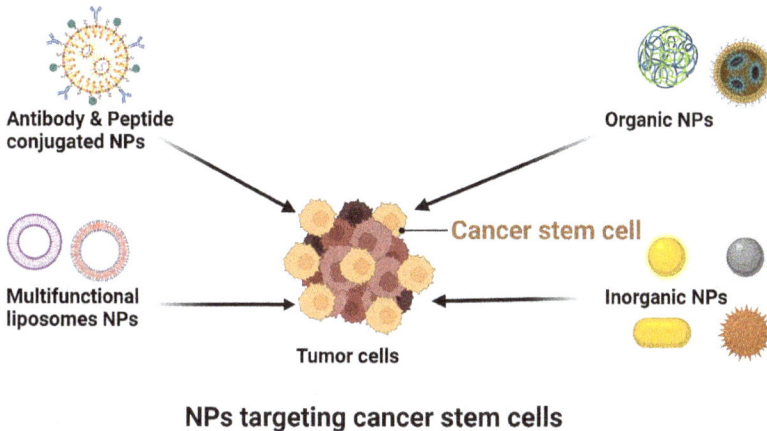

NPs targeting cancer stem cells

Figure 3: Schematic representation of different types of NPs targeting specifically cancer stem cells (CSCs) in the bulk of abnormal transformed cells.

4. Mechanistic action of NPs reprogramming the tumor microenvironment (TME)

The primary obstacles to effective cancer treatment are the tumor microenvironment (TME) of tumors. The TME consists of a variety of stromal cell types, including immune cells, macrophages, fibroblasts, lymphocytes and other secreted proteins [134]. The tumor microenvironment (TME) has been linked to tumor development, spread, and persistence. Both innate and adaptive immunity are the result of the actions of immune cells. TME is largely composed of fibroblasts, which have a role in all phases of cancer [134]. Fibroblasts that have been activated by cancer cells are known as CAFs. By producing various kind of cytokines which in turn stimulate TME remodelling. CAFs inhibit cytotoxic T cells activity and attracts lymphocytes that contribute to tumor growth by causing inflammation [135]. CAFs can also use stroma-derived factors to revert tumor-suppressing settings to ones that promote metastasis. For instance, fibrinogen-like protein 2 can promote tumorigenic CAF activity, which in turn promotes tumor suppression by myeloid-derived suppressor cells (MDSCs) via CXCL12. These faulty signaling pathways contribute to a hypoxic aberrant vasculature that is characteristic of the tumorigenic TME. Changes in TME activity, increased tumorigenesis, and worse treatment efficacy have all been linked to hypoxic conditions. Due to its ability to alter medication action and distribution, the tumor microenvironment (TME) can contribute to drug resistance by shielding pro-proliferation factors inside the TME and tumor mass [136]. Because of this dynamic network of connections tumor growth, response to therapy and the development of drug resistance are all always in flux. Targeting aberrant TME components has become a viable method for treating aggressive and resistant malignancies including gastrointestinal (GI) cancers.

In addition, macrophages are also essential for a number of processes, including immunity, homeostasis, and the healing and regeneration of wounds. They are specialized cells within the phagocyte system that facilitate phagocytosis by engulfing non-self material. Macrophages undergo distinct M1 and M2 differentiation after being exposed to signals. Cancer development is slowed when macrophages are polarized into conventional M1 macrophages by LPS and IFN-α. By secreting IL-10 and promoting tissue repair, macrophages exposed to IL-4 and IL-13 aid in tumor formation. Macrophages are seen as "double-edged swords" due to their ability to promote tumor growth in some immunological contexts while suppressing tumor growth in others. During the early stages of tumor development, TAMs promote anti-cancer immunity, but in later stages, they become centers of immune suppression and angiogenesis. The adaptability of macrophages is to blame for this disparity in behaviour. Changes in the tumor microenvironment (TME) during tumor progression trigger a phenotypic switch in TAMs from M1 to M2-like states. Therefore, TAM polarization reprogramming may have an impact on TAM function and lead to enhanced anti-tumor actions. Thus, nano-based therapies that selectively target TAMs may offer a viable cancer immunotherapy technique if they are able to modulate their polarization.

It's fascinating to learn that several NPs have been synthesized with the goal of controlling TAM polarization. For instance, Zanganeh et al. [137] found that IO NPs like ferumoxytol, which are currently licensed for treating iron deficiency, can provoke a switch from M2 to pro-inflammatory M1 phenotypes in solid tumor tissues, hence reducing tumor development. Therefore, ferumoxytol may be used to improve immunotherapies that regulate macrophages. A further iron-based nanostrategy involved creating nanotraps by modifying S-dot-bound proteins that target TAMs [138]. The nanotraps were able to adsorb, collect, and selectively transport iron to TAMs because they included multiple oxygen-containing groups on their surfaces. Internalization of the nanotraps within lysosomes resulted in iron release and oxidative pressure, which switched TAMs from M2 to M1 subtypes. In order to repolarize and retrain themselves, activated M1 released H_2O_2. In addition, ultra-small nanotraps penetrated tumors more deeply, polarized TAMs more effectively, and enhanced immune responses. Increased production of NO, TNF-α, and IL-10 was found following expression of M2 (CD206) and M1 (CD86) macrophage markers. This research demonstrates that natural chemicals already present in the body may be used to improve immunotherapy. The tumor microenvironment (TME) can be reprogrammed with the use of an anti-tumor vaccination developed using cutting-edge nano-based techniques. Antigenic microparticles formed from tumors (T-MPs) have the same markers and biological properties as the cancer cells from whence they were derived. Through a process of self-recognition, the contents of their cavities may be transported and delivered to parental cancerous cells [139,140]. Furthermore, T-MPs preserve and freely expose cancer-specific antigens that are highly immunogenic.

Interestingly, biomimetic anticancer compound designed by Zhang et al. [141] to reprogram M2 macrophages and boost antitumor immunity. The team synthesized a B16-OVA membrane to encapsulate the PNP@R surfaces (PNP@R@M) and coated it with resiquimod (R848), a Tolllike receptor 7/8 agonist and potent driver of macrophage polarization. This allowed the team to reprogram M2 subtypes. The NPs are shielded from RES clearance by this membrane, which expresses high amounts of CD47. In addition, they used synthetic PEG to alter the membrane, which lessens the interactions between NPs and blood proteins. Thereby protects the NPs from phagocytosis and aggregation. Successful selective reprogramming of M2 phenotypes led to increased lymphocyte infiltration and the immunity of the host [141]. Therefore, this engineered technique, which targets Toll-like receptors for their therapeutic potential via an enhanced formulation, may offer a useful method of effectively and selectively repolarizing M2 macrophages. Bacterial treatment is only one of the many methods that researchers have looked into to enhance macrophage reprogramming.

These results imply that tailored small-molecule NP-based solutions may decrease clearance, increase circulation duration, and enhance absorption by particular targets in the TME, resulting in better immunotherapeutic effects. Therefore, NPs provide promising avenues for engineering that circumvent the common challenges associated with immunotherapies targeting TAMs, such as non-specific delivery and clearance from the body. In addition, these NPs may improve the efficiency of immunotherapies in combination with chemotherapies and radiation treatments. Additional NPs that alter TAM programming are enlisted in **Table 10**.

Table 10 List of synthesized NPs that target TAMs and modulates its function in the TME.

NPs type	Model type	Benefit	Reference
AuNPs and AgNPs	TAMs isolated from murine fibrosarcoma	1. Promotes antitumorigenic response in TAMs through ROS and RNS generation 2. Protumorigenic nature of TAMs is altered	[142]
Fe3O4@Chitosan	Target TAMs present in Lewis lung carcinoma-bearing mouse	The expression of CD86 and secretion of TNF-α by TAMs was increased and inhibited cancer metastasis	[143]
CHO-PLGA-RA NPs + Anti-PD-L1	Target TAMs in allographic colorectal cancer	Blocked the polarization of M2-TAMs and supress EMT in colorectal cancer	[144]
biomimetic NPs-HPA/AS/CQ@Man-EM	Orthotopic colorectal mouse model	Promotes NPs accumulation in the tumor tissue and repolarizes TAMs	[145]
XAV-NPs	In vitro macrophage co-culture assays with melanoma cells	Inhibited β-catenin activation in TAMs and induces anti-tumor effects	[146]
DOX-PBCA NPs	TAMs-liver cancer cells co-culture model	Showed greater cytotoxicity in the developed co-culture model	[147]

Fibroblasts, which make up the tumor stroma, interact with tumor cells in a two-way conversation that contributes to tumor growth. The ECM is deposited by fibroblasts, which then promotes tissue healing and alters the tissue's normal organization and activity. They are called CAFs and they are permanently active in cancerous tissue. By increasing VEGF, SMA and various pro-angiogenic signals, CAFs impede effective medication delivery and induce treatment resistance. Therefore, focusing on CAFs is a vital tactic for changing the therapeutic TME, increasing drug uptake and neutralizing evasion mechanisms.

Hu et al. [148] created an injectable hydrogel NP solution containing losartan. In orthotopic mouse model, the hydrogel was able to inhibit collagen and CAFs development in TME for up to 9 days. The effectiveness of Pegylated and DOX-loaded liposomes was also enhanced by losartan hydrogels. Results from this study suggest that losartan hydrogels may be useful for improving the efficacy of pre-operative treatment in various types of solid tumors by altering the stromal cells. Likewise, one of the studies reported the development of pH-sensitive nano system consisted of a monophosphorylated gemcitabine (p-GEM) and paclitaxel NPs. The study's conclusions demonstrated that the innovative formulation (T-RKP) significantly reduced the expression of α-SMA in tumor cells and killed the tumor cells without harming healthy stromal cells. This nanosystem thereby permitted precise stromal targeting and increased medication effectiveness [149]. **Table 11** represents the list of other NPs targeting CAFs in TME.

Table 11 List of synthesized NPs that target CAFs and modulates its function in the TME.

NPs type	Model type	Benefit	Reference
LPPR NPs + αPD-L1	4T1^{luc}/CAFs tumor spheres in vitro and 4T1 tumor bearing-mouse	Reshaped immune microenvironment and led to facilitate antitumor effect	[150]
Res-NPs	H-357 Oral cancer cells incubated with CAFs-CM and xenograft mouse model	Inhibited the secretion of CXCL-12 and IL-6 in TME	[151]
GNPs	ovarian CAFs	Decreases multiple fibroblast activation proteins, such as TGF-β1, PDGF, uPA and TSP1	[152]
HSA-PTX@CAP-ITSL + NIR	Pan 02 cell line co-culture with mouse embryonic fibroblast and orthotopic tumor mouse	Accumulate in the tumor site and release HSA-PTX via cleavable CAP responsive to FAP-α in the TME	[135]

5.   NPs as new tools for inhibition of cancer angiogenesis

Solid tumor growth and development depend on the presence of tumor vasculature, which are primarily made up of blood vessels. Blood vessels are necessary for a tumor's uptake of nutrients, oxygen, and waste products from metabolism. When tumor tissue is bigger than 2 mm^3, it needs to produce blood vessels in order to get the nutrients it needs to survive and grow [153]. Chemotherapeutic drug delivery and penetration into tumor tissue are restricted by poor perfusion and elevated IFP, whereas chemotherapy, radiation, and immunotherapy are less effective due to aberrant TME. As a result, aberrant tumor vasculature are acknowledged as a significant cause of tumor therapeutic failure. Hematogenous diffusion, or the spread of cancer cells through endothelial gap junction in the blood vessels, is another significant mechanism in the metastatic cascade. The goal of the vascular disruption technique is to damage the endothelial cells (ECs) that currently line tumor blood vessels, cause thrombosis, and obstruct the tumor's ability to receive blood. By suppressing pro-angiogenic pathways such as VEGF pathway and the Ang 2 pathway, anti-angiogenic strategies seek to reduce tumor angiogenesis [153]. Using a coagulation response or gel phase change to restrict tumor blood vessel is known as a vascular blockade method. The goal of the vascular normalization approach is to enhance the transport and effectiveness of chemotherapeutic medications by correcting the tumor vasculature aberrant status and restoring their normal function[154].

Therapeutic effectiveness of small drug compounds like vascular disrupting agents (VDAs) is reduced due to their high distribution volumes and short half-lives following systemic administration. Effective accumulation of VDAs at tumor locations, continuous release, extended half-life, and increased activity in disrupting tumor vasculature are all made possible by NPs. The tubulin-binding agent VDA combretastatin A4 (CA4). Microtubule depolymerization and subsequent shape alterations in endothelial cells (ECs) upon binding contribute to vascular damage. Liu developed poly (L-glutamic acid)-CA4 conjugate NPs to improve CA4's activity against tumors. Increased disruption time and considerably enhanced therapeutic efficacy resulted from the NPs ability to boost tumor retention and accumulation of CA4 via EPR action and sustain a higher expression of CA4 in the transformed cells [155,156].

To selectively transport CA4 to tumor tissues, Liu engineered a GSH-responsive self-assembled NPs. The NPs was well-distributed in the body, highly responsive to GSH stimulation, stable in a typical environment, and degradable when exposed to GSH. Through the manipulation of targeting ligands, the NPs are also able to provide active targeted delivery of CA4 to cancer cells [157]. VDAs are able to produce greater necrosis in the cancer cells but in tumors with a high number of blood vessel density, some tumor cells will always survive. The normal blood vessels in the nearby healthy tissue are usually unaffected by VDAs, which may explain why the remaining external tumor cells are able to receive oxygen and nutrients from them. These lingering marginal cells have a role in tumor progression and recurrence. As a result, VDAs coupled with chemotherapy and other treatment strategies delivered by NPs may produce higher anti-cancer impact.

Overexpression of VEGF is common in tumor tissue and has been characterized as a crucial pro-angiogenic agent. Bevacizumab (BVZ) is a monoclonal antibody that blocks the binding of VEGF to VEGFR. Vandetanib, sorafenib (SFN), and other tyrosine kinase inhibitors may block VEGFR kinase activity and so limit receptor activation. The PI3K/Akt/mTOR pathway is just one of several downstream intracellular signaling pathways that may be activated by VEGF binding to VEGFR. Consequently, rapamycin and other inhibitors of the PI3K/Akt/mTOR pathways can likewise

Advances in Healthcare and Nanoparticle Toxicology   Materials Research Forum LLC
Materials Research Foundations 171 (2024) 22-75   https://doi.org/10.21741/9781644903339-2

suppress the VEGF signaling pathways. Nanocarrier medication delivery improves treatment efficacy and reduces adverse effects [158].

For the purpose of passive targeted distribution of BVZ, Battaglia et al. synthesized solid lipid NPs (SLNs) from palmitic acid and stearic acid. When compared to unloaded BVZ, the performance of BVZ loaded in SLNs is around 100-200 times greater [159]. Similarly, Wang et al. created a nanocarrier for co-delivery of MTX and sorafenib employing amine-terminated thermally carbonized porous silicon (PSi) loaded with MTX on the surface to operate as the targeted ligand via chemical bonding and loaded SFN inside to take advantage of the hydrophobic effect. Sorafenib's solubility was greatly enhanced by the dual-drug co-delivery nano-system, leading to enhanced suppression of tumor angiogenesis. In addition, it improves the capacity to destroy tumor cells by increasing MTX absorption and regulated release [160]. PEG-PLA amphiphilic block copolymer self-assembled NPs were also made by Wang et al., which were then employed to load vandetanib internally through hydrophobic effect and change a tumor-targeting peptide iRGD on the surface. The NPs may improve vandetanib's therapeutic and targeted effectiveness [161].

Numerous additional routes, including as Ang 2 and other pro-angiogenic signaling pathways, are also implicated in angiogenesis. Pericytes are also linked to tumor angiogenesis. The goal of using NMs to limit angiogenesis can also be accomplished by focusing on these locations. Another essential pro-angiogenic signaling implicated in tumor angiogenesis is Ang/Tie2. Multiple studies have demonstrated that blocking the Ang 2 pathway inhibits tumor angiogenesis and that it works in concert with VEFG inhibition [162]. To inhibit the expression of Ang 2 in tumor tissues, Shan et al. [163] created chitosan based magnetic NPs (CMNPs) carrying plasmids producing SiRNA for Ang 2 (Ang 2-CMNPs). The NPs shown excellent accumulation in tumor cells in the presence of a steady external magnetic field. The NPs showed potent Ang2 expression and tumor growth suppression, and they were able to trigger tumor cell death through the mitochondrial route. The Tie-2 receptor can also be blocked as a strategy. Considering the amphiphilic peptide mPEG1000-K(DEAP)-AAN-T4, Zhang et al. [164] created passively targeted, dual-responsive NPs. A small peptide called T4 inhibits the Tie2 receptor. Diethylaminopropyl isothiocyanate (DEAP) was protonated in TME and exposed the AAN-T4 to legumain, which is often overexpressed in TME. This caused the AAN to be cleaved and the T4 to be released, which was then followed by the blocking of the Tie2 receptor. Due to a localized blockade of Tie2 signaling, the P-T4 dramatically lowered vascular density.

Numerous studies have proven that p53 gene mutation is linked to angiogenesis. Overexpression of pro-angiogenic proteins and downregulation of angiogenesis inhibitors, two hallmarks of angiogenesis, can be developed in cancer cells with a p53 gene mutation. Angiogenesis is also prevented in cancer cells by restoring p53 activity. For the passive delivery of p53 plasmid DNA, Prabha [165] produced PLGA NPs. In order to sustain continuous gene transfection and reduce cell growth, the NPs may continuously release DNA plasmids into tumor cells. Following therapy, tumor tissue displayed overexpression of the endogenous angiogenic inhibitor thrombospondin-1 (TSP-1), as well as a decrease in the density of microvessels and an increase in cell death.

Pericytes may be a target for anti-angiogenic treatment since, according to the research, they are also linked to angiogenesis and may efficiently maintain the integrity of tumor vasculature. PEG-PLA NPs modified with the TH10 peptide were developed by Guan et al. [166] and loaded with docetaxel via the hydrophobic effect. Through its interaction with NG2 receptors, the TH10 peptide increased cellular absorption of the NMs in pericytes. Vascular pericytes may be precisely

Advances in Healthcare and Nanoparticle Toxicology
Materials Research Foundations 171 (2024) 22-75

Materials Research Forum LLC
https://doi.org/10.21741/9781644903339-2

targeted by NPs, and pericytes might be killed off by DTX. The improved anti-tumor activity was linked to the reduction in pericytes and micro-vessels, and the NPs could greatly extend the survival of tumor-bearing animals without having any noticeable hazardous side effects.

NPs create a practical platform for the concomitant delivery of anti-angiogenic drugs that target several signalling pathways, with the potential to boost therapy efficacy and reduce drug resistance. Additionally, certain nanomaterials exhibit anti-angiogenic properties on their own. AuNPs have been shown by Mukherjee et al. [167] to have anti-angiogenic ability. Growth factors that bind to heparin, such VEGF165 and bFGF, can be inhibited by the presence of AuNPs. Chitosan NPs, as reported by Xu et al. [168], also exhibit an anti-angiogenic action that is linked to the down-regulation of VEGFR2 expression. **Figure 4** illustrates the numerous approaches taken to designed NPs to target the tumor stromal cells in the TME.

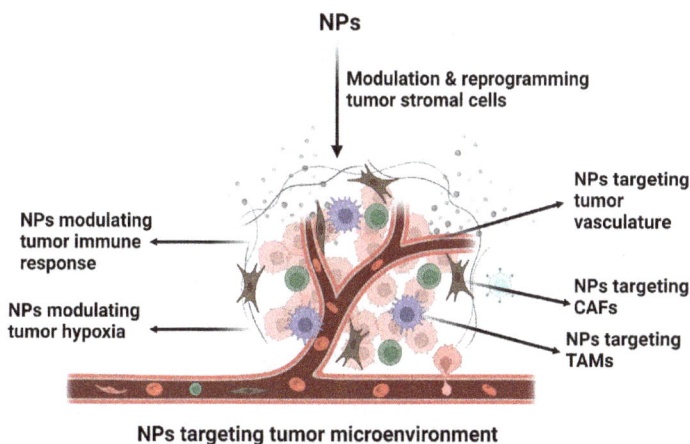

Figure 4: Strategies of various engineered NPs targeting various factors in the tumor microenvironment.

6.   NPs in overcoming drug resistance property in cancer

The development of drug resistance, which is unavoidable, is the main obstacle to cancer therapy. According to the underlying reasons, drug resistance may be categorized into two categories: primary and acquired. Primary resistance refers to elements that diminish the effectiveness of chemotherapy in the tumor tissue prior to treatment [169]. During the first few weeks of treatment, the body can change in a number of ways that can lead to acquired resistance. These include upregulation of therapeutic targets and stimulation of alternative signalling pathways. Some cancers have the traditional characteristics of continuously active proliferative signals, inactive growth suppressors, and active metastasis factors, which make them resistant to treatment. As the understanding of the mechanisms behind drug resistance in chemotherapy increased, it became clear that tumor heterogeneity, the TME and CSCs were all strongly related to these events. Multiple drug resistance (MDR) and other genetic drug transport factors are suspected to have a potential role in the generation of drug resistance alongside the epithelial-mesenchymal transition (EMT) in cancer cells [170].

The development of resistance to treatment is exceedingly subtle at the level of tumor tissue. Tumor heterogeneity in the TME, MDR and cancer stem cells (CSCs) all well-known contributors to the non-specific target. Drug efflux mediated by drug transporters reduces the therapeutic concentration of chemotherapeutic agents that are delivered to cells. Additional factors that may improve the survival compensation effect include a low pH, a hypoxic TME, and anti-apoptotic agents. In addition to the aforementioned five reasons for drug resistance, there are many others. These include mutations and instability in the genome, epigenetic alterations including DNA methylation and acetylation, suppression of apoptotic response and higher expression of anti-apoptotic proteins [171]. It has been shown in clinical practice that high levels of resistance to chemotherapeutic medications greatly reduce their therapeutic efficacy. As a result, new drugs development from a comprehensive molecular knowledge of drug resistance, leading to more effective and long-lasting cancer therapies.

Multiple studies have shown that hypoxia in the TME plays a critical role in the development of clinical resistance to chemotherapy through a variety of signal transduction pathways and molecular alterations. Therefore, creating efficient ways of delivering NPs remains a crucial challenge in the quest to overcome the resistance of hypoxic malignancies. There are three broad categories into which the tactics explored for modifying hypoxia using NPs: preventing hypoxia, adapting to hypoxia, and capitalizing on hypoxia. Hypoxia-reduction tactics are a tried and true targeted technique. Previous research by Alsaab et al., for instance, developed a multifunctional tumor hypoxia-directed NPs loaded with Acetazolamide (ATZ) to reverse Sorafenib resistance in renal cell carcinoma by targeting the marker of tumor hypoxia, carbonic anhydrase IX (CA IX) [172]. Taking use of perfluorocarbon's (PFC) biocompatibility and oxygen dissolving capacity, Song et al. developed a novel nano-PFC based on oxygen shuttle for ultrasound-triggered tumor-targeted administration of oxygen. Overcoming hypoxia-related medication resistance in cancer treatment with this strategy has great promise [173].

Hypoxia may be responsible for the development of acquired resistance of cancer cells to cisplatin treatment by regulating the expression of DNA self-repairing proteins such xeroderma pigmentosum F (XPF) [174]. A new nano-chemotherapy technique including the construction of a liposome NPs comprising the combination of glucose oxidase (GOx), tirapazamine (TPZ) and platinum (IV) prodrug was used to overcome cisplatin resistance. To maximize TPZ drug action, the nanodrug can be used to its maximum potential and even exacerbate the intracellular hypoxia of cisplatin-resistant malignant tumor cells. By efficiently suppressing XPF protein expression, which promotes DNA repair in tumor cells, activated TPZ enabled synergistic enhancement of anticancer treatment with platinum medication [175].

As the field of nanoscience progressed, more and more nanostructures were identified. Due to its exceptional optical characteristics and surface plasmon resonance effect, AuNPs are finding widespread application as a vehicle for the delivery of small anti-cancer drug molecules. Chemotherapy medications stability and blood-flow duration can be improved by attaching PEG to AuNPs. Furthermore, altering the form of AuNPs from spheres to rods would alter the NPs surface plasmon band (SPR). AuNPs combined with near-infrared (NIR) irradiation showed promising results. One of the study carried out by Vishwakarma et al. [176] which created a stable colloidal suspension of sorafenib-gold nanoconjugate (SF-GNP) for reversing drug resistance in HCC cells in a 3D tumor model, is one noteworthy piece of work. After being administered intraperitoneally to mice, the NPs greatly decreased the development of resistant tumor cells with very few negative effects. **Table 12** represents the list of other NPs that overcome from the drug resistance. **Figure 5** is a graphical depiction of new approaches to cancer chemotherapy that rely on NPs to overcome drug resistance.

Table 12 Enlisted other NPs that are able to circumvent drug resistance in cancer cells.

NPs type	Model type	Benefit	Reference
[DD]NpH-T	mouse orthotopic breast cancer	Effectively target multiple drug-resistance pathways	[177]
CSP-NPs	human GBM cell lines	CSP-NPs significantly enhanced accumulations of anticancer drugs and reversed the MDR transporters in human GBM cells.	[178]
SV/DOX@TPGS2k-PLGA NPs	Drug-resistant colon cancer SW620/AD300 cells	Enhanced the effectiveness of DOX in tumors by decreasing P-gp expression and inducing mitochondrial dysfunction	[179]
pH-sensitive pullulan-based NPs	MCF-7/ADR cells	Promotes drug internalization and showed higher drug toxicity	[180]
QZO-DER NPs	DOX-resistant human cancer cell lines like HCT-116, MCF-7, MDA-231, and A549	Reverse the drug resistance	[181]
vitamin E-TPGS NPs	drug resistant MCF-7/ADR cancer cell line	Improved tumor cell inhibitory efficiency and decreased the expression of P-gp	[182]
FA-RES + DTX-NP	DTX resistant PCa cells	Decreased cancer cells' propensity for developing drug resistance	[183]
HA-SLN NPs	MCF7, MDA-MB-231, and MCF7/ADR cells	Cellular uptake and cytotoxicity of HA-SLN were higher in MCF7/ADR cells	[184]

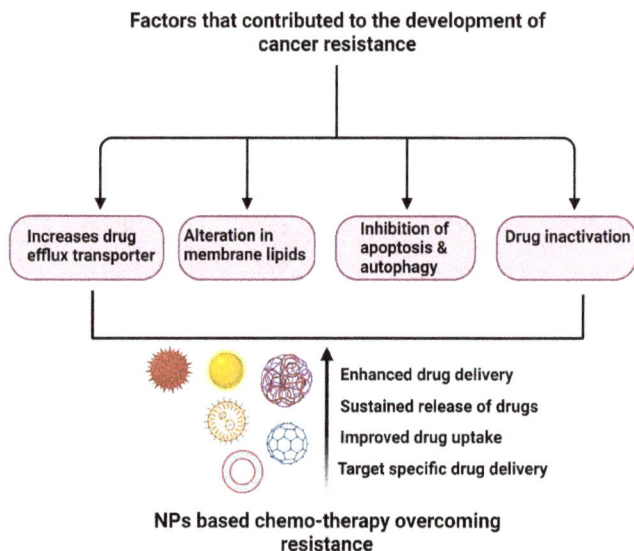

Figure 5: Graphical representation showing emerging NPs based strategies overcoming drug resistance in cancer chemotherapy.

7. List of Approved NPs for cancer treatment

Since the early 1990s, NPs have been commercially available for cancer treatment in clinics, with the polymer-protein conjugate Zinostatin stimalamer approved in Japan for the treatment of liver cancer and liposome based Doxil® commercially available in the United States for ovarian cancer treatment. Multiple organizations across the globe have approved numerous nano-formulations for medical use including metal, liposomes, metal oxide, polymeric micelles and lipid NPs (**Table 13**).

Table 13 *Clinically-approved lists of NPs for cancer therapy.*

Product	Drug encapsulated	NPs type	Cancer type	Benefit	Approval year
Zinostatin stimalamer	Styrene maleic anhydride neocarzinostatin (SMANCS)	Polymer protein conjugate	Primary unresectable hepatocellular carcinoma	Greater accumulation and enhanced EPR effect	1994 (Japan)
Doxil (Caelyx)	Doxorubicin hydrochloride	Pegylated liposome	Ovarian cancer and AIDS-related Kaposi's sarcoma	Prolonged drug circulation time, drug loading and active tumor targeting	1995 (FDA)
Myocet	Doxorubicin	Liposome	Breast cancer	Reduced cardiotoxicity with no loss of antitumor efficacy	2000 (EMA)
Depocyt	Cytarabine	Liposome	Lymphomatous malignant meningitis	Facilitate transport while decreasing toxicity	1999 (FDA)
Nanoxel	Docetaxel	Polymeric micelle	Breast and ovarian cancers, NSCLC, and AIDS-related Kaposi's sarcoma	High therapeutic effectiveness and favorable pharmacokinetics with low adverse effects.	2006 (India)
Marqibo	Vincristine	Liposome	Leukemia	Sustained release of drug	2012 (FDA)
Onivyde	Irinotecan	Liposome	Pancreatic cancer	Improved drug delivery and reduced systemic toxicity	2015 (FDA)
Vyxeos	Daunorubicin and cytarabine	Liposome	Acute myeloid leukemia (AML)	Improved overall survival	2017 (EMA)
Oncaspar	L-Asparaginase	PEGylated conjugate	Acute lymphoblastic leukemia	Drug load stability is enhanced.	2006 (FDA)
Eligard	Leuprolide acetate	Polymeric NPs	Advanced prostate cancer	Sustained release of drug	2002 (FDA)
Abraxane	Paclitaxel	Protein carrier	Various cancers including metastatic and pancreatic cancers HER2+ breast cancer	Overcomes paclitaxel's very poor solubility	2005 (FDA)

8. A novel strategy for cancer treatment using photothermal modality

Chemotherapy, radiation, and surgery are the gold standards of clinical treatment, and they are also the most widely used. Patients undergoing conventional treatment, however, may be at risk for therapeutic failure or long-term secondary side effects. Photothermal therapy (PTT) is a new cancer treatment approach that uses the photosensitizers to kill the tumor with intense heat [185]. Due to its non-invasive nature, better penetration depth, shorter treatment duration, and quicker recovery, PTT is of considerable research value. It's known that the objective of cancer treatment approach is to eliminate malignant cells while minimizing collateral damage to healthy tissue. Photosensitizers are administered to the tumor area and the area is subsequently irradiated with NIR to increase the temperature. Therefore, the simplest approach is to increase the concentration of photosensitizers at the tumor site, creating a concentration differential between normal and tumor tissue that causes the tumor site temperature to rise preferentially. Intratumoral injection is now the most accessible and widely used treatment option. By specifically increasing the temperature at the tumor location, photosensitizers can be concentrated in the tumor tissue in high amounts. Based on the literature, it appears that targeted hyperthermia can increase tumor blood flow and the percentage of the tumor that is perfused. Improved vascular tumor permeability, higher oxygenation, decreased fluid interstitial pressure and restoration of normal physiological pH conditions are only some of the key physiological changes that can emerge from alterations in blood circulation. These alterations in tumor physiology may benefit the therapeutically active TME as well as the accumulation and distribution of nanomedicine. Several mechanisms involved in cancer cell survival can be disrupted by hyperthermia, which also improves chemotherapeutic effectiveness. Hyperthermic stimulation of cancer cells inhibits DNA repair and cell survival signalling leading to the final induction of apoptosis [185]. Simultaneously, the release of antigenic peptides from the died cancer cells activates the immune cells which kills the localized and metastatic cancer cells; hence, enabling long-term immunological memory retention thereby enhancing immune system activation.

PTT and chemotherapy are often used together for better cancer treatment. For example, Zhang et al. were able to add functions to molybdenum telluride nanosheets (MoTe2-PEG-cRGD) and then load them with doxorubicin (DOX), a potent anti-cancer drug. The MoTe2-PEG-cRGD/DOX that was synthesized has a good potentiality to turn light into heat and kill malignant cancer cells upon NIR irradiation. The cyclic arginine-glycine-aspartate (cRGD) motif targets tumors specifically, MoTe2-PEG-cRGD/DOX binds well to malignant cancer cells and has a strong anti-tumor effect [186].

In the same way, Lam et al. used the self-assembly method to make a nanosystem with high photothermal conversion efficiency, good packing and the ability to control its release. When used to treat orthotopic xenotransplantation mice model, the nanosystem with DOX can be used as a solution for both photothermal and chemotherapy. This nanosystem has a good effect on a mouse breast cancer model. In particular, the combined treatment not only gets rid of the primary tumor, but it also gets the immune system of the whole body to work against the secondary tumors. Since PTT can make the body's cancer immune system work better, it can be used to treat tumors by making the body's antitumor immune system work better [187]. Figure demonstrating PTT stimulating the immune system, which destroys primary and secondary tumor cells and builds an immunological memory to prevent cancer recurrence has been represented graphically in **Figure 6**. Various photosensitizer NPs that have been used in combination with chemotherapy for cancer treatment have been listed in **Table 14**.

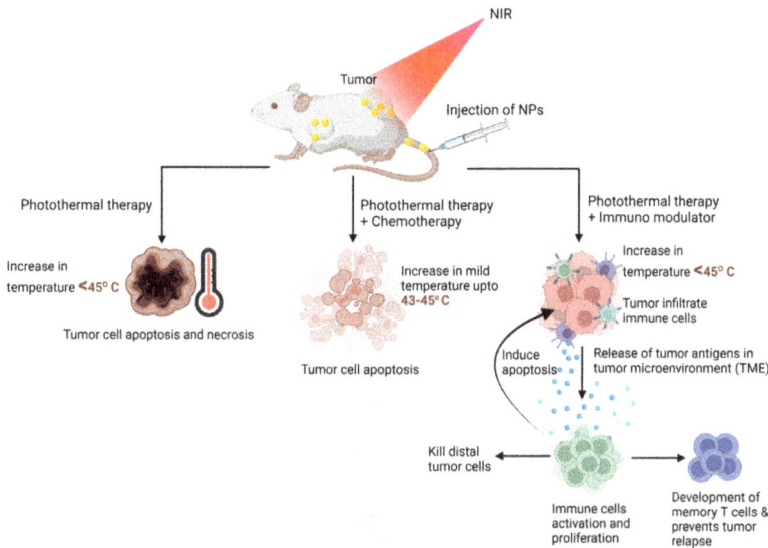

Figure 6: Figure showing PTT's role in activating the immune system, which kills both primary and secondary tumor cells and creates an immunological memory that protects against cancer recurrence.

Table 14 Represents list of other photosensitizers that has been used in combination with other treatment procedure for effective cancer treatment.

NPs type	Model type	Benefit	Reference
AMNDs-LHRH nano-system + NIR	Metastatic prostate tumorigenic mice model	Enhanced the accumulation of the nano-system at the tumor site and improved prostate cancer targeting specificity	[188]
solid lipid NPs co-loaded with gold nanorods and mitoxantrone and functionalized with folic acid + NIR	MCF-7 breast cancer cells	Greater accumulation of NPs and improves light induced drug release	[189]
TiO2−x nanosheets + NIR-II	cervical cancer	NPs exhibited enhanced absorption and photothermal conversion properties	[190]

SA/DOX@hydrogel + NIR	Hela, 4T1 cells and heterotopic transplanted tumor model	Mice main organs are less likely to be damaged and biocompatibility is improved	[191]
QAuNP + NIR	SCC-9 oral CSCs and xenograft mice model	Combined treatment induces apoptosis in OSCC-CSCs by deregulating HSP-70 in the DISC	[132]
QAuNP + NIR	Patient-derived breast CSCs and PDX mice model	Reduced TGF-β-mediated angiogenesis	[133]

9.   Potential risk of NPs to environment

The harmful consequences of NPs are not commonly acknowledged, despite their widespread usage in everyday life, from cancer research to detection and therapy. The beneficial benefits of NPs are emphasized in the scientific literature, whereas their potential drawbacks are mostly ignored. Excellent penetration efficiencies, which vary with NPs size and surface charge, are a result of the NPs diminutive dimensions. Thanks to recent developments in nanotechnology, NPs are now accepted as a reliable drug transporter. While NPs conjugated with the anti-cancer drugs can preferentially target cancer cells, it also has non-intended effects on other healthy tissues and cells [192]. Although NPs are generally well considered for their advantages, some people are concerned about the dangers they may pose in the future. The detrimental influence of NPs on society is a cause for worry, despite the fact that their use has a far more positive effect on societal progress. NPs are discharged into the environment and pollute air, water, and soil having an adverse effect on plants, animals and microbes. This is why the problem of nanowaste is important to address. Therefore, the release of NPs into the environment has negative consequences for plant, microbe, animal, and human health, leading to an overall disruption of the ecosystem.

To combat cancer, organic NPs have been developed as a vaccine platform thanks to their superior biocompatibility, less systemic toxicity, and biodegradability. Their additional benefits include adaptability to a variety of delivery methods, chemical compositions, physical dimensions, and the functionalization of their surfaces with targeted antigens and adjuvants to elicit an immune response. However, these organic NPs have a number of undesirable consequences, including neurotoxicity, organ toxicity, and oxidative stress [192]. Likewise, inorganic NPs are mostly used by the cancer research community for imaging and photothermal therapy application. Inorganic NPs are superior to organic ones for antigen administration because of their smaller particle size, greater stability, controlled tunability, higher permeability, bigger drug loadings, and triggered release. The accumulation of NPs in the liver, kidney, and spleen can cause severe toxicity, making inorganic NPs a double-edged sword [192]. Similar to what was seen with chronic inflammation. Consequently, new approaches need to be created in order to address the aforementioned issues with NPs. Used and unused NPs-containing items should be taken to designated facilities for the proper disposal. The government should introduce new laws to properly clean hospital and pharmaceutical industries effluents before discharging them into waterways. One of the methods for removing nanowaste from soil and water is called phytoremediation, which is facilitated by planting trees of specified species close to the place of NPs disposal. The NPs are easily absorbed by the plant, where they are metabolized after being broken down into its smaller components. Before releasing NPs into the market, the government should implement new laws and procedures to ensure that they have been properly and effectively assessed for any type of risks.

Conclusion and Future perspectives

In recent years, nanotechnology has seen widespread application throughout science & engineering disciplines, and significant research and development have been made in this domain. Many cancer nanomedicine researchers have looked at methods that are mostly constant such as synthesis, characterization and preclinical investigation of its anti-cancer efficacy. Validation efforts are hindered by poor prediction performance and low throughput. The development of cancer nanomedicine from its conception to the present state of the art has been discussed in this review. Advances in nanotechnology have been seen in every discipline, including the medical sector. Numerous pre-clinical and clinical research are continuing, and some of the formulations have received approval. It was preferable to the traditional approach due to its longer circulation period, adequate size range, lower toxicity, and excellent drug accumulation potentiality. Through passive and active targeting, NPs can target cancer cell. NPs may readily penetrate tumor cells because of the EPR effect and their tiny size, and they can keep them in the tumor bed because of the extended circulation time. Additionally, the greatest strategy to slow tumor development is through targeting TME. For specific tumor targeting, theranostic and drug administration, several NPs are being studied. To tackle the present difficulties, new strategies or technologies are always being considered.

Due to the composite structure of NPs, toxicity assessment is very difficult, and there isn't a sufficiently validated model to assess nano-bio interactions, especially nano-immuno interactions. Reproducibility and openness are the other two barriers to approval. In addition, the increased expense and complexity have placed further limitations on the success of the translation. The development of nano-medicines travels a very long and complicated path from the production stage to the approval stage. The development of a medicinal product also requires advancements in equipment and characterisation techniques. Additionally, consideration should be given to the choice of drugs with a combinatorial effect for the particular illness. Biological barriers frequently lead to an unbalance in endo/lysosomal escape, targeted delivery, permeability and penetration. Large-scale manufacture appears to be a difficult undertaking owing to constraints in the sophisticated experimental set-up and a lack of adequate knowledge on the scale-up technologies. Additionally, negative consequences are experienced as a result of the challenges faced in repeating the preparation process and scaling up operations. FDA and EMA often examine the developed nanomedicines, however the lack of a standard operating process for the evaluation of NPs makes the work more difficult. Regarding safety and biocompatibility, it appears essential to monitor the medications for any changes during each stage of the clinical studies. Additionally, the beginning of the clinical trials depends on the drug's features and repeatability being well specified. Combination medicines are now often employed in cancer studies to boost effectiveness. Clinical studies for nanomedicines are mostly intended to assess monotherapy, which might impede their advancement. It is necessary to enhance cancer therapy so that it can target and regulate immune system components. The application of a well-designed NPs as a potential tool for tumor detection is possible. To enhance drug delivery, it is necessary to consider each stage, including internalization, circulation, penetration into the TME, binding to the target, and tumor cell death. Researchers have recently been concentrating on computer algorithms for effective medicine delivery. The two sides of a coin are opportunities and difficulties. To reconcile preclinical and in vitro investigations, further clinical trials are required. These need to contribute to the thorough understanding of how NPs interact with cells and what results. For anticancer administration, the three major drug delivery methods are oral, intravenous, and subcutaneous. New delivery methods include lung delivery, rectal delivery, and inhalation delivery. However, the substantial toxicity brought on by the combined effects of drug deposition and high therapeutic efficacy limits the adoption of these approaches. Due to medication loss in the pulmonary route, substantial therapeutic dosages are further required. Low absorption is another significant issue with rectal administration. Due to mitochondria's role in the development and growth of tumors, mitochondria-based targeting has recently demonstrated considerable potential as a cancer therapeutic. The main difficulties encountered during mitochondrial-targeted treatment include difficulties in clinical trials and bio-safety issues since NPs with positive charges are harmful to normal cells. According to a different research, contact to blood proteins causes liposomal formulations to lose their function over time. Because of their non-uniform properties, the main disadvantage of carbon-based NPs is the difficulty in standardizing; this is why phototherapy is linked with a lack of targeting. Cancer nanomedicine therapy will undoubtedly overcome all of its difficulties in the future and play a significant part in all facets of cancer treatment, including diagnostics, bio-imaging, and drug administration.

Acknowledgement

We want to thank DST-INSPIRE, the Government of India, for providing a research fellowship to Somya Ranjan Dash.

References

[1]C. Pucci, C. Martinelli, G. Ciofani, Innovative approaches for cancer treatment: current perspectives and new challenges, Ecancermedicalscience 13 (2019) 961. https://doi.org/10.3332/ecancer.2019.961

[2]P. Krzyszczyk, A. Acevedo, E.J. Davidoff, L.M. Timmins, I. Marrero-Berrios, M. Patel, C. White, C. Lowe, J.J. Sherba, C. Hartmanshenn, K.M. O'Neill, M.L. Balter, Z.R. Fritz, I.P. Androulakis, R.S. Schloss, M.L. Yarmush, The growing role of precision and personalized medicine for cancer treatment, Technology (Singap World Sci) 6 (2018) 79–100. https://doi.org/10.1142/S2339547818300020

[3]M. Zhang, S. Gao, D. Yang, Y. Fang, X. Lin, X. Jin, Y. Liu, X. Liu, K. Su, K. Shi, Influencing factors and strategies of enhancing nanoparticles into tumors in vivo, Acta Pharmaceutica Sinica B 11 (2021) 2265–2285. https://doi.org/10.1016/j.apsb.2021.03.033

[4]M. Overchuk, R.A. Weersink, B.C. Wilson, G. Zheng, Photodynamic and Photothermal Therapies: Synergy Opportunities for Nanomedicine, ACS Nano 17 (2023) 7979–8003. https://doi.org/10.1021/acsnano.3c00891

[5]B. Aslan, B. Ozpolat, A.K. Sood, G. Lopez-Berestein, NANOTECHNOLOGY IN CANCER THERAPY, J Drug Target 21 (2013) 904–913. https://doi.org/10.3109/1061186X.2013.837469

[6] V.S. Nagtode, C. Cardoza, H.K.A. Yasin, S.N. Mali, S.M. Tambe, P. Roy, K. Singh, A. Goel, P.D. Amin, B.R. Thorat, J.N. Cruz, A.P. Pratap, Green Surfactants (Biosurfactants): A Petroleum-Free Substitute for Sustainability—Comparison, Applications, Market, and Future Prospects, ACS Omega 8 (2023) 11674–11699. https://doi.org/10.1021/acsomega.3c00591

[7] D. Chenthamara, S. Subramaniam, S.G. Ramakrishnan, S. Krishnaswamy, M.M. Essa, F.-H. Lin, M.W. Qoronfleh, Therapeutic efficacy of nanoparticles and routes of administration, Biomater Res 23 (2019) 20. https://doi.org/10.1186/s40824-019-0166-x

[8] S. Palazzolo, S. Bayda, M. Hadla, I. Caligiuri, G. Corona, G. Toffoli, F. Rizzolio, The Clinical Translation of Organic Nanomaterials for Cancer Therapy: A Focus on Polymeric Nanoparticles, Micelles, Liposomes and Exosomes, Curr Med Chem 25 (2018) 4224–4268. https://doi.org/10.2174/0929867324666170830113755

[9] M.A. Subhan, S.S.K. Yalamarty, N. Filipczak, S. Parveen, V.P. Torchilin, Recent Advances in Tumor Targeting via EPR Effect for Cancer Treatment, J Pers Med 11 (2021) 571. https://doi.org/10.3390/jpm11060571

[10] E. Paszko, M.O. Senge, Immunoliposomes, Curr Med Chem 19 (2012) 5239–5277. https://doi.org/10.2174/092986712803833362

[11] W. Jn, M. Rl, Y. Mb, Z. Ds, Liposomes and local hyperthermia: selective delivery of methotrexate to heated tumors, Science (New York, N.Y.) 204 (1979). https://doi.org/10.1126/science.432641

[12] J.N. Cruz, S. Muzammil, A. Ashraf, M.U. Ijaz, M.H. Siddique, R. Abbas, M. Sadia, Saba, S. Hayat, R.R. Lima, A review on mycogenic metallic nanoparticles and their potential role as antioxidant, antibiofilm and quorum quenching agents, Heliyon 10 (2024). https://doi.org/10.1016/j.heliyon.2024.e29500

[13] M.B. Yatvin, J.N. Weinstein, W.H. Dennis, R. Blumenthal, Design of liposomes for enhanced local release of drugs by hyperthermia, Science 202 (1978) 1290–1293. https://doi.org/10.1126/science.364652

[14] T.O.B. Olusanya, R.R. Haj Ahmad, D.M. Ibegbu, J.R. Smith, A.A. Elkordy, Liposomal Drug Delivery Systems and Anticancer Drugs, Molecules 23 (2018) 907. https://doi.org/10.3390/molecules23040907

[15] R.A. Schwendener, Liposomes as vaccine delivery systems: a review of the recent advances, Ther Adv Vaccines 2 (2014) 159–182. https://doi.org/10.1177/2051013614541440

[16] I.N. de F. Ramos, M.F. da Silva, J.M.S. Lopes, J.N. Cruz, F.S. Alves, J. de A.R. do Rego, M.L. da Costa, P.P. de Assumpção, D. do S. Barros Brasil, A.S. Khayat, Extraction, Characterization, and Evaluation of the Cytotoxic Activity of Piperine in Its Isolated form and in Combination with Chemotherapeutics against Gastric Cancer, Molecules 28 (2023) 5587. https://doi.org/10.3390/molecules28145587

[17] G. Ricciuti, E. Finolezzi, S. Luciani, E. Ranucci, M. Federico, M. Di Nicola, I.A.L. Zecca, F. Angrilli, Combination of rituximab and nonpegylated liposomal doxorubicin (R-NPLD) as front-line therapy for aggressive non-Hodgkin lymphoma (NHL) in patients 80 years of age or older: a single-center retrospective study, Hematol Oncol 36 (2018) 44–48. https://doi.org/10.1002/hon.2386

[18]   J. Zhou, W.-Y. Zhao, X. Ma, R.-J. Ju, X.-Y. Li, N. Li, M.-G. Sun, J.-F. Shi, C.-X. Zhang, W.-L. Lu, The anticancer efficacy of paclitaxel liposomes modified with mitochondrial targeting conjugate in resistant lung cancer, Biomaterials 34 (2013) 3626–3638. https://doi.org/10.1016/j.biomaterials.2013.01.078

[19]   M. Legut, D. Lipka, N. Filipczak, A. Piwoni, A. Kozubek, J. Gubernator, Anacardic acid enhances the anticancer activity of liposomal mitoxantrone towards melanoma cell lines - in vitro studies, Int J Nanomedicine 9 (2014) 653–668. https://doi.org/10.2147/IJN.S54911

[20]   A. Hardiansyah, L.-Y. Huang, M.-C. Yang, T.-Y. Liu, S.-C. Tsai, C.-Y. Yang, C.-Y. Kuo, T.-Y. Chan, H.-M. Zou, W.-N. Lian, C.-H. Lin, Magnetic liposomes for colorectal cancer cells therapy by high-frequency magnetic field treatment, Nanoscale Res Lett 9 (2014) 497. https://doi.org/10.1186/1556-276X-9-497

[21]   J.N. Mock, L.J. Costyn, S.L. Wilding, R.D. Arnold, B.S. Cummings, Evidence for distinct mechanisms of uptake and antitumor activity of secretory phospholipase A2 responsive liposome in prostate cancer, Integr Biol (Camb) 5 (2013) 172–182. https://doi.org/10.1039/c2ib20108a

[22]   V. Bensa, E. Calarco, E. Giusto, P. Perri, M.V. Corrias, M. Ponzoni, C. Brignole, F. Pastorino, Retinoids Delivery Systems in Cancer: Liposomal Fenretinide for Neuroectodermal-Derived Tumors, Pharmaceuticals (Basel) 14 (2021) 854. https://doi.org/10.3390/ph14090854

[23]   C. Oerlemans, W. Bult, M. Bos, G. Storm, J.F.W. Nijsen, W.E. Hennink, Polymeric Micelles in Anticancer Therapy: Targeting, Imaging and Triggered Release, Pharm Res 27 (2010) 2569–2589. https://doi.org/10.1007/s11095-010-0233-4

[24]   J.K. Patra, G. Das, L.F. Fraceto, E.V.R. Campos, M. del P. Rodriguez-Torres, L.S. Acosta-Torres, L.A. Diaz-Torres, R. Grillo, M.K. Swamy, S. Sharma, S. Habtemariam, H.-S. Shin, Nano based drug delivery systems: recent developments and future prospects, Journal of Nanobiotechnology 16 (2018) 71. https://doi.org/10.1186/s12951-018-0392-8

[25]   M.H. Sarfraz, M. Zubair, B. Aslam, A. Ashraf, M.H. Siddique, S. Hayat, J.N. Cruz, S. Muzammil, M. Khurshid, M.F. Sarfraz, A. Hashem, T.M. Dawoud, G.D. Avila-Quezada, E.F. Abd_Allah, Comparative analysis of phyto-fabricated chitosan, copper oxide, and chitosan-based CuO nanoparticles: antibacterial potential against Acinetobacter baumannii isolates and anticancer activity against HepG2 cell lines, Frontiers in Microbiology 14 (2023) 1188743. https://doi.org/10.3389/fmicb.2023.1188743

[26]   Y. Bae, T.A. Diezi, A. Zhao, G.S. Kwon, Mixed polymeric micelles for combination cancer chemotherapy through the concurrent delivery of multiple chemotherapeutic agents, J Control Release 122 (2007) 324–330. https://doi.org/10.1016/j.jconrel.2007.05.038

[27]   J.R. Hasenstein, H.-C. Shin, K. Kasmerchak, D. Buehler, G.S. Kwon, K.R. Kozak, Anti-tumor Activity of Triolimus: A Novel Multi-Drug Loaded Micelle Containing Paclitaxel, Rapamycin and 17-AAG, Mol Cancer Ther 11 (2012) 2233–2242. https://doi.org/10.1158/1535-7163.MCT-11-0987

[28]   H.S. Na, Y.K. Lim, Y.-I. Jeong, H.S. Lee, Y.J. Lim, M.S. Kang, C.-S. Cho, H.C. Lee, Combination antitumor effects of micelle-loaded anticancer drugs in a CT-26 murine

colorectal carcinoma model, Int J Pharm 383 (2010) 192–200.
https://doi.org/10.1016/j.ijpharm.2009.08.041

[29]   H. Wang, Y. Zhao, Y. Wu, Y. Hu, K. Nan, G. Nie, H. Chen, Enhanced anti-tumor efficacy by co-delivery of doxorubicin and paclitaxel with amphiphilic methoxy PEG-PLGA copolymer nanoparticles, Biomaterials 32 (2011) 8281–8290.
https://doi.org/10.1016/j.biomaterials.2011.07.032

[30]   Y. Han, Z. He, A. Schulz, T.K. Bronich, R. Jordan, R. Luxenhofer, A.V. Kabanov, Synergistic Combinations of Multiple Chemotherapeutic Agents in High Capacity Poly(2-oxazoline) Micelles, Mol Pharm 9 (2012) 2302–2313. https://doi.org/10.1021/mp300159u

[31]   U. Katragadda, Q. Teng, B.M. Rayaprolu, T. Chandran, C. Tan, Multi-drug delivery to tumor cells via micellar nanocarriers, Int J Pharm 419 (2011) 281–286.
https://doi.org/10.1016/j.ijpharm.2011.07.033

[32]   S.S. Desale, S.M. Cohen, Y. Zhao, A.V. Kabanov, T.K. Bronich, Biodegradable hybrid polymer micelles for combination drug therapy in ovarian cancer, J Control Release 171 (2013) 339–348. https://doi.org/10.1016/j.jconrel.2013.04.026

[33]   H. Cho, G.S. Kwon, Thermosensitive poly-(d,l-lactide-co-glycolide)-block-poly(ethylene glycol)-block-poly-(d,l-lactide-co-glycolide) hydrogels for multi-drug delivery, J Drug Target 22 (2014) 669–677. https://doi.org/10.3109/1061186X.2014.931406

[34]   A. Pugazhendhi, T.N.J.I. Edison, I. Karuppusamy, B. Kathirvel, Inorganic nanoparticles: a potential cancer therapy for human welfare, International Journal of Pharmaceutics 539 (2018) 104–111.

[35]   S. Pushkar, A. Philip, K. Pathak, D. Pathak, Dendrimers: Nanotechnology derived novel polymers in drug delivery, Indian Journal of Pharmaceutical Education and Research 40 (2006) 153.

[36]   L. Palmerston Mendes, J. Pan, V.P. Torchilin, Dendrimers as Nanocarriers for Nucleic Acid and Drug Delivery in Cancer Therapy, Molecules 22 (2017) 1401.
https://doi.org/10.3390/molecules22091401

[37]   Z. Bober, D. Bartusik-Aebisher, D. Aebisher, Application of Dendrimers in Anticancer Diagnostics and Therapy, Molecules 27 (2022) 3237.
https://doi.org/10.3390/molecules27103237

[38]   H.L. Crampton, E.E. Simanek, Dendrimers as drug delivery vehicles: non-covalent interactions of bioactive compounds with dendrimers, Polym Int 56 (2007) 489–496.
https://doi.org/10.1002/pi.2230

[39]   M.A. Fuertes, C. Alonso, J.M. Pérez, Biochemical modulation of cisplatin mechanisms of action: enhancement of antitumor activity and circumvention of drug resistance, Chemical Reviews 103 (2003) 645–662

[40]   D.B. Longley, D.P. Harkin, P.G. Johnston, 5-fluorouracil: mechanisms of action and clinical strategies, Nature Reviews Cancer 3 (2003) 330–338.

[41] P.-S. Lai, P.-J. Lou, C.-L. Peng, C.-L. Pai, W.-N. Yen, M.-Y. Huang, T.-H. Young, M.-J. Shieh, Doxorubicin delivery by polyamidoamine dendrimer conjugation and photochemical internalization for cancer therapy, Journal of Controlled Release 122 (2007) 39–46.

[42] P. Kozub, M. Simaljakova, Systemic therapy of psoriasis: methotrexate, Bratisl Lek Listy 112 (2011) 390–394.

[43] K. Wróbel, A. Deręgowska, G. Betlej, M. Walczak, M. Wnuk, A. Lewińska, S. Wołowiec, Cytarabine and dexamethasone-PAMAM dendrimer di-conjugate sensitizes human acute myeloid leukemia cells to apoptotic cell death, Journal of Drug Delivery Science and Technology 81 (2023) 104242. https://doi.org/10.1016/j.jddst.2023.104242

[44] F. Hassani, A. Heydarinasab, H. Ahmad Panahi, E. Moniri, Surface modification of tungsten disulfide nanosheets with pH/Thermosensitive polymer and polyethylenimine dendrimer for near-infrared triggered drug delivery of letrozole, Journal of Molecular Liquids 371 (2023) 121058. https://doi.org/10.1016/j.molliq.2022.121058

[45] C. Ni, Z. Ouyang, G. Li, J. Liu, X. Cao, L. Zheng, X. Shi, R. Guo, A tumor microenvironment-responsive core-shell tecto dendrimer nanoplatform for magnetic resonance imaging-guided and cuproptosis-promoted chemo-chemodynamic therapy, Acta Biomaterialia (2023). https://doi.org/10.1016/j.actbio.2023.04.003

[46] S. Michlewska, M. Maly, D. Wójkowska, K. Karolczak, E. Skiba, M. Hołota, M. Kubczak, P. Ortega, C. Watala, F. Javier de la Mata, M. Bryszewska, M. Ionov, Carbosilane ruthenium metallodendrimer as alternative anti-cancer drug carrier in triple negative breast cancer mouse model: A preliminary study, International Journal of Pharmaceutics 636 (2023) 122784. https://doi.org/10.1016/j.ijpharm.2023.122784

[47] S. Shen, Y. Gao, Z. Ouyang, B. Jia, M. Shen, X. Shi, Photothermal-triggered dendrimer nanovaccines boost systemic antitumor immunity, Journal of Controlled Release 355 (2023) 171–183. https://doi.org/10.1016/j.jconrel.2023.01.076

[48] S. Huang, X. Huang, H. Yan, Peptide dendrimers as potentiators of conventional chemotherapy in the treatment of pancreatic cancer in a mouse model, European Journal of Pharmaceutics and Biopharmaceutics 170 (2022) 121–132. https://doi.org/10.1016/j.ejpb.2021.11.005

[49] A. Lewińska, K. Wróbel, D. Błoniarz, J. Adamczyk-Grochala, S. Wołowiec, M. Wnuk, Lapatinib- and fulvestrant-PAMAM dendrimer conjugates promote apoptosis in chemotherapy-induced senescent breast cancer cells with different receptor status, Biomaterials Advances 140 (2022) 213047. https://doi.org/10.1016/j.bioadv.2022.213047

[50] D.E. Ybarra, M.N. Calienni, L.F.B. Ramirez, E.T.A. Frias, C. Lillo, S. del V. Alonso, J. Montanari, F.C. Alvira, Vismodegib in PAMAM-dendrimers for potential theragnosis in skin cancer, OpenNano 7 (2022) 100053. https://doi.org/10.1016/j.onano.2022.100053

[51] M. Gonçalves, V. Kairys, J. Rodrigues, H. Tomás, Polyester Dendrimers Based on Bis-MPA for Doxorubicin Delivery, Biomacromolecules 23 (2021) 20–33.

[52] F. Zhu, L. Xu, X. Li, Z. Li, J. Wang, H. Chen, X. Li, Y. Gao, Co-delivery of gefitinib and hematoporphyrin by aptamer-modified fluorinated dendrimer for hypoxia alleviation and

enhanced synergistic chemo-photodynamic therapy of NSCLC, European Journal of Pharmaceutical Sciences 167 (2021) 106004. https://doi.org/10.1016/j.ejps.2021.106004

[53]   G. Ma, X. Du, J. Zhu, F. Xu, H. Yu, J. Li, Multi-functionalized dendrimers for targeted co-delivery of sorafenib and paclitaxel in liver cancers, Journal of Drug Delivery Science and Technology 63 (2021) 102493. https://doi.org/10.1016/j.jddst.2021.102493

[54]   T. Barrett, G. Ravizzini, P.L. Choyke, H. Kobayashi, Dendrimers in medical nanotechnology, IEEE Eng Med Biol Mag 28 (2009) 12–22. https://doi.org/10.1109/MEMB.2008.931012

[55]   S. Bhattacharyya, R.A. Kudgus, R. Bhattacharya, P. Mukherjee, Inorganic nanoparticles in cancer therapy, Pharm Res 28 (2011) 237–259. https://doi.org/10.1007/s11095-010-0318-0

[56]   D. Kovács, N. Igaz, M.K. Gopisetty, M. Kiricsi, Cancer Therapy by Silver Nanoparticles: Fiction or Reality?, Int J Mol Sci 23 (2022) 839. https://doi.org/10.3390/ijms23020839

[57]   G.M. Vlăsceanu, Ş. Marin, R.E. Ţiplea, I.R. Bucur, M. Lemnaru, M.M. Marin, A.M. Grumezescu, E. Andronescu, Chapter 2 - Silver nanoparticles in cancer therapy, in: A.M. Grumezescu (Ed.), Nanobiomaterials in Cancer Therapy, William Andrew Publishing, 2016: pp. 29–56. https://doi.org/10.1016/B978-0-323-42863-7.00002-5

[58]   X.-F. Zhang, W. Shen, S. Gurunathan, Silver Nanoparticle-Mediated Cellular Responses in Various Cell Lines: An in Vitro Model, Int J Mol Sci 17 (2016) 1603. https://doi.org/10.3390/ijms17101603

[59]   G. Nadar Rajivgandhi, G. Chackaravarthi, G. Ramachandran, C. Kanisha Chelliah, M. Maruthupandy, M.S. Alharbi, N.S. Alharbi, J.M. Khaled, W.-J. Li, Morphological damage and increased ROS production of biosynthesized silver nanoparticle against MCF-7 breast cancer cells through in vitro approaches, Journal of King Saud University - Science 34 (2022) 101795. https://doi.org/10.1016/j.jksus.2021.101795

[60]   T.T.H. Le, T.H. Ngo, T.H. Nguyen, V.H. Hoang, V.H. Nguyen, P.H. Nguyen, Anti-cancer activity of green synthesized silver nanoparticles using Ardisia gigantifolia leaf extract against gastric cancer cells, Biochemical and Biophysical Research Communications 661 (2023) 99–107. https://doi.org/10.1016/j.bbrc.2023.04.037

[61]   Y. Peng, W. Ni, T. Ni, P. Xu, C. Gu, W. Yu, A. Xie, M. Yao, Silver nanoparticles induce cytotoxicity by releasing Ag + from the lysosome and increasing lysosomal membrane permeabilit, (2023). https://doi.org/10.21203/rs.3.rs-2938573/v1

[62]   L. Fageria, V. Pareek, R.V. Dilip, A. Bhargava, S.S. Pasha, I.R. Laskar, H. Saini, S. Dash, R. Chowdhury, J. Panwar, Biosynthesized protein-capped silver nanoparticles induce ros-dependent proapoptotic signals and prosurvival autophagy in cancer cells, ACS Omega 2 (2017) 1489–1504

[63]   F.S. Alves, J.N. Cruz, I.N. de Farias Ramos, D.L. do Nascimento Brandão, R.N. Queiroz, G.V. da Silva, G.V. da Silva, M.F. Dolabela, M.L. da Costa, A.S. Khayat, J. de Arimatéia Rodrigues do Rego, D. do Socorro Barros Brasil, Evaluation of Antimicrobial Activity and Cytotoxicity Effects of Extracts of Piper nigrum L. and Piperine, Separations 10 (2023) 21. https://doi.org/10.3390/separations10010021

[64]   K.C. Hembram, S. Chatterjee, C. Sethy, D. Nayak, R. Pradhan, S. Molla, B.K. Bindhani, C.N. Kundu, Comparative and mechanistic study on the anticancer activity of quinacrine-based silver and gold hybrid nanoparticles in head and neck cancer, Molecular Pharmaceutics 16 (2019) 3011–3023.

[65]   E. Locatelli, M. Naddaka, C. Uboldi, G. Loudos, E. Fragogeorgi, V. Molinari, A. Pucci, T. Tsotakos, D. Psimadas, J. Ponti, Targeted delivery of silver nanoparticles and alisertib: in vitro and in vivo synergistic effect against glioblastoma, Nanomedicine 9 (2014) 839–849.

[66]   A. Espinosa, A. Curcio, S. Cabana, G. Radtke, M. Bugnet, J. Kolosnjaj-Tabi, C. Péchoux, C. Alvarez-Lorenzo, G.A. Botton, A.K. Silva, Intracellular biodegradation of Ag nanoparticles, storage in ferritin, and protection by a Au shell for enhanced photothermal therapy, ACS Nano 12 (2018) 6523–6535.

[67]   Z. Liu, H. Tan, X. Zhang, F. Chen, Z. Zhou, X. Hu, S. Chang, P. Liu, H. Zhang, Enhancement of radiotherapy efficacy by silver nanoparticles in hypoxic glioma cells, Artificial Cells, Nanomedicine, and Biotechnology 46 (2018) 922–930.

[68]   J. Liu, Y. Zhao, Q. Guo, Z. Wang, H. Wang, Y. Yang, Y. Huang, TAT-modified nanosilver for combating multidrug-resistant cancer, Biomaterials 33 (2012) 6155–6161.

[69]   E. Liu, M. Zhang, H. Cui, J. Gong, Y. Huang, J. Wang, Y. Cui, W. Dong, L. Sun, H. He, Tat-functionalized Ag-Fe3O4 nano-composites as tissue-penetrating vehicles for tumor magnetic targeting and drug delivery, Acta Pharmaceutica Sinica B 8 (2018) 956–968.

[70]   Y. Tang, J. Liang, A. Wu, Y. Chen, P. Zhao, T. Lin, M. Zhang, Q. Xu, J. Wang, Y. Huang, Co-delivery of trichosanthin and albendazole by nano-self-assembly for overcoming tumor multidrug-resistance and metastasis, ACS Applied Materials & Interfaces 9 (2017) 26648–26664.

[71]   J. Zhao, P. Liu, J. Ma, D. Li, H. Yang, W. Chen, Y. Jiang, Enhancement of radiosensitization by silver nanoparticles functionalized with polyethylene glycol and aptamer As1411 for glioma irradiation therapy, International Journal of Nanomedicine (2019) 9483–9496.

[72]   C. Nie, P. Du, H. Zhao, H. Xie, Y. Li, L. Yao, Y. Shi, L. Hu, S. Si, M. Zhang, Ag@TiO2 nanoprisms with highly efficient near-infrared photothermal conversion for melanoma therapy, Chemistry–An Asian Journal 15 (2020) 148–155.

[73]   K. Habiba, K. Aziz, K. Sanders, C.M. Santiago, L.S.K. Mahadevan, V. Makarov, B.R. Weiner, G. Morell, S. Krishnan, Enhancing colorectal cancer radiation therapy efficacy using silver nanoprisms decorated with graphene as radiosensitizers, Scientific Reports 9 (2019) 1–9.

[74]   K. Sztandera, M. Gorzkiewicz, B. Klajnert-Maculewicz, Gold Nanoparticles in Cancer Treatment, Mol. Pharmaceutics 16 (2019) 1–23. https://doi.org/10.1021/acs.molpharmaceut.8b00810

[75]   G.F. Paciotti, L. Myer, D. Weinreich, D. Goia, N. Pavel, R.E. McLaughlin, L. Tamarkin, Colloidal gold: a novel nanoparticle vector for tumor directed drug delivery, Drug Deliv 11 (2004) 169–183. https://doi.org/10.1080/10717540490433895

[76]   Y. Yang, X. Zheng, L. Chen, X. Gong, H. Yang, X. Duan, Y. Zhu, Multifunctional Gold Nanoparticles in Cancer Diagnosis and Treatment, Int J Nanomedicine 17 (2022) 2041–2067. https://doi.org/10.2147/IJN.S355142

[77]   T.A. Salamone, L. Rutigliano, B. Pennacchi, S. Cerra, R. Matassa, S. Nottola, F. Sciubba, C. Battocchio, M. Marsotto, A. Del Giudice, A. Chumakov, A. Davydok, S. Grigorian, G. Canettieri, E. Agostinelli, I. Fratoddi, Thiol functionalised gold nanoparticles loaded with methotrexate for cancer treatment: from synthesis to in vitro studies on neuroblastoma cell lines, Journal of Colloid and Interface Science (2023). https://doi.org/10.1016/j.jcis.2023.06.078

[78]   G. Niu, L. Zhao, Y. Wang, Y. Jiang, PDA/gold nanorod-based nanoparticles for synergistic genetic and photothermal combination therapy for cancer treatment, ChemPhysMater 2 (2023) 83–89. https://doi.org/10.1016/j.chphma.2022.07.001

[79]   Z. Khademi, P. Lavaee, M. Ramezani, M. Alibolandi, K. Abnous, S.M. Taghdisi, Co-delivery of doxorubicin and aptamer against Forkhead box M1 using chitosan-gold nanoparticles coated with nucleolin aptamer for synergistic treatment of cancer cells, Carbohydrate Polymers 248 (2020) 116735. https://doi.org/10.1016/j.carbpol.2020.116735

[80]   S. Yu, J. Zhang, S. Liu, Z. Ma, H. Sun, Z. Liu, L. Wang, Self-assembly synthesis of flower-like gold nanoparticles for photothermal treatment of cancer, Colloids and Surfaces A: Physicochemical and Engineering Aspects 647 (2022) 129163. https://doi.org/10.1016/j.colsurfa.2022.129163

[81]   J. Wan, X. Ma, D. Xu, B. Yang, S. Yang, S. Han, Docetaxel-decorated anticancer drug and gold nanoparticles encapsulated apatite carrier for the treatment of liver cancer, Journal of Photochemistry and Photobiology B: Biology 185 (2018) 73–79. https://doi.org/10.1016/j.jphotobiol.2018.05.021

[82]   C.S. Kumar, M.D. Raja, D.S. Sundar, M. Gover Antoniraj, K. Ruckmani, Hyaluronic acid co-functionalized gold nanoparticle complex for the targeted delivery of metformin in the treatment of liver cancer (HepG2 cells), Carbohydrate Polymers 128 (2015) 63–74. https://doi.org/10.1016/j.carbpol.2015.04.010

[83]   D.-S. Hsieh, H. Wang, S.-W. Tan, Y.-H. Huang, C.-Y. Tsai, M.-K. Yeh, C.-J. Wu, The treatment of bladder cancer in a mouse model by epigallocatechin-3-gallate-gold nanoparticles, Biomaterials 32 (2011) 7633–7640. https://doi.org/10.1016/j.biomaterials.2011.06.073

[84]   C.U. Tunç, D.Y. Öztaş, D. Uzunoğlu, Ö.F. Bayrak, M. Çulha, Silencing breast cancer genes using morpholino embedded DNA-tile-AuNPs nanostructures, Human Gene Therapy 30 (2019) 1547–1558.

[85]   Y. Liu, M. Xu, Y. Zhao, X. Chen, X. Zhu, C. Wei, S. Zhao, J. Liu, X. Qin, Flower-like gold nanoparticles for enhanced photothermal anticancer therapy by the delivery of pooled siRNA to inhibit heat shock stress response, Journal of Materials Chemistry B 7 (2019) 586–597.

[86]   S. Huang, Y. Liu, X. Xu, M. Ji, Y. Li, C. Song, S. Duan, Y. Hu, Triple therapy of hepatocellular carcinoma with microRNA-122 and doxorubicin co-loaded functionalized gold nanocages, Journal of Materials Chemistry B 6 (2018) 2217–2229.

Materials Research Forum LLC
https://doi.org/10.21741/9781644903339-2

[87]   P. Thevenot, J. Cho, D. Wavhal, R.B. Timmons, L. Tang, Surface chemistry influences cancer killing effect of TiO2 nanoparticles, Nanomedicine: Nanotechnology, Biology and Medicine 4 (2008) 226–236. https://doi.org/10.1016/j.nano.2008.04.001

[88]   R.K. Kawassaki, M. Romano, N. Dietrich, K. Araki, Titanium and Iron Oxide Nanoparticles for Cancer Therapy: Surface Chemistry and Biological Implications, Frontiers in Nanotechnology 3 (2021). https://www.frontiersin.org/articles/10.3389/fnano.2021.735434 (accessed July 4, 2023).

[89]   A.M. Itoo, M. Paul, S.G. Padaga, B. Ghosh, S. Biswas, Nanotherapeutic Intervention in Photodynamic Therapy for Cancer, ACS Omega 7 (2022) 45882–45909. https://doi.org/10.1021/acsomega.2c05852

[90]   D. Ziental, B. Czarczynska-Goslinska, D.T. Mlynarczyk, A. Glowacka-Sobotta, B. Stanisz, T. Goslinski, L. Sobotta, Titanium Dioxide Nanoparticles: Prospects and Applications in Medicine, Nanomaterials (Basel) 10 (2020) 387. https://doi.org/10.3390/nano10020387

[91]   S. Senapati, A.K. Mahanta, S. Kumar, P. Maiti, Controlled drug delivery vehicles for cancer treatment and their performance, Signal Transduct Target Ther 3 (2018) 7. https://doi.org/10.1038/s41392-017-0004-3

[92]   J. Violet Mary, C. Pragathiswaran, N. Anusuya, Photocatalytic, degradation, sensing of Pb2+ using titanium nanoparticles synthesized via plant extract of Cissusquadrangularis: In-vitroanalysis of microbial and anti-cancer activities, Journal of Molecular Structure 1236 (2021) 130144. https://doi.org/10.1016/j.molstruc.2021.130144

[93]   J. Shen, J. Karges, K. Xiong, Y. Chen, L. Ji, H. Chao, Cancer cell membrane camouflaged iridium complexes functionalized black-titanium nanoparticles for hierarchical-targeted synergistic NIR-II photothermal and sonodynamic therapy, Biomaterials 275 (2021) 120979. https://doi.org/10.1016/j.biomaterials.2021.120979

[94]   H. Kim, D. Jeon, S. Oh, K. Nam, S. Son, M.C. Gye, I. Shin, Titanium dioxide nanoparticles induce apoptosis by interfering with EGFR signaling in human breast cancer cells, Environmental Research 175 (2019) 117–123. https://doi.org/10.1016/j.envres.2019.05.001

[95]   M. Pourmadadi, A. Tajiki, M. Abdouss, A green approach for preparation of polyacrylic acid/starch incorporated with titanium dioxide nanocomposite as a biocompatible platform for curcumin delivery to breast cancer cells, International Journal of Biological Macromolecules 242 (2023) 124785. https://doi.org/10.1016/j.ijbiomac.2023.124785

[96]   S.A. Ferraro, M.G. Domingo, A. Etcheverrito, D.G. Olmedo, D.R. Tasat, Neurotoxicity mediated by oxidative stress caused by titanium dioxide nanoparticles in human neuroblastoma (SH-SY5Y) cells, Journal of Trace Elements in Medicine and Biology 57 (2020) 126413. https://doi.org/10.1016/j.jtemb.2019.126413

[97]   Q. Feng, Y. Liu, J. Huang, K. Chen, J. Huang, K. Xiao, Uptake, distribution, clearance, and toxicity of iron oxide nanoparticles with different sizes and coatings, Scientific Reports 8 (2018) 1–13.

[98]   A.J. Giustini, A.A. Petryk, S.M. Cassim, J.A. Tate, I. Baker, P.J. Hoopes, Magnetic Nanoparticle Hyperthermia in Cancer Treatment, Nano Life 1 (2010) 10.1142/S1793984410000067. https://doi.org/10.1142/S1793984410000067

[99]   K. Maier-Hauff, F. Ulrich, D. Nestler, H. Niehoff, P. Wust, B. Thiesen, H. Orawa, V. Budach, A. Jordan, Efficacy and safety of intratumoral thermotherapy using magnetic iron-oxide nanoparticles combined with external beam radiotherapy on patients with recurrent glioblastoma multiforme, J Neurooncol 103 (2011) 317–324. https://doi.org/10.1007/s11060-010-0389-0

[100]   Y. Pu, H. Ke, C. Wu, S. Xu, Y. Xiao, L. Han, G. Lyu, S. Li, Superparamaetic iron oxide nanoparticles target BxPC-3 cells and silence MUC4 for theranostics of pancreatic cancer, Biochimica et Biophysica Acta (BBA) - General Subjects 1867 (2023) 130383. https://doi.org/10.1016/j.bbagen.2023.130383

[101]   C. Nascimento, F. Castro, M. Domingues, A. Lage, É. Alves, R. de Oliveira, C. de Melo, C. Eduardo Calzavara-Silva, B. Sarmento, Reprogramming of tumor-associated macrophages by polyaniline-coated iron oxide nanoparticles applied to treatment of breast cancer, International Journal of Pharmaceutics 636 (2023) 122866. https://doi.org/10.1016/j.ijpharm.2023.122866

[102]   T.-L. Ho, C. Mutalik, L. Rethi, H.-N.T. Nguyen, P.-R. Jheng, C.-C. Wong, T.-S. Yang, T.T. Nguyen, B.W. Mansel, C.-A. Wang, E.-Y. Chuang, Cancer-targeted fucoidan-iron oxide nanoparticles for synergistic chemotherapy/chemodynamic theranostics through amplification of P-selectin and oxidative stress, International Journal of Biological Macromolecules 235 (2023) 123821. https://doi.org/10.1016/j.ijbiomac.2023.123821

[103]   Z. Wang, Y. Wang, H. Li, Y. Lan, Z. Zeng, J. Yao, M. Li, H. Xia, Fabrication of Etoposide-loaded superparamagnetic iron oxide nanoparticles (SPIONs) induced apoptosis in glioma cancer cells, Process Biochemistry 128 (2023) 126–136. https://doi.org/10.1016/j.procbio.2023.02.026

[104]   B. Freis, M.D.L.Á. Ramírez, S. Furgiuele, F. Journe, C. Cheignon, L.J. Charbonnière, C. Henoumont, C. Kiefer, D. Mertz, C. Affolter-Zbaraszczuk, F. Meyer, S. Saussez, S. Laurent, M. Tasso, S. Bégin-Colin, Bioconjugation studies of an EGF-R targeting ligand on dendronized iron oxide nanoparticles to target head and neck cancer cells, International Journal of Pharmaceutics 635 (2023) 122654. https://doi.org/10.1016/j.ijpharm.2023.122654

[105]   Y. Li, X. Wang, B. Ding, C. He, C. Zhang, J. Li, H. Wang, Z. Li, G. Wang, Y. Wang, H. Chen, P. Ma, B. Sun, Synergistic Apoptosis-Ferroptosis: Oxaliplatin loaded amorphous iron oxide nanoparticles for High-efficiency therapy of orthotopic pancreatic cancer and CA19-9 levels decrease, Chemical Engineering Journal 464 (2023) 142690. https://doi.org/10.1016/j.cej.2023.142690

[106]   S. Majeed, N.A.B. Mohd Rozi, M. Danish, M.N. Mohamad Ibrahim, E.L. Joel, In vitro apoptosis and molecular response of engineered green iron oxide nanoparticles with l-arginine in MDA-MB-231 breast cancer cells, Journal of Drug Delivery Science and Technology 80 (2023) 104185. https://doi.org/10.1016/j.jddst.2023.104185

[107] J. Nayak, K.S. Prajapati, S. Kumar, V.K. Vashistha, S.K. Sahoo, R. Kumar, Thiolated β-cyclodextrin modified iron oxide nanoparticles for effective targeted cancer therapy, Materials Today Communications 33 (2022) 104644. https://doi.org/10.1016/j.mtcomm.2022.104644

[108] S. Bhal, C.N. Kundu, Targeting crosstalk of signaling pathways in cancer stem cells: a promising approach for development of novel anti-cancer therapeutics, Med Oncol 40 (2023) 82. https://doi.org/10.1007/s12032-022-01905-7

[109] S. Muzammil, J. Neves Cruz, R. Mumtaz, I. Rasul, S. Hayat, M.A. Khan, A.M. Khan, M.U. Ijaz, R.R. Lima, M. Zubair, Effects of Drying Temperature and Solvents on In Vitro Diabetic Wound Healing Potential of Moringa oleifera Leaf Extracts, Molecules 28 (2023) 710. https://doi.org/10.3390/molecules28020710

[110] B. Lu, X. Huang, J. Mo, W. Zhao, Drug Delivery Using Nanoparticles for Cancer Stem-Like Cell Targeting, Frontiers in Pharmacology 7 (2016). https://www.frontiersin.org/articles/10.3389/fphar.2016.00084 (accessed July 4, 2023).

[111] M.F. Attia, N. Anton, J. Wallyn, Z. Omran, T.F. Vandamme, An overview of active and passive targeting strategies to improve the nanocarriers efficiency to tumour sites, Journal of Pharmacy and Pharmacology 71 (2019) 1185–1198. https://doi.org/10.1111/jphp.13098

[112] Z. Yang, N. Sun, R. Cheng, C. Zhao, J. Liu, Z. Tian, Hybrid nanoparticles coated with hyaluronic acid lipoid for targeted co-delivery of paclitaxel and curcumin to synergistically eliminate breast cancer stem cells, Journal of Materials Chemistry B 5 (2017) 6762–6775.

[113] J. Li, W. Xu, X. Yuan, H. Chen, H. Song, B. Wang, J. Han, Polymer–lipid hybrid anti-HER2 nanoparticles for targeted salinomycin delivery to HER2-positive breast cancer stem cells and cancer cells, International Journal of Nanomedicine (2017) 6909–6921.

[114] A. Abou-ElNaga, G. Mutawa, I.M. El-Sherbiny, H. Abd-ElGhaffar, A.A. Allam, J. Ajarem, S.A. Mousa, Novel nano-therapeutic approach actively targets human ovarian cancer stem cells after xenograft into nude mice, International Journal of Molecular Sciences 18 (2017) 813.

[115] R.K. Verma, W. Yu, S.P. Singh, S. Shankar, R.K. Srivastava, Anthothecol-encapsulated PLGA nanoparticles inhibit pancreatic cancer stem cell growth by modulating sonic hedgehog pathway, Nanomedicine: Nanotechnology, Biology and Medicine 11 (2015) 2061–2070.

[116] N. Sun, C. Zhao, R. Cheng, Z. Liu, X. Li, A. Lu, Z. Tian, Z. Yang, Cargo-free nanomedicine with pH sensitivity for codelivery of DOX conjugated prodrug with SN38 to synergistically eradicate breast cancer stem cells, Molecular Pharmaceutics 15 (2018) 3343–3355.

[117] M. Wang, F. Xie, X. Wen, H. Chen, H. Zhang, J. Liu, H. Zhang, H. Zou, Y. Yu, Y. Chen, Therapeutic PEG-ceramide nanomicelles synergize with salinomycin to target both liver cancer cells and cancer stem cells, Nanomedicine 12 (2017) 1025–1042.

[118] Y. Mi, Y. Huang, J. Deng, The enhanced delivery of salinomycin to CD133+ ovarian cancer stem cells through CD133 antibody conjugation with poly (lactic-co-glycolic acid)-poly (ethylene glycol) nanoparticles, Oncology Letters 15 (2018) 6611–6621.

[119] D. Chen, X. Pan, F. Xie, Y. Lu, H. Zou, C. Yin, Y. Zhang, J. Gao, Codelivery of doxorubicin and elacridar to target both liver cancer cells and stem cells by polylactide-co-

glycolide/d-alpha-tocopherol polyethylene glycol 1000 succinate nanoparticles, International Journal of Nanomedicine (2018) 6855–6870.

[120] Y. Zhang, Q. Zhang, J. Sun, H. Liu, Q. Li, The combination therapy of salinomycin and gefitinib using poly (d, l-lactic-co-glycolic acid)-poly (ethylene glycol) nanoparticles for targeting both lung cancer stem cells and cancer cells, OncoTargets and Therapy (2017) 5653–5666.

[121] A.L.C. de Souza, A. do Rego Pires, C.A.F. Moraes, C.H.C. de Matos, K.I.P. dos Santos, R.C. e Silva, S.P.C. Acuña, S. dos Santos Araújo, Chromatographic methods for separation and identification of bioactive compounds, in: J.N. Cruz (Ed.), Drug Discovery and Design Using Natural Products, Springer Nature Switzerland, Cham, 2023: pp. 153–176. https://doi.org/10.1007/978-3-031-35205-8_6

[122] V. Ejigah, O. Owoseni, P. Bataille-Backer, O.D. Ogundipe, F.A. Fisusi, S.K. Adesina, Approaches to Improve Macromolecule and Nanoparticle Accumulation in the Tumor Microenvironment by the Enhanced Permeability and Retention Effect, Polymers (Basel) 14 (2022) 2601. https://doi.org/10.3390/polym14132601

[123] M.A. Dheyab, A.A. Aziz, P. Moradi Khaniabadi, M.S. Jameel, N. Oladzadabbasabadi, S.A. Mohammed, R.S. Abdullah, B. Mehrdel, Monodisperse Gold Nanoparticles: A Review on Synthesis and Their Application in Modern Medicine, Int J Mol Sci 23 (2022) 7400. https://doi.org/10.3390/ijms23137400

[124] Y. Zhao, W. Zhao, Y.C. Lim, T. Liu, Salinomycin-Loaded Gold Nanoparticles for Treating Cancer Stem Cells by Ferroptosis-Induced Cell Death, Mol Pharm 16 (2019) 2532–2539. https://doi.org/10.1021/acs.molpharmaceut.9b00132

[125] C. Hu, M. Niestroj, D. Yuan, S. Chang, J. Chen, Treating cancer stem cells and cancer metastasis using glucose-coated gold nanoparticles, Int J Nanomedicine 10 (2015) 2065–2077. https://doi.org/10.2147/IJN.S72144

[126] H.J. Kim, H. Takemoto, Y. Yi, M. Zheng, Y. Maeda, H. Chaya, K. Hayashi, P. Mi, F. Pittella, R.J. Christie, Precise engineering of siRNA delivery vehicles to tumors using polyion complexes and gold nanoparticles, ACS Nano 8 (2014) 8979–8991

[127] Y. Yi, H.J. Kim, P. Mi, M. Zheng, H. Takemoto, K. Toh, B.S. Kim, K. Hayashi, M. Naito, Y. Matsumoto, Targeted systemic delivery of siRNA to cervical cancer model using cyclic RGD-installed unimer polyion complex-assembled gold nanoparticles, Journal of Controlled Release 244 (2016) 247–256.

[128] K. Cao, Y. Du, X. Bao, M. Han, R. Su, J. Pang, S. Liu, Z. Shi, F. Yan, S. Feng, Glutathione-Bioimprinted Nanoparticles Targeting of N6-methyladenosine FTO Demethylase as a Strategy against Leukemic Stem Cells, Small 18 (2022) e2106558. https://doi.org/10.1002/smll.202106558

[129] S.-T. Ning, S.-Y. Lee, M.-F. Wei, C.-L. Peng, S.Y.-F. Lin, M.-H. Tsai, P.-C. Lee, Y.-H. Shih, C.-Y. Lin, T.-Y. Luo, M.-J. Shieh, Targeting Colorectal Cancer Stem-Like Cells with Anti-CD133 Antibody-Conjugated SN-38 Nanoparticles, ACS Appl Mater Interfaces 8 (2016) 17793–17804. https://doi.org/10.1021/acsami.6b04403

Materials Research Forum LLC
https://doi.org/10.21741/9781644903339-2

[130] L.M. Lazer, Y. Kesavan, R. Gor, I. Ramachandran, S. Pathak, S. Narayan, M. Anbalagan, S. Ramalingam, Targeting colon cancer stem cells using novel doublecortin like kinase 1 antibody functionalized folic acid conjugated hesperetin encapsulated chitosan nanoparticles, Colloids Surf B Biointerfaces 217 (2022) 112612. https://doi.org/10.1016/j.colsurfb.2022.112612

[131] E. Espinosa-Cano, M. Huerta-Madroñal, P. Cámara-Sánchez, J. Seras-Franzoso, S. Schwartz, I. Abasolo, J. San Román, M.R. Aguilar, Hyaluronic acid (HA)-coated naproxen-nanoparticles selectively target breast cancer stem cells through COX-independent pathways, Mater Sci Eng C Mater Biol Appl 124 (2021) 112024. https://doi.org/10.1016/j.msec.2021.112024

[132] S.R. Dash, S. Chatterjee, S. Sinha, B. Das, S. Paul, R. Pradhan, C. Sethy, R. Panda, J. Tripathy, C.N. Kundu, NIR irradiation enhances the apoptotic potentiality of quinacrine-gold hybrid nanoparticles by modulation of HSP-70 in oral cancer stem cells, Nanomedicine 40 (2022) 102502. https://doi.org/10.1016/j.nano.2021.102502

[133] S.R. Dash, B. Das, C. Das, S. Sinha, S. Paul, R. Pradhan, C.N. Kundu, Near-infrared enhances antiangiogenic potentiality of quinacrine-gold hybrid nanoparticles in breast cancer stem cells via deregulation of HSP-70/TGF-β, Nanomedicine (Lond) 18 (2023) 19–33. https://doi.org/10.2217/nnm-2022-0243

[134] Y. Xiao, D. Yu, Tumor microenvironment as a therapeutic target in cancer, Pharmacol Ther 221 (2021) 107753. https://doi.org/10.1016/j.pharmthera.2020.107753

[135] Q. Yu, Y. Qiu, J. Li, X. Tang, X. Wang, X. Cun, S. Xu, Y. Liu, M. Li, Z. Zhang, Q. He, Targeting cancer-associated fibroblasts by dual-responsive lipid-albumin nanoparticles to enhance drug perfusion for pancreatic tumor therapy, Journal of Controlled Release 321 (2020) 564–575. https://doi.org/10.1016/j.jconrel.2020.02.040

[136] F. Xing, J. Saidou, K. Watabe, Cancer associated fibroblasts (CAFs) in tumor microenvironment, Front Biosci 15 (2010) 166–179.

[137] S. Zanganeh, G. Hutter, R. Spitler, O. Lenkov, M. Mahmoudi, A. Shaw, J.S. Pajarinen, H. Nejadnik, S. Goodman, M. Moseley, Iron oxide nanoparticles inhibit tumour growth by inducing pro-inflammatory macrophage polarization in tumour tissues, Nature Nanotechnology 11 (2016) 986–994.

[138] Y. Sang, Q. Deng, F. Cao, Z. Liu, Y. You, H. Liu, J. Ren, X. Qu, Remodeling macrophages by an iron nanotrap for tumor growth suppression, ACS Nano 15 (2021) 19298–19309.

[139] H. Zhao, B. Zhao, L. Wu, H. Xiao, K. Ding, C. Zheng, Q. Song, L. Sun, L. Wang, Z. Zhang, Amplified cancer immunotherapy of a surface-engineered antigenic microparticle vaccine by synergistically modulating tumor microenvironment, ACS Nano 13 (2019) 12553–12566.

[140] S.R. Ali, S. Kumari, S.K. Prasad, R.S. Prasad, S.K. Sinha, A. Shakya, Drug development projects guided by ethnobotany and ethnopharmacology studies, in: J.N. Cruz (Ed.), Drug Discovery and Design Using Natural Products, Springer Nature Switzerland, Cham, 2023: pp. 3–21. https://doi.org/10.1007/978-3-031-35205-8_1

[141]   Y. Zhang, Y. Chen, J. Li, X. Zhu, Y. Liu, X. Wang, H. Wang, Y. Yao, Y. Gao, Z. Chen, Development of toll-like receptor agonist-loaded nanoparticles as precision immunotherapy for reprogramming tumor-associated macrophages, ACS Applied Materials & Interfaces 13 (2021) 24442–24452.

[142]   R. Pal, B. Chakraborty, A. Nath, L.M. Singh, M. Ali, D.S. Rahman, S.K. Ghosh, A. Basu, S. Bhattacharya, R. Baral, M. Sengupta, Noble metal nanoparticle-induced oxidative stress modulates tumor associated macrophages (TAMs) from an M2 to M1 phenotype: An in vitro approach, International Immunopharmacology 38 (2016) 332–341. https://doi.org/10.1016/j.intimp.2016.06.006

[143]   L. Jiang, Z. Wang, Y. Wang, S. Liu, Y. Xu, C. Zhang, L. Li, S. Si, B. Yao, W. Dai, H. Li, Re-exposure of chitosan by an inhalable microsphere providing the re-education of TAMs for lung cancer treatment with assistant from sustained H2S generation, International Journal of Pharmaceutics (2023) 123142. https://doi.org/10.1016/j.ijpharm.2023.123142

[144]   R.F. de A. Júnior, G.A. Lira, T. Schomann, R.S. Cavalcante, N.F. Vilar, R.C.M. de Paula, R.F. Gomes, C.K. Chung, C. Jorquera-Cordero, O. Vepris, A.B. Chan, L.J. Cruz, Retinoic acid-loaded PLGA nanocarriers targeting cell cholesterol potentialize the antitumour effect of PD-L1 antibody by preventing epithelial-mesenchymal transition mediated by M2-TAM in colorectal cancer, Translational Oncology 31 (2023) 101647. https://doi.org/10.1016/j.tranon.2023.101647

[145]   J. Peng, J. Zhou, R. Sun, Y. Chen, D. Pan, Q. Wang, Y. Chen, Z. Gong, Q. Du, Dual-targeting of artesunate and chloroquine to tumor cells and tumor-associated macrophages by a biomimetic PLGA nanoparticle for colorectal cancer treatment, International Journal of Biological Macromolecules 244 (2023) 125163. https://doi.org/10.1016/j.ijbiomac.2023.125163

[146]   C. Pundkar, F. Antony, X. Kang, A. Mishra, R.J. Babu, P. Chen, F. Li, A. Suryawanshi, Targeting Wnt/β-catenin signaling using XAV939 nanoparticles in tumor microenvironment-conditioned macrophages promote immunogenicity, Heliyon 9 (2023) e16688. https://doi.org/10.1016/j.heliyon.2023.e16688

[147]   M. Mahmoudian, A. Namdar, P. Zakeri-Milani, H. Valizadeh, S. Elahi, A.M. Darwesh, J.M. Seubert, A.G. Siraki, W.H. Roa, N.B. Chacra, R. Löbenberg, Interaction of M2 macrophages with hepatocellular carcinoma co-culture system in the presence of doxorubicin-loaded nanoparticles, Journal of Drug Delivery Science and Technology 73 (2022) 103487. https://doi.org/10.1016/j.jddst.2022.103487

[148]   C. Hu, X. Liu, W. Ran, J. Meng, Y. Zhai, P. Zhang, Q. Yin, H. Yu, Z. Zhang, Y. Li, Regulating cancer associated fibroblasts with losartan-loaded injectable peptide hydrogel to potentiate chemotherapy in inhibiting growth and lung metastasis of triple negative breast cancer, Biomaterials 144 (2017) 60–72.

[149]   X. Chen, W. Zhou, C. Liang, S. Shi, X. Yu, Q. Chen, T. Sun, Y. Lu, Y. Zhang, Q. Guo, Codelivery nanosystem targeting the deep microenvironment of pancreatic cancer, Nano Letters 19 (2019) 3527–3534.

[150]   H. Zhang, L. Chen, Y. Zhao, N. Luo, J. Shi, S. Xu, L. Ma, M. Wang, M. Gu, C. Mu, Y. Xiong, Relaxin-encapsulated polymeric metformin nanoparticles remodel tumor immune

microenvironment by reducing CAFs for efficient triple-negative breast cancer immunotherapy, Asian Journal of Pharmaceutical Sciences 18 (2023) 100796. https://doi.org/10.1016/j.ajps.2023.100796

[151] R. Pradhan, S. Paul, B. Das, S. Sinha, S.R. Dash, M. Mandal, C.N. Kundu, Resveratrol nanoparticle attenuates metastasis and angiogenesis by deregulating inflammatory cytokines through inhibition of CAFs in oral cancer by CXCL-12/IL-6-dependent pathway, The Journal of Nutritional Biochemistry 113 (2023) 109257. https://doi.org/10.1016/j.jnutbio.2022.109257

[152] Y. Zhang, C.K. Elechalawar, M.N. Hossen, E.R. Francek, A. Dey, S. Wilhelm, R. Bhattacharya, P. Mukherjee, Gold nanoparticles inhibit activation of cancer-associated fibroblasts by disrupting communication from tumor and microenvironmental cells, Bioactive Materials 6 (2021) 326–332. https://doi.org/10.1016/j.bioactmat.2020.08.009

[153] N. Nishida, H. Yano, T. Nishida, T. Kamura, M. Kojiro, Angiogenesis in Cancer, Vasc Health Risk Manag 2 (2006) 213–219.

[154] I. Zuazo-Gaztelu, O. Casanovas, Unraveling the Role of Angiogenesis in Cancer Ecosystems, Frontiers in Oncology 8 (2018). https://www.frontiersin.org/articles/10.3389/fonc.2018.00248 (accessed July 6, 2023).

[155] T. Liu, D. Zhang, W. Song, Z. Tang, J. Zhu, Z. Ma, X. Wang, X. Chen, T. Tong, A poly (l-glutamic acid)-combretastatin A4 conjugate for solid tumor therapy: Markedly improved therapeutic efficiency through its low tissue penetration in solid tumor, Acta Biomaterialia 53 (2017) 179–189.

[156] A.V. Samrot, M.S. Sree, D. Rajalakshmi, L.N.R. Prakash, P. Prakash, Natural biopolymers as scaffold, in: J.N. Cruz (Ed.), Drug Discovery and Design Using Natural Products, Springer Nature Switzerland, Cham, 2023: pp. 23–36. https://doi.org/10.1007/978-3-031-35205-8_20.

[157] Z. Liu, N. Shen, Z. Tang, D. Zhang, L. Ma, C. Yang, X. Chen, An eximious and affordable GSH stimulus-responsive poly (α-lipoic acid) nanocarrier bonding combretastatin A4 for tumor therapy, Biomaterials Science 7 (2019) 2803–2811.

[158] D. Zhu, Y. Li, Z. Zhang, Z. Xue, Z. Hua, X. Luo, T. Zhao, C. Lu, Y. Liu, Recent advances of nanotechnology-based tumor vessel-targeting strategies, J Nanobiotechnology 19 (2021) 435. https://doi.org/10.1186/s12951-021-01190-y

[159] L. Battaglia, M. Gallarate, E. Peira, D. Chirio, I. Solazzi, S.M.A. Giordano, C.L. Gigliotti, C. Riganti, C. Dianzani, Bevacizumab loaded solid lipid nanoparticles prepared by the coacervation technique: preliminary in vitro studies, Nanotechnology 26 (2015) 255102.

[160] C.-F. Wang, E.M. Mäkilä, M.H. Kaasalainen, M.V. Hagström, J.J. Salonen, J.T. Hirvonen, H.A. Santos, Dual-drug delivery by porous silicon nanoparticles for improved cellular uptake, sustained release, and combination therapy, Acta Biomaterialia 16 (2015) 206–214.

[161] J. Wang, H. Wang, J. Li, Z. Liu, H. Xie, X. Wei, D. Lu, R. Zhuang, X. Xu, S. Zheng, iRGD-decorated polymeric nanoparticles for the efficient delivery of vandetanib to

hepatocellular carcinoma: preparation and in vitro and in vivo evaluation, ACS Applied Materials & Interfaces 8 (2016) 19228–19237.

[162] M. Shibuya, Vascular Endothelial Growth Factor (VEGF) and Its Receptor (VEGFR) Signaling in Angiogenesis, Genes Cancer 2 (2011) 1097–1105. https://doi.org/10.1177/1947601911423031

[163] X. Shan, W. Yu, X. Ni, T. Xu, C. Lei, Z. Liu, X. Hu, Y. Zhang, B. Cai, B. Wang, Effect of chitosan magnetic nanoparticles loaded with Ang2-siRNA plasmids on the growth of melanoma xenografts in nude mice, Cancer Management and Research (2020) 7475–7485.

[164] L. Zhang, Y. Qi, H. Min, C. Ni, F. Wang, B. Wang, H. Qin, Y. Zhang, G. Liu, Y. Qin, Cooperatively responsive peptide nanotherapeutic that regulates angiopoietin receptor Tie2 activity in tumor microenvironment to prevent breast tumor relapse after chemotherapy, ACS Nano 13 (2019) 5091–5102.

[165] S. Prabha, B. Sharma, V. Labhasetwar, Inhibition of tumor angiogenesis and growth by nanoparticle-mediated p53 gene therapy in mice, Cancer Gene Therapy 19 (2012) 530–537.

[166] Y.-Y. Guan, X. Luan, J.-R. Xu, Y.-R. Liu, Q. Lu, C. Wang, H.-J. Liu, Y.-G. Gao, H.-Z. Chen, C. Fang, Selective eradication of tumor vascular pericytes by peptide-conjugated nanoparticles for antiangiogenic therapy of melanoma lung metastasis, Biomaterials 35 (2014) 3060–3070.

[167] P. Mukherjee, R. Bhattacharya, P. Wang, L. Wang, S. Basu, J.A. Nagy, A. Atala, D. Mukhopadhyay, S. Soker, Antiangiogenic properties of gold nanoparticles, Clinical Cancer Research 11 (2005) 3530–3534.

[168] Y. Xu, Z. Wen, Z. Xu, Chitosan nanoparticles inhibit the growth of human hepatocellular carcinoma xenografts through an antiangiogenic mechanism, Anticancer Research 29 (2009) 5103–5109.

[169] G. Housman, S. Byler, S. Heerboth, K. Lapinska, M. Longacre, N. Snyder, S. Sarkar, Drug Resistance in Cancer: An Overview, Cancers (Basel) 6 (2014) 1769–1792. https://doi.org/10.3390/cancers6031769

[170] N. Vasan, J. Baselga, D.M. Hyman, A view on drug resistance in cancer, Nature 575 (2019) 299–309. https://doi.org/10.1038/s41586-019-1730-1

[171] T.B. Emran, A. Shahriar, A.R. Mahmud, T. Rahman, M.H. Abir, Mohd.F.-R. Siddiquee, H. Ahmed, N. Rahman, F. Nainu, E. Wahyudin, S. Mitra, K. Dhama, M.M. Habiballah, S. Haque, A. Islam, M.M. Hassan, Multidrug Resistance in Cancer: Understanding Molecular Mechanisms, Immunoprevention and Therapeutic Approaches, Frontiers in Oncology 12 (2022). https://www.frontiersin.org/articles/10.3389/fonc.2022.891652 (accessed July 7, 2023).

[172] H.O. Alsaab, S. Sau, R.M. Alzhrani, V.T. Cheriyan, L.A. Polin, U. Vaishampayan, A.K. Rishi, A.K. Iyer, Tumor hypoxia directed multimodal nanotherapy for overcoming drug resistance in renal cell carcinoma and reprogramming macrophages, Biomaterials 183 (2018) 280–294.

[173]　X. Song, L. Feng, C. Liang, K. Yang, Z. Liu, Ultrasound triggered tumor oxygenation with oxygen-shuttle nanoperfluorocarbon to overcome hypoxia-associated resistance in cancer therapies, Nano Letters 16 (2016) 6145–6153.

[174]　Z. Wang, Z. Xu, G. Zhu, A Platinum(IV) Anticancer Prodrug Targeting Nucleotide Excision Repair To Overcome Cisplatin Resistance, Angew Chem Int Ed Engl 55 (2016) 15564–15568. https://doi.org/10.1002/anie.201608936

[175]　J. Chen, X. Wang, Y. Yuan, H. Chen, L. Zhang, H. Xiao, J. Chen, Y. Zhao, J. Chang, W. Guo, X.-J. Liang, Exploiting the acquired vulnerability of cisplatin-resistant tumors with a hypoxia-amplifying DNA repair–inhibiting (HYDRI) nanomedicine, Sci Adv 7 (2021) eabc5267. https://doi.org/10.1126/sciadv.abc5267

[176]　S.K. Vishwakarma, P. Sharmila, A. Bardia, L. Chandrakala, N. Raju, G. Sravani, B.V.S. Sastry, M.A. Habeeb, A.A. Khan, M. Dhayal, Use of biocompatible sorafenib-gold nanoconjugates for reversal of drug resistance in human hepatoblatoma cells, Scientific Reports 7 (2017) 8539.

[177]　K.L. Swetha, M. Paul, K.S. Maravajjala, S. Kumbham, S. Biswas, A. Roy, Overcoming drug resistance with a docetaxel and disulfiram loaded pH-sensitive nanoparticle, Journal of Controlled Release 356 (2023) 93–114. https://doi.org/10.1016/j.jconrel.2023.02.023

[178]　N. Maliyakkal, A. Appadath Beeran, N. Udupa, Nanoparticles of cisplatin augment drug accumulations and inhibit multidrug resistance transporters in human glioblastoma cells, Saudi Pharmaceutical Journal 29 (2021) 857–873. https://doi.org/10.1016/j.jsps.2021.07.001

[179]　B. Du, W. Zhu, L. Yu, Y. Wang, M. Zheng, J. Huang, G. Shen, J. Zhou, H. Yao, TPGS2k-PLGA composite nanoparticles by depleting lipid rafts in colon cancer cells for overcoming drug resistance, Nanomedicine: Nanotechnology, Biology and Medicine 35 (2021) 102307. https://doi.org/10.1016/j.nano.2020.102307

[180]　H. Guo, Y. Liu, Y. Wang, J. Wu, X. Yang, R. Li, Y. Wang, N. Zhang, pH-sensitive pullulan-based nanoparticle carrier for adriamycin to overcome drug-resistance of cancer cells, Carbohydrate Polymers 111 (2014) 908–917. https://doi.org/10.1016/j.carbpol.2014.05.057

[181]　S.F. El-Menshawe, O.M. Sayed, H.A. Abou Taleb, M.A. Saweris, D.M. Zaher, H.A. Omar, The use of new quinazolinone derivative and doxorubicin loaded solid lipid nanoparticles in reversing drug resistance in experimental cancer cell lines: A systematic study, Journal of Drug Delivery Science and Technology 56 (2020) 101569. https://doi.org/10.1016/j.jddst.2020.101569

[182]　B.-Y. Liu, C. Wu, X.-Y. He, R.-X. Zhuo, S.-X. Cheng, Multi-drug loaded vitamin E-TPGS nanoparticles for synergistic drug delivery to overcome drug resistance in tumor treatment, Science Bulletin 61 (2016) 552–560. https://doi.org/10.1007/s11434-016-1039-5

[183]　S.K. Singh, J.W. Lillard, R. Singh, Reversal of drug resistance by planetary ball milled (PBM) nanoparticle loaded with resveratrol and docetaxel in prostate cancer, Cancer Letters 427 (2018) 49–62. https://doi.org/10.1016/j.canlet.2018.04.017

[184]　S.-E. Lee, C.D. Lee, J.B. Ahn, D.-H. Kim, J.K. Lee, J.-Y. Lee, J.-S. Choi, J.-S. Park, Hyaluronic acid-coated solid lipid nanoparticles to overcome drug-resistance in tumor cells,

Journal of Drug Delivery Science and Technology 50 (2019) 365–371. https://doi.org/10.1016/j.jddst.2019.01.042

[185] S.R. Dash, C.N. Kundu, Photothermal Therapy: A New Approach to Eradicate Cancer, Current Nanoscience 18 (n.d.) 31–47.

[186] N. Ma, M.-K. Zhang, X.-S. Wang, L. Zhang, J. Feng, X.-Z. Zhang, NIR Light-Triggered Degradable MoTe2 Nanosheets for Combined Photothermal and Chemotherapy of Cancer, Advanced Functional Materials 28 (2018) 1801139. https://doi.org/10.1002/adfm.201801139

[187] L. Zhang, D. Jing, L. Wang, Y. Sun, J.J. Li, B. Hill, F. Yang, Y. Li, K.S. Lam, Unique Photochemo-Immuno-Nanoplatform against Orthotopic Xenograft Oral Cancer and Metastatic Syngeneic Breast Cancer, Nano Lett. 18 (2018) 7092–7103. https://doi.org/10.1021/acs.nanolett.8b03096

[188] Z. Wang, H. Xing, A. Liu, L. Guan, X. Li, L. He, Y. Sun, A.V. Zvyagin, B. Yang, Q. Lin, Multifunctional nano-system for multi-mode targeted imaging and enhanced photothermal therapy of metastatic prostate cancer, Acta Biomaterialia 166 (2023) 581–592. https://doi.org/10.1016/j.actbio.2023.05.014

[189] A. Granja, R. Lima-Sousa, C.G. Alves, D. de Melo-Diogo, C. Nunes, C.T. Sousa, I.J. Correia, S. Reis, Multifunctional targeted solid lipid nanoparticles for combined photothermal therapy and chemotherapy of breast cancer, Biomaterials Advances 151 (2023) 213443. https://doi.org/10.1016/j.bioadv.2023.213443

[190] L. Bian, N. Wang, K. Tuersong, A. Kaidierdan, J. Li, J. Gong, Oxygen vacancy engineering of TiO2 nanosheets for enhanced photothermal therapy against cervical cancer in the second near-infrared window, Colloids and Surfaces B: Biointerfaces 229 (2023) 113427. https://doi.org/10.1016/j.colsurfb.2023.113427

[191] J. Yang, Z. Sun, Q. Dou, S. Hui, P. Zhang, R. Liu, D. Wang, S. Jiang, NIR-light-responsive chemo-photothermal hydrogel system with controlled DOX release and photothermal effect for cancer therapy, Colloids and Surfaces A: Physicochemical and Engineering Aspects 667 (2023) 131407. https://doi.org/10.1016/j.colsurfa.2023.131407

[192] S.R. Dash, C.N. Kundu, Promising opportunities and potential risk of nanoparticle on the society, IET Nanobiotechnol 14 (2020) 253–260. https://doi.org/10.1049/iet-nbt.2019.0303

Advances in Healthcare and Nanoparticle Toxicology　　　　　Materials Research Forum LLC
Materials Research Foundations 171 (2024) 76-92　　　　https://doi.org/10.21741/9781644903339-3

Chapter 3

Nanoparticles Interaction with Bacteria and Viruses

Farwa Farooq[1], Sumreen Hayat[1], Asma Ashraf[2], Bilal Aslam[1], Muhmmad Hussnain Siddique[3], Muhammad Umar Ijaz[4], Zeeshan Taj[1], Azalfah Ibrar[1], Saima Muzammil[1]*

[1]Institute of Microbiology, Government college University Faisalabad, Pakistan

[2]Department of Zoology, Government college University Faisalabad, Pakistan

[3]Department of Bioinformatics and Biotechnology, Government college University Faisalabad, Pakistan

[4]Department of Zoology, Wildlife and Fisheries, University of Agriculture, Faisalabad, Pakistan

*saimamuzammil83@gmail.com

Abstract

Nanotechnology is growing rapidly because of its widespread uses in different fields of science. History of nanoparticles ranges back to ancient Egyptians times when the people used Pbs nanoparticles in their hair dyes; approximately 4000 years ago. Nanoparticles are tiny particles that vary in size from 1 to 100 nanometers. They are classified into three distinct types according to their origin. Nanoparticles can be produced by various chemical and biological methods and from various organisms as well, such as Bacteria, Fungi, Algae, viruses and Yeast. Viral nanoparticles can be designed by altering the genetic and chemical makeup of viruses to enhance their functionality. This has led us to successfully manufacture numerous antibiotics, as antibiotic resistance has significant impact in global healthcare. They have enabled us to deal with the microbial mechanism of antibiotic resistance with the help of Beta-Lactam ring that is present on their surface. Moreover, nanoparticles play a pivotal role in drug delivery, gene therapy and nano therapy in case of bacteria and fungi. Although they have uncountable positive impacts, but they also exhibit some of the negative effects too, mostly in plants.

Keywords

Nanoparticle Synthesis, Viral Nanoparticles Design, Antibiotic Mechanisms, Drug Therapy

Contents

1.   Introduction

Development of nano materials along with their uses and synthesis are the focus of the newly developing scientific discipline known is nanotechnology. Nanotechnology and nanomaterials are increasingly coming into attention because of their widespread uses in different fields of science such as medicine, industries, energy, pharmaceuticals, space industries, electronics and even in cosmetics, clothes and construction industries. [1] The term "nano" comes from a Greek word "nanos" that means dwarf. Nano describes the measurement on the scale of one-billionth (10^9) of a meter [2]. The size variety of nanoparticles is from 1 to 100 nanometers and they exist as solid ultrafine particles [3]. They have dimensions lower than the wavelength of light. This property of nanoparticles makes them transparent and because of which they are very useful in packaging, cosmetics and coatings [4].

However, surface area to volume ration is also a significant feature of nanoparticles because it allows nanoparticles to interact efficiently with other particles [5]. One of the important fields of research in nanotechnology is the synthesis of nanoparticles with the help of ecofriendly, nontoxic, clean and reliable resources. As a result, a number of options like enzymes, polysaccharides, vitamins, biodegradable polymers and even microorganisms have been used to achieve this ultimate objective [6]. Some metallic nanoparticles are also produced from certain metals e.g., silver, copper, magnesium, gold, titanium, alginate and zinc [7]. In recent years, nanoparticles synthesized from certain biodegradable polymers that are particularly coated with poly (ethylene glycol) (PEG) have been successfully used as drug delivery devices. This is because they are able to target a particular organ as carriers of DNA in gene therapy and they can circulate for extended

Advances in Healthcare and Nanoparticle Toxicology   Materials Research Forum LLC
Materials Research Foundations 171 (2024) 76-92   https://doi.org/10.21741/9781644903339-3

periods of time. Using nanoparticles in drug delivery system has several benefits. For instance, their size can easily be manipulated and therefore, we can attain effective passive and active drug targeting. Most importantly we can achieve site-specific targeting.

Nanoparticles have been prepared most commonly by three methods: 1) polymerization of monomers 2) coacervation of hydrophilic polymers and 3) dispersion of preformed polymers ,other than these; a number of other methods like wet chemical precipitation, sol-gel technique, vapor phase condensation and sputtering have also been developed for the synthesis of nanoparticles [8]. Apart from these methods certain microorganisms have been frequently used in biosynthesis of nanoparticles. Bacteria are one the best candidates because of their ability to reduce heavy metals [9]. For example, T. thiooxidans and Sulfolobus acidocaldarius reduces ferric oxide to ferrous state as an energy source when it grows on elemental sulfur. Similarly, Enterobacter cloacae, Desulfovibrio desulfuricans and Rhodospirillum rubrum can reduce selenite to selenium [4].

2.   Historical perspective of nanoparticles

One of the earliest examples of nanomaterials could be found 4000 years ago at the time of Ancient Egyptians when they synthesized approximately 5nm Pbs NPs for hair dyes. The synthesis of metallic nanoparticles dates back to 13th and 14th century when Egyptians and Mesopotamians manufactured glass with the help of metals via chemical methods [10].The first scientific description about NPs was reported in 1857 when Michael Faraday synthesized a colloidal Au NP Solution. (Abbas Mohajerani) Richard Zsigmonday and Henry Siedentopf developed an ultramicroscope which aided in the detection of structures smaller than 4nm in 1902. These small structures were detected in ruby glasses. Years later the idea of nanotechnology was introduced in 1959 by a famous physicist Richard Feynman. At the California Institute of Technology, he spoke to an American Physical Society Meeting on "There's Plenty of Room at the Bottom''. When Norio Taniguchi highlighted the term "Nanotechnology'' in 1947 at the International Conference on Production Engineering, people began to take notice of Feynman's proposal [11].

In later years a book titled "The Engine of Creation: Coming Era of Nanotechnology" was published in 1986 by Kim Eric Drexler. Kim presented the concept of Molecular Nanotechnology (MNT). By suggesting the synthesis of materials with the help of both molecular and atomic components-molecule by molecule and atom by atom he advanced the concept of nanotechnology (Abbas Mohajerani). Following these methods a great deal of study and efforts were done, leading to numerous innovations and discoveries in this field of science and significant growth was seen in the number of publications on Nanotechnology.

3.   Classification of nanoparticles

According to their origin, a three-fold classification system may be used to classify nanoparticles and nanomaterials.

3.1   Incidental nanoparticles

In daily life, inadvertent nanoparticle synthesis is not always evident, but it might become evident if it is recognized and appropriately handled. Diverse pathways lead to the production of incidental nanoparticles. These are inadvertently created by anthropogenic processes or direct or direct human impacts [12]. In a similar vein, NPs are also produced when animals and plants shed their

Advances in Healthcare and Nanoparticle Toxicology   Materials Research Forum LLC
Materials Research Foundations 171 (2024) 76-92   https://doi.org/10.21741/9781644903339-3

skin and hair. In dessert and terrestrial regions, sandstorms are the primary source of particulate matter (NPs0, which makes asthma and emphysema two of the biggest issues that people deal with. Moreover, the production of reactive oxygen species by metal-containing dust nanoparticles might cause significant harm to lung tissues. Since a single eruption may release up to 30×10^6 tons of NPs in the form of ash, volcanic eruptions are also one of the main generators of NPs. Along with these natural events that can produce nanoparticles there are some human activities too that generate NPs. These activities include transportation, charcoal burning, industrial operations and exhaust emissions. Human activity is responsible for around 10% of the aerosols created in the environment; the remaining 90% are produced mostly by natural processes [13].

3.2   Engineered nanoparticles

Engineered nanoparticles are produced from simple combustion of fuel, oil and coal for power generation, airplane engines, from vehicles, smelting and even from cooking. Certain nanoparticles are purposefully created via top-down or bottom-up synthesis techniques. Top-down process involves reducing the size of larger structures into nano size particles and bottom-up process involves the synthesis of nanoparticles from individual atoms or molecules [13]. The first ever nanomaterial was engineered in 1940s from aerosols (fumed silica) [11]. Then later in 1960s first silica nanosphere was synthesized from aqueous solutions. Furthermore, classification can be done on the basis of synthesis mechanisms that are used to engineer NPs. This involves three categories 1) chemical process, 2) physical process and 3) mechanical process [14].

3.3   Natural nanoparticles

Naturally occurring nanoparticles are found in living things, including people, animals, plants, birds and microorganisms including bacteria, algae and viruses. These are produced from natural biogeochemical or mechanical processes that do not have any involvement of anthropogenic activities [10]. Through a mechanism of evolution, insects create nanostructures that enables them to withstand hostile environments. In order to flourish, plants also need the nutrients found in soil and water. Plants store these biominerals in nanoforms. In a similar vein, several human organs such as the bones are composed of nanostructures. Furthermore, substances in the nanometer size range include DNA, antibodies, enzymes and certain fluids that are critical to the healthy operation of human body [15].

4.   Methods for production of nanoparticles

There are various methods for the production of nanoparticles. Some widely used and the most common methods are as follows:

4.1   Physical and chemical methods

Nanoparticles can be produced from various different methods which include chemical methods, physical methods, biological methods and hybrid methods. If we talk about physical method; there are a number of approaches like aerosol technologies, laser ablation, lithography, ultraviolet irradiation and ultrasonic fields [16]. Whereas in chemical method, nanoparticles are grown in a liquid medium with the help of reactants. This technique is a characteristic of sol-gel approach and is also used in production of quantum dots [4].

Advances in Healthcare and Nanoparticle Toxicology   Materials Research Forum LLC
Materials Research Foundations 171 (2024) 76-92   https://doi.org/10.21741/9781644903339-3

However, use of toxic chemical agents, hazardous commodities and high vigor consumptions are the major problems that arise with use of these methods. Both methods are very costly as physical methods require high energy for the synthesis of nanoparticles, while on the other hand in chemical methods waste disposal issue is a major problem as it requires different capping and toxin reducing agents [16].

There are two methods for producing nanoparticles: "top-down" and "bottom-up". In "top-down" method, there is mechanical crushing of the source material occurs via a milling process. However, the "bottom-up" strategy describes the synthesis of structures by chemical process. The required features and chemical composition of the nanoparticles dictate the best process to choose. While performing mechanical production, micro-particles are crushed via milling. This method is used to create metallic and ceramic nanoparticles. Milling is an energy-intensive process that includes thermal stress. Process times longer than necessary might potentially erode the grinding media and contaminate the particles. Reactive milling, in which the milling process is followed by chemical or chemo-physical reactions, can be used in conjunction with pure mechanical milling.

In "Bottom-up" method, atoms or molecules are carefully chosen which are to be used. This method creates structures of complicated size, shape and range. It goes through various procedures which include solgel process, precipitation reactions and aerosol processes [17]

4.2   Biological approaches for NPs production

The drawbacks of physical and chemical methods have led us to ultimately find clean, non-hazardous and biocompatible methods for NPs production. Since a few years ago, biological agents such as bacteria, yeast, algae, fungus, plants and viruses have been employed to create nanoparticles by biological processes. A biological approach is preferred over other methods because of it being eco-friendly and not very energy intensive. Secondly, the nanoparticles which are synthesized using biological agents have high catalytic activity, increased biomass, greater surface and size uniformity [16].

4.3   Nanoparticle synthesis by fungi, algae and yeast

The use of fungi for synthesis of nanoparticles has received significant importance as it has certain advantages over the use of bacteria. This is because fungi secrets higher amounts of protein in comparison to bacteria so this ultimately leads to higher rate of nanoparticles production. Fungi produce nanoparticles both intracellularly as well as extracellularly and they are excellent candidates for metal and metal sulfide nanoparticle synthesis [18]. Intracellular production involves production of nanoparticles in the presence of enzymes through transport of ions into microbial cells. While extracellular production involves secretory components that play a role in capping and reduction of nanoparticles [7].

Several studies have been conducted with Fusarium oxysporum and Aspergillus fumigatus; in attempt to produce silver nanoparticles. Fusarium has also been shown to produce cadmium sulfide, zinc sulfide, lead sulfide and molybdenum sulfide nanoparticles when certain salts are added to the growth medium [18].

Advances in Healthcare and Nanoparticle Toxicology   Materials Research Forum LLC
Materials Research Foundations 171 (2024) 76-92   https://doi.org/10.21741/9781644903339-3

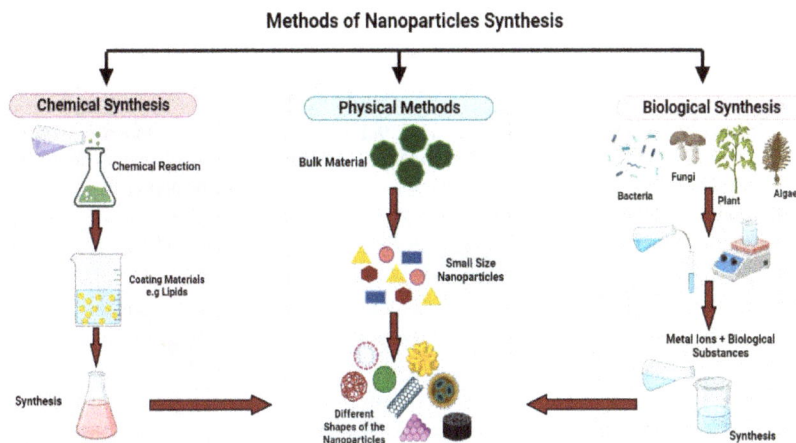

Fig 1. Methods for Synthesis of Nanoparticles

5. Nano-organisms

Biosynthesis of nanoparticles is a green route for NPs production by microorganisms. Similarly, these microorganisms are also called nano-organisms because they are naturally occurring nanomaterials, they can synthesize nanoparticles in their bodies [19]. Nano-organisms are a wide range of organisms that are present all around us even inside human bodies. These include yeast, fungi, algae, viruses and nanobacteria. Certain species of algae such as Chlorella vulgaris also supports the formation of Ag NPs. But some yeast like Candida glabrata, Torulopsis sp., Schizosaccharomyces pombe and MKY3 were also used in the synthesis of NPs [20].

5.1 Magneto tactic bacteria

Magneto tactic bacteria are very useful in the production of magnetic oxide NPs. These NPs are helpful in biomedicine and biological separation because of their high coercive force and micro configuration. Magneto tactic bacteria are also involved in synthesis of iron oxide, iron sulfides, biocompatible magnetite (Fe_3O_4) and maghemite (Fe_2O_3) [21]. These help in targeted treatment of cancer through DNA analysis, magnetic resonance imaging (MRI), hyperthermia and gene therapy. Similarly, a number of other NPs were also produced via magneto tactic bacteria which include modified iron NPs, superparamagnetic NPs, surface-distributed magnetic iron-sulfide particles and 12nm octahedral NPs [22].

5.2 Nanobacteria

Metallic nanoparticles are produced by bacteria known as nanobacteria which attach easily to heavy, soluble, hazardous metals and precipitate them onto the outside of the cell. The term nanobacteria was patented by Dr. Olavi kajender. In 1998, Kajender and Neva Ciftcioglu announced their discovery of nano bacteria in a seminal paper [23] .Nanobacteria are also named as Calcifying Nano-particles (CNPs). They are the smallest, self-replicating, gram-negative bacteria. It requires aerobic conditions, i.e., 5% CO2 and 95% air for its growth. Its metabolic rate is almost 1000 times slower than E.coli with a doubling time of about three days [24]. Nanobacteria are immensely useful in the production of low toxicity NPs. The first ever bacteria that was used to produce Ag NPs, was Pseudomonas stutzeri A259. Following that, several more bacterial strains were employed to create a range of metal nanoparticles, including alloys, nonmagnetic oxide and gold nanoparticles. Nanobacteria are of great importance in nanomedicine applications as they are helpful in reducing potential cellular toxicity. However, there are some drawbacks also. These NPs are difficult to filter and require longer production time and they also produce low yield as compared to the chemical synthesis [25].

Table 1: Nanoparticles synthesized by various Bacteria.

Name of bacteria	Nanoparticles produced	Size (nm)	References
Bacillus cereus	Silver	20-40	[26]
Klebsiella pneumoniae	Silver	~52.2	[27]
Escherichia coli	Cadmium sulfide Silver Gold	2-5 8-9 >10-50	[27]
Desulfovibrio desulfuricans	Palladium	–	[28]
Lactobacillus strains	Gold Silver Silver- Gold alloys Titanium	20-50 and above 100 15-500 100-300 40-60	[29]
Rhodopseudomonas capsulata	Gold	10-20	[29]
Cornybacterium sp	Silver	10-15	[30]
Planomicrobium sp.	Titanium dioxide	8.89	[31]
Enterobacter cloacae	Silver selenium	–	[27]
Marinobacter pelagius	Gold	10	[32]
Vibrio alginolyticus	Silver	50-100	[33]
Shewanella strains	Platinum Gold Magnitite	5 Size change with pH 10-50	[34]
Alteromonas macleodii	Silver	70	[33]
Pseudomonas strains	Gold Silver	15-30 ~70	[35]

5.3   Viral nanoparticles

Viral nanoparticles are tiny structures which are made from viral proteins. They offer a broad spectrum of possible uses which include imaging, medication delivery and vaccine development. We can use them safely on humans because of their biocompatibility and biodegradation. We can engineer and designed them by using both genetic and chemical protocols in order to provide more functionality to the nanoparticles [36].

Moreover, viral nanoparticles consist of one important class "bio-inspired" nanocarriers which have the ability of carrying biologically active molecules to the target site. Bioinspired system consists of lipids, polymers and biomaterials. VNPs are mostly plant viruses that consist of protein shells and contain nucleic acid that is necessary element for viral replication in plants. The role of their hollow protein shells is to allow encapsulation of high concentration of therapeutics. As they are safe to use for humans, so they are also non-infectious to plants [37].

However, the icosahedral Cowpea Mosaic Virus (CPMV) is a plant virus that is a member of the picornavidae family. It is 30nm in diameter. Its capsid naturally encases two positive sense RNA molecules and is composed of 60 copies of both tiny and big proteins. The application of nanomedicines in the study of CPMV and its empty virus-like particles (eVLPs) is being investigated. CPMV is expressed on the surface of both cancerous and endothelial tissues and it binds vimentin receptors preferentially. They have potential to deliver agents that can either help visualize or treat cancer to that specific tissue. Moreover, CPMV has also major role in vaccinology as it acts structure to showcase different immunogenic epitopes that can fight animal pathogens. eVlps have also been used drugs as well as cancer immunotherapeutic agents. [38].

The L and S proteins of CPMV are not digestible when treated with a set amount of pepsin and pancreatic enzymes. CPMV was able to penetrate the mice's circulation when given orally, but it was 10-100 times less abundant in the spleen when given orally than when it was injected [39].

6.   Viral nanoparticle approaches against bacterial infection

Various antibiotics are successfully made against infectious diseases. Despite of this success, bacterial infections are very challenging for the global healthcare. Antibiotic resistance has had a significant impact and threatens to roll back the advancements made against an extensive variety of bacterial infections to the pre-antibiotic era for some pathogens like tuberculosis which have complex mechanisms for undermining their host defenses as well as delivery challenges that prevent antibiotics from reaching infection sites. For these challenges, there must be some alternative and effective antimicrobial strategies [40].

In recent few decades, nanoparticles played a vital role for drug delivery in medicines. Nanoparticle delivery technology has additional benefits such as enhanced drug absorption, precise drug targeting, long-term and precise release. As a result, drug delivery systems that employ nanoparticles are approved for use in clinical settings to treat a range of infectious disorders. In addition to this, several antibacterial nanoparticle compositions are undergoing numerous early-stage and human clinical studies right now. Scientists are finding more ways to use nanoparticles to fight against bacterial infections. They are getting better at making multifunctional nanoparticles, which makes them even more effective [41].

Bacterial infections can make it easier for drugs to passively targeted and reach the affected area by increasing the permeability of blood vessels. At site of infection, various inflammatory

Advances in Healthcare and Nanoparticle Toxicology
Materials Research Foundations 171 (2024) 76-92

Materials Research Forum LLC
https://doi.org/10.21741/9781644903339-3

mediators are triggered by the release and accumulation of bacterial components which include bacterial protease and lipopolysaccharide from gram negative bacteria and lipoteichoic acid from gram positive bacteria [42]. When immune cells are activated by bacterial components, they release different inflammatory and vascular mediators that interact with the vascular endothelium. This interaction can cause gaps to widen, barriers to break down and eventually increase blood vessel permeability. Moreover, bacterial infections can also cause lymphatic drainage to not work properly. This can help nanoparticles accumulate at the site of infection. Nanoparticles can use the characteristics of bacterial infections to their advantage. They can use the enhanced permeation and retention (EPR) effect to deliver antibiotics to the targeted sites [43].

6.1   Targeted delivery of antibiotics in bacterial infections

Under normal physiological conditions, pathogenic bacteria have a surface charge that is negative. It is studied that cationic nanoparticles that can bind to bacteria through electrostatic interactions. This research is aimed at finding more effective ways to target bacteria. Its multivalent effect and ability to target polymicrobial infection make this strategy successful. Incorporating a wide variety of bactericidal polymers and peptides into different nanoparticle designs has been done for antibacterial purpose [44]. The local charge and mass densities can be enhanced by nanoparticles formulation which results in improving therapeutic index. Moreover, if we upgrade the biodegradability of nanoparticles, it results in lessening positive charged related toxicity. According to this viewpoint, it has been demonstrated that cationic nanoparticles which are self-assembled from polycarbonate-based block polymers with excellent ability of biodegradability, may kill bacteria without causing systemic hemolysis. Another method to target bacteria is "active targeting" which involves adding pathogen-binding ligands directly to the surface of nanoparticles [45]. For instance, iron oxide nanoparticles, gold nanoparticles and porous silica nanoparticles have small molecules like vancomycin conjugated to their surfaces, resulting in the preferential binding of nanoparticles to Gram-positive bacteria [46]. Molecular orientation, surface as temperature, electric field magnetic field and ultrasound are the factors which really affect the attacking efficiency of small molecules. There are also some other factors on which targeting efficiency of small molecules depend which can be chemical signals like pH, ionic strength and some enzymatic activities. Among these, pH has been an important factor that is used to design approachable nanoparticles for antibiotic delivery. For site-specific antibiotic administration, nanoparticles have been developed at the organ level to respond to the pH gradient throughput the gastrointestinal tract and the acidic environment of human skin. For triggered medication release, nanoparticles have been designed to react to acidic pH inside end lysosomal compartments at intracellular level. In addition to pH gradient, the release of antimicrobial compounds to stop the development of the target bacteria has also been triggered by bacterial enzymatic activities, including those of released toxins[37].

Lately, there has been an addition of new delivery strategy of environment-responsive in which small, charged nanoparticles are attached onto liposome surface for stabilization of liposome and it triggers antimicrobial delivery. The steric repulsion caused by non-specific adsorption of charged nanoparticles onto phospholipids bilayers prevented the fusing of liposomes[47].

7.   Using viruses as nano medicine

For medical therapy to be successful and efficient, it is essential to be able to manage and target drug delivery. However, one of the biggest challenges in contemporary medicine is the inability to control site-specific localization and therefore, bioavailability at the targeted location. Particularly, medication solubility is a significant problem with approximately half of recently discovered and already marketed pharmaceuticals having poor water solubility [48]. Because a small portion of the medicine accumulates in the intended location due to their lack of solubility, they work much worse overall. More importantly, the loss after delivery frequently results in unfavorable side effects. Using salt or excipients such as organic solvents, oils or surfactants are two solutions for low solubility. The logical design of effective, targetable carrier systems, including nanoparticles, nanocontainers and biomaterials has attracted a lot of interest lately. The physical characteristics of the nanocarriers exterior such as its size, shape and charge, directly influence cellular uptake, intracellular distribution and accumulation, retention and excretion periods when it comes to the design of effective carrier system. Although viruses have evolved to enter and infect host cells to efficiently transport genetic material, current advances in nanotechnology, have transformed viruses into secure delivery methods [49].

One of the biggest obstacles to treating cancer is metastasis, the unchecked spread of cancer cells from a primary tumor to the other regions of the body. Surgery has thus far been shown to be very useless in treating metastatic cancer; instead, contemporary therapy frequently uses extremely powerful cytotoxic drugs, which can have serious adverse effects. A number of promising plant viruses, including the bacteriophages Hong Kong 97 (HK97) and M13, the red clover necrotic mosaic virus (RCNMV) the hibiscus chlorotic singspot virus (HCRRSV), and the cowpea mosaic virus (CPMV) have emerged for the applications for cancer treatment, including molecular detection, targeting and trafficking. Even while some viruses such as CPMV and canine and parvovirus (CPV), naturally prefer certain types of cells, cellular absorption is frequently ineffective. The virus capsid's outer surface can be made accessible for genetic engineering and chemical modification techniques like N-hydroxy succinimide coupling, Michael addition to maleimides and carbodiimide activation by making use of the reactive lysine, cysteine, and aspartic acid and glutamic acid groups. This has been widely used for application in cell imaging and targeting because it enables molecular cargo or prosthetic groups such as aptamers, proteins, antibodies, carbohydrates, fluorescent dyes and medicines to be functionalized. Using conventional chemical conjugation techniques, fluorescent moieties can be added to the surface of wild-type virus particles. For applications in medicines, it is crucial to take into account the unique capability of directly targeting individual cells or cellular areas [50].

8.   Nanoparticles as antimicrobial agents

Over the past few decades, nanoparticles have been used in various fields of science. Most importantly vigorous efforts have been put into research for nanoparticles due to their applications in medical science and drug delivery. In recent times infectious diseases have become a burden on the world's economy. As these diseases are treated by antibiotics but due to increasing microbial resistance the treatment for infectious diseases is becoming difficult day by day. To overcome this microbial assistance; nanoparticles conjugated with antibiotics is the best approach to deal with infectious diseases as there are less chances of resistance from microorganisms [51].

9. Mechanism of action of nanoparticles

Apparently, the antimicrobial activity of nanoparticles is due to the presence of B-lactam ring that is present on the surface of nanoparticles. In the case of bacteria and fungi nano-therapy can cause the inhibition of biofilm formation and it ultimately has bactericidal effect. Reportedly, biogenic selenium nanoparticles can inhibit the biofilm formation of *Pseudomonas aeruginosa*. Similarly, TiO2s nanoparticles can impose inhibitory effect on fungal biofilms. Studies have shown that certain features of nanoparticles like size, surface area and shape contribute in antimicrobial activity. For instance, it has been proven that the size of nanoparticles confers antimicrobial activity whereas the large surface area helps the nanoparticles in attachment and rapid penetration into the cell surface. Similarly, the shapes of nanoparticles play important role in determining the level of interaction with the microbial membrane or enzymes. There are not proper studies which can explain the blockage mechanism by nanoparticles [52].

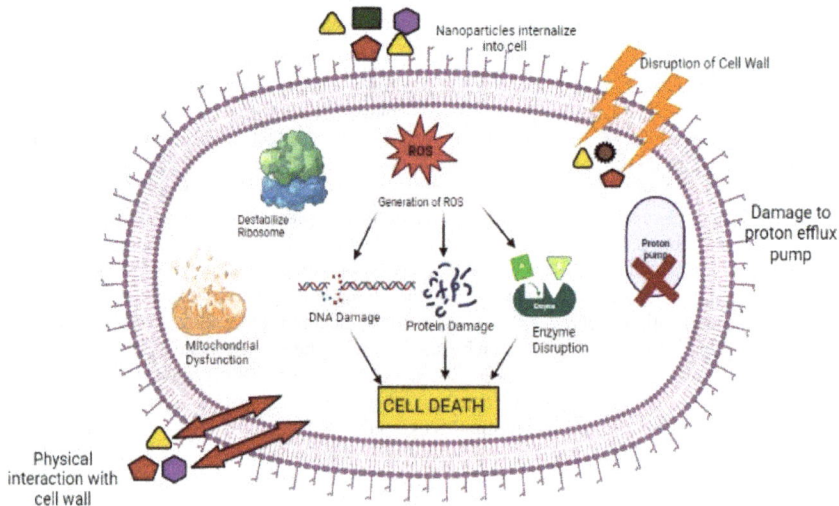

Fig 2. Anti-bacterial mechanism of nanoparticles

In addition, there are some fundamental mechanisms which include cell wall penetration, destruction of DNA, damage to proteins and deprivation of cell immortality. There are some stages to be followed; penetration of nanoparticles in the cell is the first step which includes adsorption or diffusion of them to the cell surface. It can be achieved by formation of a bond between nanoparticles and functional groups of proteins which possess negative charge on them. There is evolvement in studies to determine antimicrobial activity of nanoparticles against bacteria, fungi and viruses. There are different aims of this study which includes the summing of all the effects

Advances in Healthcare and Nanoparticle Toxicology   Materials Research Forum LLC
Materials Research Foundations 171 (2024) 76-92   https://doi.org/10.21741/9781644903339-3

of nanoparticles against pathogenic infections along with study of their mechanism of actions. It also encourages researchers to study potential benefits of nanoparticles in this particular field [53].

10. Applications of nanoparticles

The disciplines of material sciences and biology are combined through the use of nanomaterials in biotechnology. With their distinct characteristics and promise for a wide range of therapeutic uses, nanoparticles offer a particularly helpful platform. Due to a number of characteristics including their comparable size to biomolecules like proteins and polynuclear acid, nanoparticles have wide range of special features and applications. Additionally, a variety of metal that contribute beneficial features like luminescence and magnetic attraction may be used to create nanoparticles. In biological field, it has its major applications in gene delivery, bio-sensing and bio-imaging. Using nanoparticles, surface enhanced Raman scattering (SERS) has been effectively used to biological sensing [54].

Drug delivery systems also play a role by increasing efficacy of free medications by increasing their solubility. Biomolecules like DNA. RNA or proteins can be transported through the cell membrane using nanoparticles as efficient transporters, preventing their decomposing. These biomolecules can deliver in a safe manner, opening the door to the use of gene therapy and protein-based treatments [55].

A promising method for influencing cellular and extracellular processes for a variety of biological applications, including transcriptional control, catalytic blocking, and transportation and sensing that is surface recognizing of biological molecules by nanoparticles acting as artificial receptors. An essential objective for medical diagnosis, forensic sciences and ecological surveillance is the sensing of biological containments, illness and harmful materials. A sensor typically has two parts: a transducer component for signaling attaching events and a recognition component for adhering to targets. These systems are intriguing possibilities for applications in sensing because of distinctive physiochemical characteristics of nanoparticles. And they built-in improvement in signal-to-noise ratio afforded by miniaturization. For example, due to the size and form, gold nanoparticles display distinctive optical and electrical characteristics [56].

Other applications include fluorescent biological labels, detection of pathogens, protein detection, analyzing of DNA structure, and bioengineering of tissues and removal of tumors with high temperature. Nano-silicon plays role in insect repellant. After being sprayed to a plant's surface, silicon kills insects physically. Poly-ethylene glycol-coated nanoparticles employed as a pesticide have insect repellent properties against adult Tribolium castaneum insects [57].

Since silver has long been recognized to have antibacterial characteristics, a variety of commercial items have been created to take advantage of these characteristics, including antimicrobial paints, pants and other household textiles. Some solutions to these problems, such as masks or sprays for surface employing silver nanoparticles and the other components or technologies, have recently been introduced to market to fight against SARS-CoV-2 epidemic e.g., FFP2 face masks. However, antiviral silver nanoparticles could be utilized in the creation of goods and tools in other industries that have high danger of viral infections [58].

11.   Negative impact of nanoparticles

Although nanoparticles have more positive effects and applications in biotechnology, pharmaceutical science, drug targeting and many other fields of study but it also causes detrimental effects in the field of agriculture. These negative effects include plant growth inhibition, suppression of chlorophyll synthesis and it also affects photosynthesis. According to plant species, the type of nanoparticles utilized and their concentration, effects of nanoparticles might be either favorable or unfavorable [59].

Some of the nanoparticles have both positive and negative influence on plant growth. They have positive effects on some plants and same particles have negative effects on other plant species. For example, AgNPs nanoparticles cause increase in growth of plants and seed sprouting in *Lolium multifolium* and *Eruca sativa* but it causes inhibition of root length when there is high concentration and there is increase in root length at low concentrations of nanoparticles. In some plants, antiviral nanoparticles decrease weight of shoots and enhance its biomass, e.g. in *Triticum aestivum* [60]. On the other hand, Nano-silicon and nano-titanium dioxide increase the ability of Glycine max seed to soak up the water and nutrients by boosting nitrate reductase which in turn boost the speed sprouting. Nano-silicon dioxide allows plants to thrive in stressful environments and increase leaf weight, proline, accumulation of amino acids and chlorophyll, adsorption of nutrients and enzyme metabolism [61].

Conclusion

The last 10 years have seen a significant amount of research on nanoparticles because on their many potential uses. They can be produced by chemical, physical and biological methods and also by some microorganisms like algae, yeast and fungi. Antibacterial administration has significantly advanced as a result of development in field of nanotechnology especially nanoparticle engineering and the growing understanding of infectious illness. The development of several nanoparticle-based delivery systems has received attention. By providing targeted, adaptable and creative antibiotic administration, efficient antibacterial vaccination and quick pathogenic detection, these nanoparticle techniques have demonstrated great results in treatment and diagnosis bacterial infections.

They have become more essential for medication delivery in pharmaceuticals and treating cancer along with other diseases. For healthcare to be successful and efficient, it is imperative that medicine distribution may be managed and targeted. One of the biggest challenges facing modern medicine is the ability to manage site-specific localization and hence availability at the intended location. Other than that, nanoparticles exert positive and negative impacts on multiple fields.

References

[1]Iravani, S., *Bacteria in nanoparticle synthesis: current status and future prospects.* International scholarly research notices, 2014. 2014.

[2]Harper, G.D., et al., Roadmap for a sustainable circular economy in lithium-ion and future battery technologies. Journal of Physics: Energy, 2023. 5(2): p. 021501.

Materials Research Forum LLC
https://doi.org/10.21741/9781644903339-3

[3] Sakthi, P., et al., Sonochemical synthesis of interconnected SnS nanocrystals for supercapacitor and solar-physical conversion applications. Optical Materials, 2022. 132: p. 112759.

[4] Trivedi, R., et al., Recent Advancements in Plant-Derived Nanomaterials Research for Biomedical Applications. Processes, 2022. 10(2): p. 338.

[5] Capeness, M.J., V. Echavarri-Bravo, and L.E. Horsfall, Production of biogenic nanoparticles for the reduction of 4-Nitrophenol and oxidative laccase-like reactions. Frontiers in microbiology, 2019. 10: p. 997.

[6] Haghighat, M., et al., Cytotoxicity properties of plant-mediated synthesized K-doped ZnO nanostructures. Bioprocess and Biosystems Engineering, 2022: p. 1-9.

[7] Ali, A., et al., Synthesis and Characterization of Silica, Silver-Silica, and Zinc Oxide-Silica Nanoparticles for Evaluation of Blood Biochemistry, Oxidative Stress, and Hepatotoxicity in Albino Rats. ACS omega, 2023.

[8] Ranjan, N. and R. Kumar, *Nanoparticles: Properties and its 3D printing applications.* Materials Today: Proceedings, 2022. 48: p. 1316-1319.

[9] Sarraf, M., et al., Mixed oxide nanotubes in nanomedicine: A dead-end or a bridge to the future? Ceramics international, 2021. 47(3): p. 2917-2948.

[10] Egbuna, C., et al., *Toxicity of nanoparticles in biomedical application: nanotoxicology.* Journal of Toxicology, 2021. 2021: p. 1-21.

[11] Sumanth, B., et al., *Fungal biogenesis of NPs and their limitations.* Microbial Nanotechnology: Green Synthesis and Applications, 2021: p. 81-101.

[12] Mohajerani, A., et al., Nanoparticles in construction materials and other applications, and implications of nanoparticle use. Materials, 2019. 12(19): p. 3052.

[13] Pan, S., et al., Therapeutic applications of metal and metal-oxide nanoparticles: Dermato-cosmetic perspectives. Frontiers in Bioengineering and Biotechnology, 2021. 9: p. 724499.

[14] Ben Amor, I., et al., Sol-gel synthesis of ZnO nanoparticles using different chitosan sources: effects on antibacterial activity and photocatalytic degradation of AZO Dye. Catalysts, 2022. 12(12): p. 1611.

[15] Barhoum, A., et al., Review on natural, incidental, bioinspired, and engineered nanomaterials: history, definitions, classifications, synthesis, properties, market, toxicities, risks, and regulations. Nanomaterials, 2022. 12(2): p. 177.

[16] Ringwal, S., et al., Citrus Medica Mediated Ag Doped Mgo Nanocomposites as Green Adsorbent and its Catalytic Performance in the Rapid Treatment of Water Contaminants. Aayasha and Bartwal, Dr, Ankit S., Citrus Medica Mediated Ag Doped Mgo Nanocomposites as Green Adsorbent and its Catalytic Performance in the Rapid Treatment of Water Contaminants.

[17] Lopes, J., et al., How to treat melanoma? The current status of innovative nanotechnological strategies and the role of minimally invasive approaches like ptt and pdt. Pharmaceutics, 2022. 14(9): p. 1817.

[18]   Balkrishna, A., R. Joshi, and A.D.a.V. Arya, Endophytic Fungi: A Comprehensive Review on Their Secondary Metabolites, Pharmacological Interventions and Host Plant Interactions. Endophytic Fungi, 2022: p. 1.

[19]   Chauhan, A., et al., Biogenic synthesis: A sustainable approach for nanoparticles synthesis mediated by fungi. Inorganic and Nano-Metal Chemistry, 2023. 53(5): p. 460-473.

[20]   Singh, K., G. Singh, and J. Singh, Sustainable synthesis of biogenic ZnO NPs for mitigation of emerging pollutants and pathogens. Environmental Research, 2023. 219: p. 114952.

[21]   Alphandéry, E., Applications of magnetotactic bacteria and magnetosome for cancer treatment: a review emphasizing on practical and mechanistic aspects. Drug Discovery Today, 2020. 25(8): p. 1444-1452.

[22]   Shimoshige, H., et al., Fundidesulfovibrio magnetotacticus sp. nov., a sulphate-reducing magnetotactic bacterium, isolated from sediments and freshwater of a pond. International Journal of Systematic and Evolutionary Microbiology, 2022. 72(9): p. 005516.

[23]   Wang, W., et al., Engineering versatile nano-bacteria hybrids for efficient tumor therapy. Coordination Chemistry Reviews, 2023. 488: p. 215178.

[24]   Alavi, M., et al., Synergistic combinations of metal, metal oxide, or metalloid nanoparticles plus antibiotics against resistant and non-resistant bacteria. Micro Nano Bio Aspects, 2022. 1(1): p. 1-9.

[25]   Kanwal, A., et al., Antimicrobial applications of nanoparticles, in Handbook of Research on Green Synthesis and Applications of Nanomaterials. 2022, IGI Global. p. 269-288.

[26]   Sunkar, S. and C.V. Nachiyar, Biogenesis of antibacterial silver nanoparticles using the endophytic bacterium Bacillus cereus isolated from Garcinia xanthochymus. Asian Pacific Journal of Tropical Biomedicine, 2012. 2(12): p. 953-959.

[27]   Kalimuthu, K., et al., Biosynthesis of silver nanocrystals by Bacillus licheniformis. Colloids and surfaces B: Biointerfaces, 2008. 65(1): p. 150-153.

[28]   Lukman, A.I., et al., Facile synthesis, stabilization, and anti-bacterial performance of discrete Ag nanoparticles using Medicago sativa seed exudates. Journal of colloid and interface science, 2011. 353(2): p. 433-444.

[29]   Shahverdi, A.R., et al., Rapid synthesis of silver nanoparticles using culture supernatants of Enterobacteria: a novel biological approach. Process Biochemistry, 2007. 42(5): p. 919-923.

[30]   Zhang, H., et al., Biosorption and bioreduction of diamine silver complex by Corynebacterium. Journal of Chemical Technology & Biotechnology: International Research in Process, Environmental & Clean Technology, 2005. 80(3): p. 285-290.

[31]   Lengke, M.F., et al., Mechanisms of gold bioaccumulation by filamentous cyanobacteria from gold (III)− chloride complex. Environmental science & technology, 2006. 40(20): p. 6304-6309.

[32]   Sharma, N., et al., Exploitation of marine bacteria for production of gold nanoparticles. Microbial cell factories, 2012. 11: p. 1-6.

Advances in Healthcare and Nanoparticle Toxicology   Materials Research Forum LLC
Materials Research Foundations 171 (2024) 76-92   https://doi.org/10.21741/9781644903339-3

[33]   Arora, D., et al., Preparation, characterization and toxicological investigation of copper loaded chitosan nanoparticles in human embryonic kidney HEK-293 cells. Materials Science and Engineering: C, 2016. 61: p. 227-234.

[34]   Malarkodi, C., et al., Bactericidal activity of bio mediated silver nanoparticles synthesized by Serratia nematodiphila. Drug invention today, 2013. 5(2): p. 119-125.

[35]   Thamilselvi, V. and K. Radha, Synthesis of silver nanoparticles from Pseudomonas putida NCIM 2650 in silver nitrate supplemented growth medium and optimization using response surface methodology. Digest Journal of Nanomaterials & Biostructures (DJNB), 2013. 8(3).

[36]   Wu, Z., et al., One-step supramolecular multifunctional coating on plant virus nanoparticles for bioimaging and therapeutic applications. ACS Applied Materials & Interfaces, 2022. 14(11): p. 13692-13702.

[37]   Azizi, M., et al., *Multifunctional plant virus nanoparticles: An emerging strategy for therapy of cancer.* Wiley Interdisciplinary Reviews: Nanomedicine and Nanobiotechnology, 2022: p. e1872.

[38]   Mellid-Carballal, R., et al., *Viral protein nanoparticles (Part 1): pharmaceutical characteristics.* European Journal of Pharmaceutical Sciences, 2023: p. 106460.

[39]   Aljabali, A.A., et al., *Protein-based nanomaterials: a new tool for targeted drug delivery.* Therapeutic delivery, 2022. 13(6): p. 321-338.

[40]   Aguilera-Correa, J.J., J. Esteban, and M. Vallet-Regí, Inorganic and polymeric nanoparticles for human viral and bacterial infections prevention and treatment. Nanomaterials, 2021. 11(1): p. 137.

[41]   Gad, S.S., et al., Selenium and silver nanoparticles: A new approach for treatment of bacterial and viral hepatic infections via modulating oxidative stress and DNA fragmentation. Journal of Biochemical and Molecular Toxicology, 2022. 36(3): p. e22972.

[42]   Patel, R.R., S.K. Singh, and M. Singh, Green synthesis of silver nanoparticles: methods, biological applications, delivery and toxicity. Materials Advances, 2023. 4(8): p. 1831-1849.

[43]   Gad, S.S., et al., Nanotechnology applications for treatment of hepatic infections via modulating Hepatic histopathological and DNA alterations. Bioorganic Chemistry, 2022. 127: p. 105927.

[44]   Wu, S., et al., Bacterial outer membrane-coated mesoporous silica nanoparticles for targeted delivery of antibiotic rifampicin against Gram-negative bacterial infection in vivo. Advanced Functional Materials, 2021. 31(35): p. 2103442.

[45]   Yeh, Y.-C., et al., Nano-based drug delivery or targeting to eradicate bacteria for infection mitigation: a review of recent advances. Frontiers in chemistry, 2020. 8: p. 286.

[46]   Abafogi, A.T., et al., Vancomycin-conjugated polydopamine-coated magnetic nanoparticles for molecular diagnostics of Gram-positive bacteria in whole blood. Journal of Nanobiotechnology, 2022. 20(1): p. 400.

[47] Jiang, L., L. Ding, and G. Liu, Nanoparticle formulations for therapeutic delivery, pathogen imaging and theranostic applications in bacterial infections. Theranostics, 2023. 13(5): p. 1545.

[48] Dash, S.R. and C.N. Kundu, Advances in nanomedicine for the treatment of infectious diseases caused by viruses. Biomaterials Science, 2023.

[49] Iravani, S. and R.S. Varma, Vault, viral, and virus-like nanoparticles for targeted cancer therapy. Materials Advances, 2023.

[50] Asr, M.H., et al., Lipid Nanoparticles as Promising Carriers for mRNA Vaccines for Viral Lung Infections. Pharmaceutics, 2023. 15(4).

[51] Sarker, S.R., et al., Functionalized concave cube gold nanoparticles as potent antimicrobial agents against pathogenic bacteria. ACS Applied Bio Materials, 2022. 5(2): p. 492-503.

[52] Racovita, A.D., *Titanium dioxide: structure, impact, and toxicity.* International Journal of Environmental Research and Public Health, 2022. 19(9): p. 5681.

[53] Zheng, Y., et al., Titanium carbide MXene-based hybrid hydrogel for chemo-photothermal combinational treatment of localized bacterial infection. Acta Biomaterialia, 2022. 142: p. 113-123.

[54] Shen, F., et al., Target-triggered Au NPs self-assembled for fluorescence-SERS dual-mode monitoring of telomerase in living cells and in vivo. Sensors and Actuators B: Chemical, 2023. 374: p. 132789.

[55] Rebelatto, E.R.L., G.S. Rauber, and T. Caon, An update of nano-based drug delivery systems for cannabinoids: Biopharmaceutical aspects & therapeutic applications. International Journal of Pharmaceutics, 2023: p. 122727.

[56] Liu, R., et al., Advances of nanoparticles as drug delivery systems for disease diagnosis and treatment. Chinese chemical letters, 2023. 34(2): p. 107518.

[57] Yeganeh, F.E., et al., Formulation and characterization of poly (ethylene glycol)-coated core-shell methionine magnetic nanoparticles as a carrier for naproxen delivery: growth inhibition of cancer cells. Cancers, 2022. 14(7): p. 1797.

[58] Sportelli, M.C., et al., *On the efficacy of ZnO nanostructures against SARS-CoV-2.* International journal of molecular sciences, 2022. 23(6): p. 3040.

[59] Bao, L., X. Cui, and C. Chen, *Toxicology for Nanotechnology*, in *Nanomedicine*. 2023, Springer. p. 157-177.

[60] Kumar, A., et al., Role of nanocomposites in sustainable crop plants' growth and production, in Engineered Nanomaterials for Sustainable Agricultural Production, Soil Improvement and Stress Management. 2023, Elsevier. p. 161-181.

[61] Bhat, K.A., et al., Silicon-and nanosilicon-mediated disease resistance in crop plants, in Silicon and Nano-silicon in Environmental Stress Management and Crop Quality Improvement. 2022, Elsevier. p. 193-205.

Advances in Healthcare and Nanoparticle Toxicology   Materials Research Forum LLC
Materials Research Foundations 171 (2024) 93-134   https://doi.org/10.21741/9781644903339-4

Chapter 4

Whispers of Healing: Navigation of Nanomaterials for Wound Restoration and Inflammation Control

Nameeta Jesudoss M.[1], Cannon Antony Fernandes[1], Jorddy Neves Cruz[2],
Vasantha Veerappa Lakshmaiah[1]*

[1]Department of Life Sciences, CHRIST (Deemed to be University), Hosur Road, Bangalore, Karnataka 560029, India

[2]Institute of Biological Sciences, Federal University of Pará, Belém, Pará, Brazil

*vasantha.vl@christuniversity.in

Abstract

Impaired healing in chronic wounds is associated with an upsurge in financial burden for treatment, high mortality rates and limited availability of medicinal resources which mounts a burden on the global healthcare system. The major challenges associated with wounds are the absence of suitable microenvironment to facilitate the migration of cells and angiogenesis, microbial infection at the wound site and aggressive inflammation. The existing treatment strategies have failed to combat these challenges and aid in faster healing of the wounds. In an effort to address these challenges, recent studies prove the use of the unique advantages of nanomaterials towards effective wound healing. Integration of nanomaterials into the current therapeutic approaches has the potential to induce the required molecular and cellular process to create an amiable microenvironment for wound healing through its anti-bacterial, anti-inflammatory properties and angiogenic effects. This review highlights the pathophysiology of wound healing and the recent findings and challenges pertaining to the use of nanomaterials towards chronic wound management.

Keywords

Chronic Wounds, Nanomaterials, Inflammation, Angiogenesis, Infection, Healing Mechanism

Contents

Graphical abstract

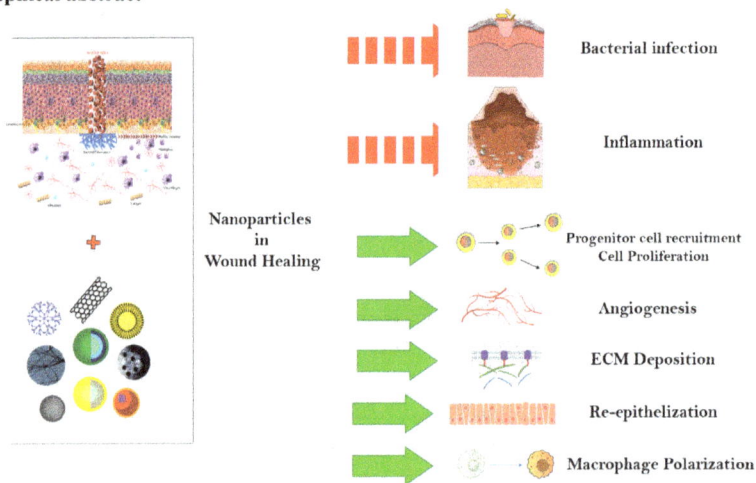

Advances in Healthcare and Nanoparticle Toxicology   Materials Research Forum LLC
Materials Research Foundations 171 (2024) 93-134   https://doi.org/10.21741/9781644903339-4

1. Introduction

Chronic wounds have been a serious life-threatening issue with the alarming surge in the mortality rate and substantial prevalence in the global population [1]. The main issues associated with the healing of chronic wounds are hindrance in the process of cell migration and proliferation, absence of microenvironment for angiogenesis, microbial infection at the site of the wound and prolonged inflammation associated with it [2]. Nanotechnology exhibits profound application in the process of wound healing in regenerative medicine. The recently described nanoparticles aid in fastening the wound healing towards recovery and serve as an effective treatment option for the injury. The unique display of the favorable properties of the nanoparticles such as decreased *in vivo* toxicity, efficient bacteriostatic and bactericidal property, biocompatibility, precise drug delivery, and decreased size of the nanoparticle which increases the surface area to volume ratio attributes towards the physiochemical properties for its best application of material science in biological process. [3,4]. The existing therapeutic approaches to combat chronic wounds have limitations but integration of nanomaterials had stimulated events at both cellular and molecular levels for hasten wound healing. This chapter will discuss the physiology of wound healing, various cellular mechanisms, application of nanomaterials in modulation of wound healing and inflammation with emerging prospects and challenges of nanomaterials in wound management.

2. Pathophysiology of wound healing

Basically, the wound healing process can be divided into four stages: hemostasis, inflammation, proliferation, and remodeling [5]. In Hemostasis, the body shows the earliest attempt to halt bleeding after a wound which is mediated by interlinked cascade of reactions to form platelet plug and coagulation by fibrin clots. There is also constriction in the blood arteries carrying the blood towards wounded region [6]. The next phase is inflammation associated with vasodilation facilitating extravasation and exudation for leukocytes and plasma fluid that aid in clearance of pathogen, cellular debris and wound healing. Inflammation is associated with cardinal signals such as redness, pain heat, swelling and loss of function. The third phase of proliferation in wound healing is to cover the surface of the wound with an epithelial layer serving as a new physical barrier followed by replenishing the lost cells by granulation tissue the fills the wounded site. Thus, new tissue is regenerated to replace lost tissue [7]. New tissue is produced during the proliferation stage to replace any lost or damaged tissue. Collagen is created by fibroblasts and serves as the framework for developing new tissue. In order to provide nutrition and oxygen to the healing tissue, new blood vessels are also produced (angiogenesis). Epithelial cells move over the wound to create a fresh barrier of defense [8]. The newly created tissue goes through alterations to improve its strength and usefulness throughout the remodeling stage. The extra collagen is broken down when the collagen fibers mature. The scarring id the natural and the final stage in wound healing which can endure for many months or even years[9]. This emphasizes the various stages wherein nanoparticles can be used for a localized, sustained and efficient wound healing.

Advances in Healthcare and Nanoparticle Toxicology   Materials Research Forum LLC
Materials Research Foundations 171 (2024) 93-134   https://doi.org/10.21741/9781644903339-4

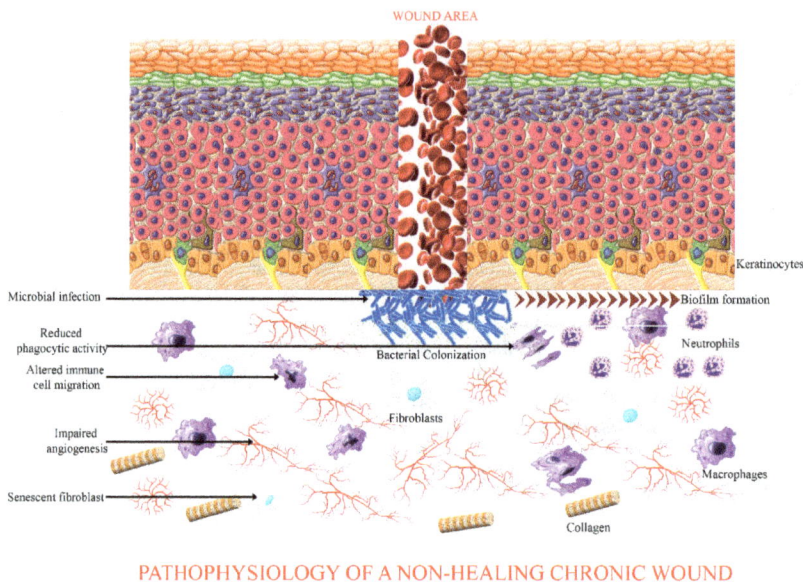

PATHOPHYSIOLOGY OF A NON-HEALING CHRONIC WOUND

Figure 1. Pathophysiology of a non-healing chronic wound

2.1 Growth factors, cytokines delivery and other endogenous compounds' roles in wound healing

Wounds can be categorized into acute and chronic types [10]. Acute wounds are generally traumatic or surgical events that heal certainly following a regular healing process. Burn wounds are another class of wounds that are caused by heat, chemicals, electricity, sunlight, radiation or friction [11].

The four complex stages of wound healing already mentioned are regulated by growth factors and cytokines that compose a complex signaling network that alters the growth, differentiation and metabolism in targeted cells[12,13] as in the fig.1. The effect of growth factors is executed by their binding to specific receptors which leads to a cascade of molecular events [14–16]. These growth factors are helpful in the wound healing of acute wounds while chronic wounds are more susceptible to infection due to the prolonged healing periods. The main distinguishing feature of chronic wounds is the persistence of inflammatory state, which is brought by an unchecked and unregulated inflammatory positive feedback loop. As a result, neutrophils persist during the entire healing phase and release significant quantities of matrix metallic proteinases (MMPs), which promote the destruction of the matrix around the wound. Additionally, the proteolytic microenvironment causes growth factors to degrade, which inhibits their activity, and the process is regulated. Additionally, fibroblasts have a diminished capacity to migrate and respond to growth factors stimuli, which slows down the healing process [17,18]. Some of the families of growth

Advances in Healthcare and Nanoparticle Toxicology   Materials Research Forum LLC
Materials Research Foundations 171 (2024) 93-134   https://doi.org/10.21741/9781644903339-4

factors and cytokines that play an active role in wound healing are, Epidermal Growth Factor (EGF), Fibroblast Growth Factor (FGF), Transforming Growth Factor-β (TGF- β), Activins, Bone Morphogenic Proteins (BMPs), Platelet Derived Growth Factor (PDGF), Vascular Endothelial Growth Factor (VEGF), Connective Tissue Growth Factor (CTGF), Granulocyte Macrophage-Colony Stimulating Factor (GM-CSF), Proinflammatory Cytokines and Chemokines [19].

These growth factors and cytokines play crucial roles in wound healing by orchestrating and regulating various cellular processes involved in the complex cascade of tissue repair. These are few crucial functions reported. Growth factors encourage keratinocytes (skin cells) and fibroblasts (connective tissue cells) to proliferate and divide. Additionally, they encourage these cells' migration to the area of the wound. This is necessary to heal the wound and restore tissue integrity [20]. Angiogenesis, the process by which new blood vessels form into a wound area, depends on growth factors, primarily vascular endothelial growth factor (VEGF). In order to promote the healing process, adequate blood flow is essential for transporting oxygen, nutrients, and immune cells to the wound site [21]. Growth factors, such as platelet-derived growth factor (PDGF) and transforming growth factor-beta (TGF-β), promote the production of the extracellular matrix (ECM). The ECM serves as a scaffold for cell migration and gives tissues structural stability required in remodeling phase. To maintain the integrity and strength of the tissue, collagen, a crucial element of the ECM, is required [22]. Growth factors direct how different cell types involved in wound healing differentiate. For instance, they promote the development of myofibroblasts, which are contractile cells involved in wound contraction and scar formation, from fibroblasts [23]. Some cytokines, such as interleukin-10 (IL-10) and transforming growth factor-beta (TGF-β) have anti-inflammatory properties. They help regulate the immune response at the wound site, preventing excessive inflammation that could hinder healing. They attract immune cells to the wound site, promoting the removal of debris and pathogens, which is essential to preventing infections [24]. In the final stages of wound healing, tissue remodeling and maturation, growth factors and cytokines play a role. They aid in controlling the ratio of ECM production to degradation, ensuring that the newly created tissue gradually acquires strength and stability. The activation of nearby stem cells can be induced by growth factors, aiding in tissue regeneration and repair. Certain growth factors, such as TGF-β, stimulate the production of collagen, which is crucial for wound strength and scar formation. Growth factors like PDGF contribute to wound contraction, which reduces the wound size and helps in the closure of the wound.

3.   Cellular interactions and extracellular matrix (ECM) remodeling

Key steps in wound healing include cellular interactions and extracellular matrix (ECM) remodeling. These procedures entail the intricate interaction of several cell types, signaling chemicals, and structural elements that cooperate to restore tissues that have been injured. The ECM has many functions in providing structure, organization, and orientation to cells and tissues; controlling morphogenesis and cellular metabolism by acting as a template for cell migration, proliferation, apoptosis, differentiation, and adhesion [25]; regulating cell activity and function via direct binding to integrins and other cell surface receptors; act as a reservoir for growth factors and regulate their bioavailability [26]. Numerous growth factors, including fibroblast growth factors, TGF-β, vascular endothelial growth factors, epidermal growth factors, and bone morphogenetic proteins, are bound by ECM proteins such as fibronectin, collagens, proteoglycans, heparin, and heparin sulphate. The local release of these growth factors from their insoluble anchoring can be induced by the proteolytic enzymes that break down ECM proteins in response to wounds, which

Advances in Healthcare and Nanoparticle Toxicology   Materials Research Forum LLC
Materials Research Foundations 171 (2024) 93-134   https://doi.org/10.21741/9781644903339-4

alters the healing process [27]. In addition, recent studies also demonstrate that ECM proteins are key components in shaping the stem cell niche to maintain stem cell homeostasis and to direct lineage commitment [28].

Collagen, laminin, elastins, and fibronectins, which are structural proteins, provide flexibility. Proteoglycans and hyaluronan, which have a high capacity to bind water, stabilize growth factors and the three-dimensional space. Integrins, which are glycoproteins, control cell adhesion and signaling between cells and the ECM. The interaction between local cells and these chemicals listed enables building the ECM niche.

3.1   Enhanced cell adhesion and proliferation

When a wound occurs, platelets and immune cells are recruited to the site. Platelets release factors that promote clot formation and stimulate the recruitment of immune cells. Fibroblasts, endothelial cells, and keratinocytes migrate to the wound site, and adheres to the extracellular matrix (ECM) for healing [29]. The ECM provides a scaffold for cell attachment and migration. Enhancing cell-ECM interactions can improve cell recruitment and stability at the wound site for regeneration. Integrins, which are cell surface receptors, mediate cell adhesion to the ECM proteins like fibronectin, collagen, and laminin [30].

Keratinocytes, the predominant cells in the epidermis, migrate from the wound edges to cover the surface during the proliferation phase. Enhanced proliferation of keratinocytes fastens wound coverage, reducing the risk of infection and dehydration [31]. Fibroblasts are responsible for synthesizing and depositing new ECM components, such as collagen and elastin. Increased fibroblast proliferation contributes to the formation of granulation tissue, which fills the wound bed and supports tissue repair [32]. New blood vessel formation (angiogenesis) is essential for supplying nutrients and oxygen to the healing tissues. Endothelial cell proliferation and migration lead to the formation of capillary networks within the wound bed [33].

3.2   ECM deposition and reorganization

ECM deposition involves the synthesis and secretion of ECM components by specialized cells such as fibroblasts, chondrocytes, and osteoblasts. These components include fibrous proteins like collagen and elastin, glycoproteins like fibronectin and laminin, and proteoglycans. Cells produce these molecules and secrete them into the extracellular space, where they assemble into a three-dimensional network [34]. The reorganization is the dynamic remodeling of the ECM after deposition. This process involves the interaction between cells and the ECM, leading to changes in the ECM's composition, structure, and mechanical properties. Cells can interact with the ECM through cell-surface receptors like integrins, which anchor them to specific ECM components. This interaction triggers signaling pathways that can influence cell behavior and ECM remodeling.

During tissue development, ECM reorganization guides cell migration, tissue morphogenesis, and differentiation. In wound healing, damaged tissue triggers ECM remodeling to create a provisional matrix that supports cell migration and tissue repair. In diseases like fibrosis, abnormal ECM reorganization can lead to excessive deposition and tissue stiffening [35]. Cell-ECM interactions are crucial for various cellular processes. Cells can sense and respond to mechanical cues from the ECM, influencing their behavior and gene expression. Integrins, as mentioned earlier, are key receptors that mediate these interactions. Cells can activate intracellular signaling pathways, such as focal adhesion kinase (FAK) and Rho GTPases, in response to ECM cues [36].

Advances in Healthcare and Nanoparticle Toxicology   Materials Research Forum LLC
Materials Research Foundations 171 (2024) 93-134   https://doi.org/10.21741/9781644903339-4

3.3   Influence on fibroblasts and keratinocytes

Fibroblasts are the primary cells responsible for synthesizing and secreting components of the ECM. They produce various types of collagens, elastin, fibronectin, proteoglycans, and other structural proteins that make up the ECM [37]. During tissue repair and wound healing, fibroblasts migrate to the wound site and deposit ECM components to form granulation tissue. This tissue provides structural support and aids in wound closure [38]. Fibroblasts are involved in ECM remodeling, which helps maintain tissue integrity and adapt to changing physiological conditions. They can degrade old or damaged ECM components and replace them with newly synthesized ones [39]. Fibroblasts interact with the ECM through cell-matrix adhesions, such as focal adhesions. These interactions influence cell behavior, migration, proliferation, and differentiation [40].

Keratinocytes are the predominant cells in the epidermis and are the outermost layer of the skin. They contribute to the formation of the epidermal barrier by producing proteins such as keratins, which provide mechanical strength and help prevent water loss [41]. Like fibroblasts, keratinocytes interact with the ECM through various cell-matrix adhesions. These interactions are essential for maintaining epidermal integrity and facilitating cell movement during wound healing [42]. They secrete factors like growth factors and cytokines that can influence fibroblast activity and ECM synthesis. For instance, keratinocytes produce transforming growth factor-beta (TGF-β), which can stimulate fibroblasts to produce collagen [43]. During wound healing, keratinocytes at the wound edges migrate to cover the wound surface, creating a temporary barrier. They also produce factors that attract immune cells and fibroblasts to the wound site, initiating the healing process [44].

4.   Nanoparticles in modulation of inflammatory responses:

Inflammatory response is an intrinsic natural phenomenon in immunity to maintain homeostasis in tissue. This is considered to be the basic defense mechanism to fight against infections, damage to cells, stress and so on. However, when its unregulated or exaggerated, it can lead to pathogenesis causing disease. Inflammation includes a series of reaction involving phagocytic cells, serum proteins, signaling cytokines and receptors to effectively mediate inflammatory response. Extensive research into the understanding of pathophysiology of inflammation had identified an unregulated inflammatory response leading to chronic diseases such as cardiovascular diseases, type 2 diabetes, neurodegenerative diseases, asthma, obesity and so on [45]. Extensive research at Harvard in Sheran's lab has provided a deeper insight towards understanding the phenomenon of inflammation as active mechanism involving the participation of endogenous activators for recovery and homeostasis. These pro-resolving cytokines such as lipoxins (e.g., LXA4), resolvins (e.g., RvD1), protectins and maresins. These can mediate few key events to end inflammation such as inhibition of leucocyte recruitment, trigger signaling pathway to induce apoptosis of leucocytes, reprogramming the macrophages to shift from proinflammatory response to resolve inflammation and thus have significant role in regulation of inflammation [46].

Thes pro-resolving mediators can be used as therapeutic agents to treat various chronic diseases. Their pharmacological application was examined in mouse model of atherosclerosis with vesicular lesion. Intravenous administration of nanomedicine comprising of polymeric nanoparticle encapsulated with annexin A1—a protein had ensured persistent release of Ac2-26 that could effectively resolve the plaque. These Col IV–Ac2-26 nano particles was able to reduce reactive oxygen species responsible for inflammation and tissue damage. Furthermore, IL-10 can serve as anti-inflammatory cytokine and suppress inflammatory response in different chronic diseased conditions [47].

IL10 can suppress the adverse effects of chronic inflammation by decreasing the antigen processing and presentation by antigen presenting cells to down regulate the immune response from T-Helper cells. It can inhibit the activation of macrophages and their infiltration to inflammation site [48]. IL-6 is a cytokine that facilitates hematopoiesis and inflammation. In inflammatory response these are responsible for senescence induced inflammation. Studies with mice models to analyze the effect of silver nanoparticles towards burnt wound healing had reported that application of Ag NPs in wound dressing had drastically reduced the production of IL-6 by keratinocytes and infiltered neutrophiles. Furthermore, regulation was by inhibiting the infiltration neutrophiles. Exposure to Ag NPs led to an upregulated expression of KGF-2 confirmed with western blotting and immunofluorescence techniques. These assist proliferation of keratinocytes at wounded sites for quick healing followed by reducing the oxidative stress in tissue and protects from ROS-induced apoptosis [49].

Cerium oxide nanoparticles (nanoceria) and their free radical scavenging activity was studies with invitro culture J774A.1 murine macrophages. This property of nanoceria was studied by treating macrophages with nanoparticles which was found to be nontoxic and biocompatible to macrophages. The ability of these nanoparticles to remove the reactive nitric oxide species had resulted in down regulation of inflammatory response. Hence nanoceria can reduce oxidative stress in macrophages and reduce the chronic effects of inflammatory response as anti-ROS and anti-inflammatory agent [50].

Fe_3O_4/CeO_2-core and shell NPs were examined for its cytotoxicity and anti-ROS and anti-inflammatory properties by performing in vitro studies with Macrophage J774A.1. Studies emphasized the nanoparticles to be a potential therapeutic agent in treating ROS associated inflammation. The cytotoxicity studies performed with Ethidium homodimer-1 fluorescence dye and fluorescent microscopy revealed nanoparticles to be biocompatible and reduce the ROS level in macrophage cells which was owing to its antioxidant properties of nanoparticles [51,52].

5.   Nanoparticles in macrophage polarization

Macrophages are immune cells in tissue defending against invading pathogens. These are cells with multiple functions associated with inflammation, allergic reactions and cancer immunity. Basically, there are two types of macrophages namely M1 and M2 cells. The major role of M1 is to serve as first line of defense against invading pathogens and pro-inflammatory response While M2 aid in cell repair and are associated with anti-inflammatory response. The physiological microenvironment drives the polarization of Macrophages with two extreme function to modulate the immune response. The process of macrophage polarization towards M1 is through signal transduction pathway mediated by various regulatory molecules such as PAMPs (Pathogen associated molecular patterns), DAMPs (Damaged associated molecular patterns). Interferon γ, LPS (lipopolysaccharides) and T-helper subtype1 produced cytokines to trigger inflammatory response. There are few regulatory molecules to promote polarization to M2 macrophages which are reported as IL-4, IL-10, IL-13 and T- helper subtype 2 cytokines. The plasticity of M1/M2 was mediated by interferon γ released by immune cells in response to TLR (Toll Like Receptor) and CLR (C-type Lectin Receptor). This channelizes signal transduction through activation of STAT1 (signal transducer and transcription activator 1) leading the polarization to M1. On other hand LPS and other microbial ligands will trigger TIRF (TIR-domain-containing adapter-inducing interferon-β) and myeloid differentiation response 88 (MyD88). These are the adapter molecules to nuclear factor kappa-B (NF-kB) pathways crucial for the expression of epigenetic transcription factors promoting polarization to M1 and induce the secretion of proinflammatory cytokines such as tumor necrosis factors (TNF), IL-1β, and IL-12.

M2 macrophages play a central role in tissue repair and are producing collagen. Stimulation of M2 macrophage polarization is triggered by IL-4. IL-10, IL-13. These cytokines bind to IL-4Rα1, IL13Rα1 & I L13Rα2 receptors to mediate the activation of JAK1 and JAK3 to mediate STAT6 signaling pathway to induce expression of M2 genes such as arginase 1 (*Arg1*), mannose receptor 1 (*Mrc1*), resistin-like α (*Retnla, Fizz1*), chitinase-like protein 3 (*Chil3, Ym1*), and the chemokine genes *Ccl17* and *Ccl24* for M2 polarization of macrophage and tissue repair[53]

Optimally formulated and administered nanoparticles can target the macrophages by changing their microenvironment, the macrophages are reprogrammed and modulated for polarization. Nanoparticles can be used to modulate and reprogramme macrophages to treat inflammation and cancer. Basically, the chemical composition, size and shape of nanoparticles can influence the process of macrophage(M0) polarization to M1 or M2. J774.A1 macrophages were treated with Au and Ag NPs of same size and monitored their pro-inflammatory cytokines such as IL-1, IL-6 and TNF-α. Au NPs were more efficient to induce the polarization of macrophages to M1 the Ag NPs. Studies confirmed that the differential cellular uptake of these nanoparticles by macrophages through receptor mediated endocytosis and pinocytosis respectively that had influenced the signaling process towards polarization [54]. Adenocarcinoma cells when co cultured with macrophages and treated with ferumoxytol nanoparticles had stimulated the reprogramming the macrophages (M2) for proinflammatory response that can inhibit the tumour growth [55] as described in the fig.2.

Folic acid formulated silver NPs (F-AgNPs) were used as therapeutic agent to treat rheumatoid arthritis (RA) was studied using RA mice model. The chronic inflammatory injury in RA is due to infiltration of M1 macrophages that induces proinflammatory response. These macrophages can effectively uptake F-AgNPs (Follic acid – Silver nanoparticles) with folic acid receptors. Released Ag ions induce ROS scavenging, apoptosis of M1 and repolarization of M1 to M2 [56]. Rebamipide (RBM) is one of the recommended drugs to treat prostatic urothelium wounds. Major limitation of this drug is poor solubility and thus poor availability for wound healing. Invitro studies with Macrophages when exposed to RBM cross linked with chitosan mediates slow and gradual release of drug for prolonged exposure to encourage wound healing. Furthermore, RBM suppress pro-inflammatory M1 macrophages polarization [57]. Large porous mesoporous silica nanoparticles (MSNs) were explored as drug delivery system for effective immunomodulation in mice models. IL4 coated on MSNs to deliver the cytokine to macrophages which directs for M2 macrophages polarization and reduce the oxidative stress and promote antiinflammation [58]. Studies with acute lung injury mice models when intratracheally injected with gold nanoparticles coated with hexapeptide had improved the stability of NPs and mediated in vivo anti-inflammatory cytokine IL-10. There was a significant increase in the number of alveolar macrophages M2 resolving the inflammation in lung tissue [59].

Osteoarthritis is a chronic disease with inflammation in joints. One of the effect means of modulating the immune response to resolve inflammation is through macrophage polarization from M1 to M2 macrophages to aid tissue repair. Nanoparticles are opsonized with IgG antibodies and compatible polymer polylysine along with anti-inflammatory agent berberine (Bb) and bilirubin as marker for oxidative stress. Invitro studies of macrophages with this designed opsonized nanoparticle (IgG/Bb@BRPL) were effectively taken up by pro-inflammatory M1 macrophage and had driven their polarization to M2 macrophages promoting the protection and repair of chondrocytes [60].

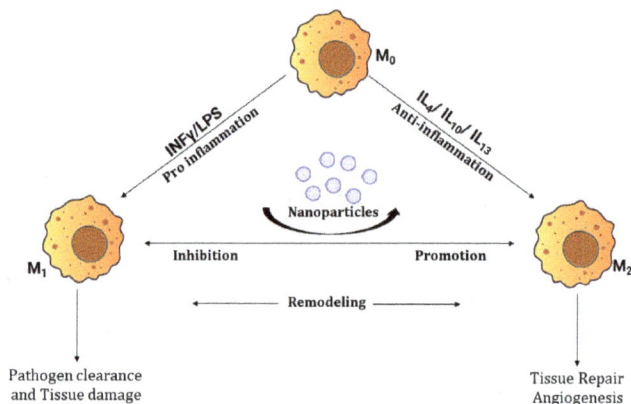

Figure 2 Nanoparticles in polarization of macrophages

6. Nanoparticles in regulation of oxidative stress

Reactive oxygen and nitrogen species are extremely reactive, strong oxidizing agents. They can help with defense mechanism as they can oxidize any biomolecules of pathogen like proteins, lipid, and even cause damage to their DNA. At physiological concentration these ROS and RNS are associated with signal transduction to regulate major cellular activities such as cell proliferation. Cell-cell interaction, differentiation and apoptosis [61]. These reactive oxygen species are the by-products of electron transport chains in mitochondria as part of cellular respiration. However, there are NADPH oxidases and nitric oxide synthase associated with phagocytic cells like macrophages, neutrophils etc. that can produce ROS & RNI during infection and inflammation to aid in immune response. The excess production of ROS/RNS can lead to cellular damage, inflammation and metabolic, neurodegenerative and chronic diseases [62]. Studies confirm the direct relation between ROS and proinflammatory cytokines. Hence there is a cross talk between them. The increase in ROS can trigger NF-κB pathway for the production of proinflammatory markers such as IL-6, IL-1, TNF-α as well as cyclooxygenase-2 (COX-2).[63] [64]

The residential macrophages have a prime role in tissue inflammatory response as they initiate, regulate and resolve the process. These macrophages can help in tissue repair by producing anti-inflammatory cytokines (IL-10 and TGF-β), growth factors (VEGF, PDGFA) and lipid meditators (resolvins, protectins, and maresins) to resolve inflammation [65]. An in vitro study was performed to determine the effect of silver nanoparticles on murine peritoneal macrophage cell line RAW 264.7 towards inflammation. When exposed to silver nanoparticles of different sizes 20 to 110nm had induced oxidative stress in cultured macrophage generating ROS. This had resulted in secretion of pro-inflammatory cytokines such as TNFα, MIPs and G-CSF. TNFα can immunomodulate the whole process of inflammatory reaction as they can induce production other cytokines participating in immune response while MIPs and G-CSF are associated with the chemoattraction and extravasation of neutrophils to inflammatory site and the growth factor mediated the development and maturation of neutrophiles respectively [66].

The role of silica nanoparticles in inflammation was studies both in vivo and in vitro studies using mice animal models and in RAW264.7 cell line. Mice treated with silica NPs intraperitonially reported the activation of macrophages. Both in vivo and in vitro studies showed enhanced NO release and pro inflammatory gene expression (IL-1, TNF- and IL-6) or other inflammatory-related enzymes (iNOS, COX-2) [67].

Different species of nanoparticles were used to study their role as mediators for inflammation. This study was done with RAW 264.7 mouse macrophages cell line wherein the cells were exposed to silver (Ag), aluminium (Al), carbon black (CB), carbon coated silver (Cag) and carbon coated gold (Cau) at a concentration of 5µg/ml. Macrophages response was ROS driven to proinflammatory cytokines produced by NF-κB signaling pathways. Major cytokines that were monitored in course of the macrophage activation were TNF-α, IL-6 and COX-2. Among the test nanoparticles, Ag ranked best as inflammation inducer followed by Al, CB and cAg [68].

Periodontitis is generally referred to as inflammation in tissue that supports the teeth. Pathogen-induced periodontitis leads to activation of tissue-based macrophages and recruited neutrophiles to produce ROS. This creates an imbalance between antioxidants and ROS leading to oxidative stress. ROS can cause pathogenesis with apoptosis of ligament stem cells, interferes with migration of ligament fibroblast and Alveolar bone resorption. Thus, eliminating the ROS can help in the recovery of periodontitis which can be easily achieved through administration of ROS scavenging nanoparticles and nanocomposites. *Larrea divaricata* Cav. mediated Chi-CMC-SiO2 nanocomposite exhibited excellent antioxidant activity to overcome the oxidative stress in inflammation [69].

Fe_3O_4 were found to mimic the enzymatic activity of horse radish peroxidase this open avenue to explore nanoparticles as nanozymes. These nanozymes have significant advantages over natural enzymes with improved stability and activity. Research in this direction had identified various nanozymes with peroxide, superoxide dismutase and catalase activity. This extended their application in medical field to maintain a healthy redox potential. There on reports on the application of these nanozymes to fight against chronic inflammatory diseases and ROS associated injury. Selenium Nanoparticles were effective in the recovery of inflammatory bowel disease that were stabilized with polysaccharides from *Ulva lactuca* served as antioxidant [70]

When CeO_2 nanoparticles are administered orally as therapeutic agent its very necessary to protect these nanoparticles from harsh acidic conditions of the digestive system Thus to increase stability and efficiency of these nanoparticle in bowel inflammation was achieved by using nanozyme carriers, montmorillonite (MMT). Studies with mice model emphasized few basic properties of inflamed colon such as the surface endothelial cell expressing high amount of negatively charged protein and high ROS. This formulated CeO_2@MMT nanozymes mimics the function as Superoxide dismutase and catalase enzymes which in turn scavenge the excess ROS release in bowel inflammation. In this formulated nanozyme the MTT is a carrier that protects CeO_2 NPs and possess a net positive charge that facilitated their delivery to epithelial cells layer with negatively charged proteins [71,72].

Manganese based nanoparticles are widely used to treat ROS injury as they tend to mimic the antioxidant activity of enzymes such as glutathione peroxidase, catalase, and superoxide dismutase. These nanoparticles are entitled to be a promising alternative to natural antioxidant enzymes to treat inflammatory diseases. Flower shaped Mn_3O_4 nanoparticles was characterized and examined for their antioxidant potential against Parkinson disease model cell line neuroblastoma-derived cell line SHSY-5Y. Studies proved that the nanoparticles were morphologically stable and nontoxic to the neurons. Furthermore, we were able to mediated the cellular repair in neuron when exposed to neurotoxin MTT [73]. Manganese-Prussian blue nano enzyme (MPBZs) was orally administered to DSS-induced colon inflammatory mouse model. These nanozymes not only function as antioxidant enzymes but can also down regulate the expression of certain chemokines such as IL-1β, IL-6, IFN-γ, TNF-α, MPO and MDA to arrest the progressive inflammatory response. These nanoparticles are also altering the signal transduction pathway to inhibit the release of pro-inflammatory cytokines [74].

7. Significance of nanoparticles in treating bacterial infection

7.1 Inorganic metal nanoparticles

Based on the classification of nanoparticles, this section gives a summary of the representatives of the reported nanomaterials that promote wound healing. Depending on the raw materials used, nanoparticles can be classified as inorganic and organic nanoparticles. Further inorganic nanoparticles can be divided into metallic and non-metallic nanoparticles.

7.2 Metallic nanoparticles

Metal-based nanoparticles find extensive use in medicine owing to their advantages, such as simplified synthesis in desired and defined shape/size, potential to alter their surface morphology i.e., facile functionalization and biocompatibility to the biological system [75]. In this regard, the commonly used metal nanoparticles in wound healing are gold, silver, copper and zinc oxide (metal oxide nanoparticles).

Silver Nanoparticles: Usage of silver-based compounds to treat wounds dates to the early 1970s through the introduction of an antibiotic supplemented with silver sulphadiazine [76,77]. Recent research validates that silver nanoparticles show increased antimicrobial activity and effectively works as anticancer agents [78]. They exhibit greater antimicrobial property due to their small size and increased binding surface area [79]. The various biomaterials used in regenerative medicine like dressings for the wound, implants, scaffolds/ controlled release carrier are supplemented with silver nanoparticles to accelerate the wound healing property [80,81]. To facilitate even distribution of the drug, bandages have silver coatings. Silver nanoparticles conjugate with biomaterials like collagen, chitosan, Polyvinyl alcohol (PVA), sodium alginate hydrogels and polycaprolactone scaffold as wound coverings display remarkable antimicrobial properties. The synergistic effect of ZnO/CuO nanocomposites along with chitosan as a dressing was tested successfully in the year 2022 [82]. Various dressings for wound healing models with the incorporation of silver nanoparticles are being tested in animal models (Table 1). In 2015, dressing material for Wistar rats made with composites containing regenerated cellulose, chitosan and silver nanoparticles embedded showed complete closure of the wound in 25 days compared to 30 days in the control [83]. Enhanced wound healing was observed in the BALB/c male mice with a dressing based on chitosan- silk fibroin impregnated with silver nanoparticles. Along with the antimicrobial activity, *in vivo* studies show increased collagen deposition, nerve repair and angiogenesis [84,85]. It was also noticed that the multicoated membranes like the electro spun polycaprolactone/gelatin/nano silver showed accelerated healing rate [86]. In the wound dressing developed by Konop et al. where keratin biomaterial was supplemented with silver nanoparticles, showed broad range spectrum of antibacterial activity, healing was faster and no toxicity was detected [80]. Compared to the commercially available wound dressings Atrauman Ag and Aquacel Ag, it was noted that silver nanoparticles have a slow, sustained release from the dressings [87]. A major limitation with the direct usage of AgNPs is their toxicity in humanoid cells. Hence, it is always preferred to be used as a composite. Some significant attempts towards incorporating silver nanoparticles to accelerate wound healing through scaffolds/ dressings from the year 2015 are described in Table 1. Examples of the biological materials used are KGM, Hyaluronic acid, alginate, g-glutamate, sericin, chitosan, hydroxyapatite, polyurethane, PVP, polyvinyl alcohol, cellulose, PEG, fumaric acid, aloe, silk fibroin, rhEGF, bFGF EGF bamboo cellulose nanocrystals [88].

Copper nanoparticles: The significant antibacterial property of copper nanoparticles against wide spectrum of strains elucidate their potential in designing wound healing dressings [89]. These release Cu^{2+} ions which disrupt the cell walls/membranes by altering the enzymes' biological functions through solidifying the constituent protein structure[90]. The catalytic property results in the cytoplasmic membrane damage, elevated ROS production and thereby kill the bacteria by peroxidation and DNA degradation.[89] Copper improves immunity through the production of interleukin 2. It stimulates angiogenesis by inducing the expression of VEGF, hypoxia inducible factor which further accelerates wound healing [91]. It is noted that at higher concentrations, Cu NPs are lethal to cells (Dose dependent). Hence, Cu NPs can be used as both direct or modified applications as mentioned in the Table 1 for improved therapeutic effect.

Gold nanoparticles: They do not show any antimicrobial activity like that of the previously discussed nanoparticles, so they are extensively used in drug delivery targeting the wound site [92,93]. Studies elucidate that owing to the above said property, it is combined with other biomolecules for effective utilization in wound healing (Table 1). The commonly adapted techniques to use them are photothermal energy utilization, drug delivery and tissue regeneration [94]. The mode of action is quite similar to the silver nanoparticles, like formation of void on cell wall thereby causing leakage of the contents[95]. Au NPs are proven to inhibit the replicatory machinery of the cell, thereby inhibiting bacterial proliferation [96]. The accumulation of ROS, inflammatory cells are prevented, thereby maintaining a sterile environment to fasten wound healing[97,98]. This improves the antibacterial property through the usage of nanoparticles. Some significant attempts towards integrating Au NP into dressing are summarized in the table 1.

Other metallic nanoparticles: Representatives of metal oxide nanoparticles are discussed under this section. Zinc oxide nanoparticles find a pivotal role in cosmetic and the fillings of therapeutic products owing to their refined advantages. These are found embedded in chitosan hydrogel, cellulose sheets etc., A composite made of ZnO/Ag/PVP/PCL nanofiber enhanced regeneration and wound healing. Other metallic oxides found to be useful are TiO_2, CeO_2, Fe_2O_3 as they emerge as potential nanoparticles for skin wound healing. Overall, all the metallic nanoparticles have a significant application in wound healing.

Table 1. List of nanoparticles with potential antimicrobial activity towards wound healing.

Composition	Therapeutic Effects	Year	Ref
Silver Nanoparticles			
Ag NP + Enoxaparin	Accelerates the healing of burn wounds. The combination was found to decrease the time for differentiation of fibroblast into myofibroblast.	2015	[99]
Regenerated cellulose + chitosan + AgNPs + Gentamicin	Composite wound dressing for accelerated wound healing	2015	[100]
ε-polylysine + AgNP nanocomposite	Eradicates antibiotic resistant bacteria and promotes wound healing by modulating CD3+ T cells and CD68+ macrophages	2016	[101]
Gentamicin + AgNP	Accelerates wound healing and reduces wound scar on *Staphylococcus aureus* infected skin wounds	2017	[102]
collagen/chitosan hybrid scaffold + AgNP	Upregulate fibroblast migration and macrophage activation	2017	[103]
Plasma treated electrospun polycaprolactone scaffold + AgNP	Prevents the removal induced damage in the wound dressing.	2018	[86]
Ag NP + Gelatin + Chitosan	Enhanced antibacterial property and retains moisture for prolonged period.	2019	[104]
Poly (vinyl alcohol)/Sodium alginate hydrogel + 5-Hydroxymethylfurfural + AgNP	Facilitates migration of human skin fibroblast and collagen production	2019	[105]
Exosomes + AgNP	Promotes angiogenesis, nerve repair, broad spectrum antimicrobial activity	2020	[84]
Fur Keratin powder + AgNP	Increased wound inflammatory response	2020	[80]
Gallocatechin + AgNP	Improved diabetic wound healing by regulating apoptosis pathway	2022	[106]
Copper Nanoparticles			
Alginate hydrogel + CuNP	Significant antibacterial activity against multidrug resistance when used as a dressing	2014	[107]

GelMA + BACA + CuNP	It would promote angiogenesis, reduce inflammation and has elevated antibacterial properties	2019	[108]
Hydrogel made of PDA + calcium silicate ceramic + CuNP	Exhibit antibacterial activity against methilin drug resistant bacteria, promotes angiogenesis and vascularization.	2020	[109]
Graphene oxide + chitosan + Hyaluronic acid + CuNP	Reduces the inflammations and accelerates wound healing	2020	[110]
Chitosan + PVA + CuNP	Inhibits the growth of gram-positive bacteria and accelerated healing	2021	[111]
Chitosan + Gallic acid + CuNP	Through the antibacterial properties and promotes wound healing	2021	[112]
PDA + Hydroxyapetite + CuNP	Accelerates angiogenesis, collagen deposition and promotes wound healing	2021	[113]
Chitosan + CDs + Graphene Oxide	Inhibits wound healing and accelerates wound healing	2021	[114]
Gold Nanoparticles			
Vancomycin + AuNP	Effective against drug resistant strains of bacteria	2003	[115]
Chitosan + AuNP	Increased free radical scavenging at the wound site, enhanced hemostasis and epithelial tissue formation for high healing rate.	2011	[116]
Pig diaphragm + AuNP	A scaffold to increase the proliferation of cells, decreased ROS and enhanced wound healing	2011	[117]
AuNP enriched photosensitizer	Enhanced the antifungal properties at the wound site	2015	[118]
Polymers/stem cells/fibroblast + AuNP	Enhanced wound healing properties	2016	[119]
Chitosan + Gelatin + AuNP	Easy usage and increased wound healing efficiency	2016	[120]
Collagen + AuNP	Reduces the inflammation at wound site, promotes granulation to establish tissue homeostasis	2016	[121]
Ag + Chitosan + AuNP	Enhanced antibacterial properties, lowered cytotoxicity and promotes accelerated wound healing	2017	[122]
Keratinocyte growth factor + gelatin + AuNP	Scaffolds keratinocyte proliferation and reepithelization	2018	[123]
Poloxamer 407 + PEG + PAH + AuNP	Enhanced collagen deposition, antimicrobial activity to promote faster wound healing	2019	[124]
Gentamicin + gelatin + AuNP	Enhanced antibacterial activity against super bacteria	2020	[125]
Chitosan + Arginine + AuNP	Promotes collagen deposition and fastens wound healing	2020	[126]

Dextran + Sericin + AuNP	Reduces the possibility of scar formation and accelerates healing	2022	[127]
Dextran + PVA + AuNP	Accelerates wound healing	2022	[128]

7.3   Inorganic non-metallic nanoparticles

Noticeable nanoparticles in this category include: SiO_2 and carbon nanoparticles. Silicon dioxide is a biocompatible, easy to obtain and does not cause any additional immune responses [129]. These possess good adsorption capacity hence can be used to carry drugs. An attempt to carry curcumin on the surface of silica nanoparticles was made [130]. A hydrogel SiO_2-PVP composite promotes wound contraction and another attempt to synthesize silica – collagen I nanogel loaded with gentamicin/rifamycin exhibits continuous drug release [131]. Depending on the form of carbon used, carbon nanoparticles can be categorized as carbon dots, carbon nanotubes and graphene. Carbon nanotubes behave in synchrony with collagen fibers and influences adhesion, proliferation and differentiation[132]. Graphene is designed for surface functionalization of drugs and also are towards cell adhesion and growth [133]. In addition to the specific properties, carbon nanoparticles have enhanced antibacterial property [134–136]. An attempt to assemble a biofilm of multiple layers of nanotubes added with polypeptides show increased effective bactericidal properties [137].

7.4   Lipid based nanoparticles

They serve as an alternative to drug delivery for therapeutic agents and offer controlled release [138–140]. An example of an antimicrobial peptide LL37 encapsulated in lipid nanoparticle would be an ideal addition at the site of wound. Encapsulating the drug in the lipid nanoparticle prevents them from degradation and aids in sustained release for prolonged duration [141]. Liposomes is a potential carrier for both hydrophobic and hydrophilic drugs [142,143]. The main advantage is the possibility to explicitly modify to enhance its functionality [144]. The negative charge on the surface enables it to attach to bacteria and thereby inhibit its growth and colonization at the wound site [145,146].

7.5   Polymeric Nanoparticles

In alliance to their advantages of feasibility in manufacture, inexpensive cost, long shelf life makes them possible to get employed in dressing materials[147]. These are used as nanospheres and nano capsules. Some examples of synthetic polymers and natural polymers used are listed in table 2 below. The natural polymers have wide prospect in wound healing as they are mixed with drugs or other materials. Since they are similar to normal tissue, they are able to mimic extracellular matrix, suitable for cell adhesion reduce possibilities for infection at the wound site, promote immune cell movement [148,149]. A significant example is the attempt to construct a calcium alginate hydrogel with chitosan nanoparticles, which has antibacterial activity and increases IL6 production. It was noticed in animal studies that this hydrogel promotes wound healing through enhanced angiogenesis [150]. The limitations are their degradability and triggering immune response. These polymers are non-toxic, biocompatible, stable with increased half-life, good infusibility and pharmacokinetics.

Table 2. List of polymers used in wound healing.

Synthetic polymers	Natural polymers	Ref
Polyethylene	DNA	
Polypropylene	Hyaluronic acid	
Polystyrene	Gelatine	
Polyvinyl chloride	Collagen	[162-164]
Polylactides	Chitosan	
Polytetrafluorethylene	Alginate	
Polymethylmethacrylate	Rosemarinic acid	
Polyamides (Nylon)		
Polysiloxanes (Silicone)		
Polyurethanes		

8. Nanoparticle-mediated drug delivery systems

In the area of wound healing, nanoparticle-mediated drug delivery methods have demonstrated considerable promise. These methods increase the efficacy and efficiency of medication administration by using nanoparticles to encapsulate and transport therapeutic ingredients directly to the wound site. To promote wound healing, nanoparticle-mediated medication delivery methods can be utilized.

8.1 Nanotechnology-based drug- delivery system for wound healing

In order to effectively release the growth factors mentioned in the earlier section, a wide variety of biomaterials have been used for developing novel drug-delivery systems, and all of them are expected to meet a series of characteristics, such as good stability, biocompatibility and biodegradability, high drug-loading, good mechanical properties, controlled or sustained release and drug protection [151]

Nanoparticles are usually preferred in drug-delivery systems for effective wound healing because nanoparticles can be incorporated into wound dressings to improve their properties. These nanoparticles can release therapeutic agents gradually, maintaining a conducive wound environment [152]. Nanoparticles can encapsulate growth factors, cytokines, and other bioactive molecules. This allows for controlled and sustained release of these agents over time, promoting cell proliferation, angiogenesis, and tissue regeneration. Nanoparticles can enhance the penetration of therapeutic agents through barriers such as the stratum corneum (outermost layer of the skin). This enables efficient delivery of drugs to deeper layers of the wound for better healing outcomes. Functionalized nanoparticles can be designed to specifically target certain cell types, such as immune cells or fibroblasts, involved in wound healing. This targeted delivery increases the effectiveness of the treatment and reduces potential side effects [153]. Nanoparticles can help reduce the systemic toxicity of certain drugs by localizing their effects on the wound site. This is particularly important for drugs that may have adverse effects when distributed throughout the body. Nanoparticles can be designed to mimic the effects of growth factors, promoting cell migration, proliferation, and wound closure. They can also enhance collagen synthesis, contributing to tissue repair [154]. Nanoparticles can be used to deliver therapeutic genes to cells at the wound site, promoting the expression of specific proteins that accelerate wound healing processes. Nanotechnology can be used to develop nanosensors that detect specific biomarkers associated with wound healing. These sensors can provide real-time information about the wound's progress and aid in personalized treatment. Nanoparticles can be engineered to promote blood clotting and hemostasis, which is essential for controlling bleeding in wounds. Nanofiber scaffolds and nanocomposite materials can be used as three-dimensional structures to support tissue regeneration and wound closure. They mimic the natural extracellular matrix and provide a framework for cell attachment and growth [155]

Due to their antibacterial qualities, *silver nanoparticles* are frequently employed. They can stop the spread of germs and infections in wounds. To keep a wound sterile, silver nanoparticles can be added to dressings or coatings [156]. Researchers have looked at the anti-inflammatory properties and wound-healing enhancement of *gold nanoparticles*. Additionally, they can be functionalized for precise medication delivery [157]. Due to their antibacterial qualities, *zinc-oxide nanoparticles* can speed up the healing of wounds by avoiding infections. They are also well-recognized for supporting tissue regeneration [158]. Chitosan, a natural polymer generated from chitin (found in crustaceans), possesses antibacterial and wound-healing qualities. Applications for regulated medication delivery and treating wounds include *chitosan nanoparticles* [159,160]. To produce a moist wound environment, encourage cell migration, and permit controlled medication release, *hydrogels* may be manufactured at the nanoscale [161]. *Nanofiber scaffolds* resemble the extracellular matrix of tissues and are frequently built from substances like polymers or collagen. They offer a three-dimensional framework that facilitates tissue regeneration and cell adherence [162]. For targeted delivery of therapeutic agents to the location of the wound, *liposomes*—nanoscale lipid vesicles—can contain medications, growth factors, or other therapeutic agents [163]. Bioactive compounds can be contained and shielded by polymeric nanoparticles, enabling their regulated and sustained release over time [164]. *Ceramic nanoparticles*, such as hydroxyapatite nanoparticles, can promote bone regeneration and wound healing in cases where bone tissue is involved [165]. *Graphene* and its derivatives possess unique mechanical and electrical properties. They have been explored for wound healing applications due to their potential to promote cell adhesion, migration, and differentiation. *Calcium Phosphate nanoparticles* are used for bone regeneration and wound healing, especially in cases where there is a need for bone tissue repair [166]. *Carbon nanotubes* have been investigated for their potential in wound healing due to their mechanical strength and potential for promoting cell growth [167]

8.2 Antibiotics and antiseptics delivery

The delivery of antibiotics and antiseptics through nanoparticles in wound healing is an innovative approach that offers several advantages. They have been explored for their potential to enhance wound healing when incorporated into ointments and topical formulations. These nanoparticles can carry therapeutic agents, such as growth factors, antimicrobial agents, anti-inflammatory compounds, and antioxidants, to the wound site. When applied as ointments, nanoparticles can provide controlled and sustained release of these agents, promoting faster and more effective wound healing [168].

Nanoparticles have been shown to be a promising and significantly efficient alternative for reducing the bacterial load and the development of biofilms in the wound site [169,170]. However, it is crucial to remember that NPs' antibacterial mechanisms rely primarily on their close interaction with cellular membranes. As a result, NPs have the potential to be toxic to both bacterial and mammalian cells, with NP cytotoxicity depending on several factors including their chemical

Advances in Healthcare and Nanoparticle Toxicology   Materials Research Forum LLC
Materials Research Foundations 171 (2024) 93-134   https://doi.org/10.21741/9781644903339-4

makeup, crystalline structure, size, shape, and concentration. Obtaining the necessary antibacterial effect while minimizing the potential damage to human cells is therefore a significant problem for the use of NPs in antibacterial medical therapies. To address this issue, significant efforts have been made, including the use of NP capping agents, to modify the size, shape, and/or release rate of NPs [171–173].

9. Nanoparticles to improve angiogenesis and blood vessel formation:

Angiogenesis can be described as the growth of new blood vessels to facilitate a continuous supply of oxygen, elements of inflammatory response and other nutrients at the wound site. This is usually active at the proliferative phase during healing to form granulation tissue. It remains a big challenge in tissue engineering. It is to note that absence of active angiogenesis during wound healing would limit to the superficial wound only. The potential of using nanomaterials to promote early angiogenesis has gained focus. Two major aspects pertaining to the use of nanomaterials in wound healing are: effective carriers to deliver factors for angiogenesis (delivery vehicle) and stimulate the existing microenvironment through chemical signals to form blood vessels locally at the wound site [174,175]. This is an effective alternative to the existing methods of using growth factors (VEGF-A or PDGF) as they are observed to cause thrombosis and fibrosis. Aiding the promotion of angiogenesis can help meet the metabolic essentials for inflammation, collagen matrix deposition and re-epithelialization. Some examples are the release of silica ions to promote angiogenesis at the wound site [176] [177], engineering an extracellular vesicle that mimics the hydrogel to deliver LncRNA-H19 to treat diabetic wounds [178]. Works suggest that gold and cerium oxide nanoparticles show increased proangiogenic properties [179–181]. It is to note that angiogenesis is mediated through a series of continuous signaling mechanisms in which nanomaterials can benefit. The possible mechanisms of action of the nanomaterials can be to promote endothelial cell migration [182], activation of redox signaling [183], regulation of the cytoskeletal remodeling, mimics the microenvironment to fasten the healing process[184], upregulation of notch signalling[185] and promotes autophagy [186]. A detailed illustration of the plausible mechanisms is described in Fig.3 . In this line a series of examples are elucidated in table 3 to highlight the application of inorganic metal-based nanoparticles in tissue regeneration.

NANOPARTICLES TO INDUCE ANGIOGENESIS AT WOUND SITE

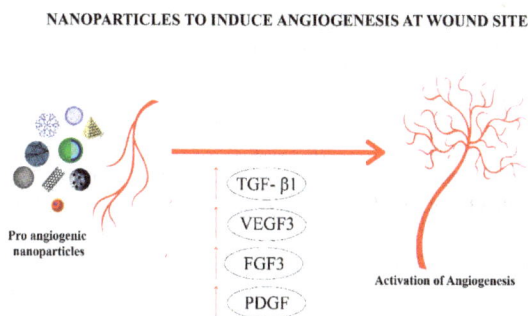

Figure 3 Nanoparticle to promote neo angiogenesis.

Table 3. Application of metallic nanoparticles for neo-angiogenesis.

Nanomaterials		Ref
Au NPs	Attempt to incorporate AuNPs onto a hydrocolloid membrane showed increased expression of angiogenesis biomarkers and was found to heal the fastest.	[187]
AuNPs	It was used as a supplement in the photo biomodulation therapy to accelerate angiogenesis.	[188]
Au nanodots	Assembly of surfactin and 1-dodecanethiol on the gold nanodots were found have new formation of collagen fibers.	[189]
SiO_2 NPs	PVP gel matrix incorporated with SiO_2 nanoparticles could heal dermal wounds with high degree of vascularization.	[190]
Cu_2S nanoflowers	Release of copper ions from the polycaprolactone and polylactide nanofibers induces the expression of VEGF to promote angiogenesis.	[191]
Cu_2O NPs	Though antiangiogenic, it works for softer and less visible scars during the process of wound healing.	[192]
Terbium (III) hydroxide as nanorods/spheres	Promotes the process of sprouting, activates P13K/AKT and exhibits proangiogenic effects.	[193]
Europium (III) hydroxide	Immunomodulatory effects by activating P13K/AKT pathway and further inhibition of TNF-a and IL-6.	[194]
Zinc oxide nanoparticles	They promote the migration of endothelial cells and by the production of nitric oxide as the product of MAPK/Akt/eNOS pathway, it enhances the formation of blood vessel.	[195]
Cerium oxide nanoparticles	Stimulates pro-angiogenesis, eases the tube structure formation and enhances vascular sprouting, VEGF	[196]
Bioactive glasses	Silica based glass embedded with SiO_2, CaO, Na_2O, P_2O_5 which shows increased dissolution of ions.	[197]

10. Nanoparticles in wound dressings and scaffolds

The inclusion of nanoparticles in the dressings is a promising factor for wound repair for their unique features of low toxicity and enhanced biocompatibility [198,199]. An ideal wound dressing should be exudate adsorbents, appreciable porosity, significant rate of water vapor transmission, antibacterial, elastic, flexible, enough capacity to load drugs, anti-inflammatory are the desired characteristics. In light of the limitations of the traditional materials, the use of nanoparticles as an alternative component to antibiotics is in the surge as they do not cause side effects, microbial resistance and are effective in inhibiting the growth of bacteria at the wound site [200]. The use of nanoparticle enables to sustain moisture at the wound site to imitate the niche of extracellular matrix [201].

The following examples are the inorganic nanoparticle embedded within a supporting polymer matrix. The trends show that largest effort of research was dedicated to dressings with silver nanoparticles. But to focus on the others, gold nanoparticles were embedded in a nano porous hydrogel made of heparin and polyvinyl alcohol which was able to inhibit bacterial growth in the burn wounds of Kunming mice [202]. Copper nanoparticle was similarly incorporated in a double layer nanofiber. Polyvinyl alcohol and chitosan making up the first layer and poly-N-vinyl pyrrolidone as the second layer. It showed accelerated healing in the albino Wistar rats [203]. Recently, zinc oxide nanoparticle containing nanofiber made of polyacrylic acid and polyallylamine hydrochloride showed effective antimicrobial properties [204]. A hydrocolloid patch with ZnO improved the healing in Sprague Dawley rats by immunomodulating the microenvironment [205]. Use of iron oxide in a nanocomposite made of chitosan, polyvinyl alcohol improved the healing in diabetics sore [206]. Cerium dioxide nanoparticles was embedded in the gelatin-caprolactone nanofibers which significantly decreased the expression of resistance genes in bacteria [207]. Another nanofiber effort was to incorporate cerium dioxide nanoparticles in poly-L-lactic acid and gelatin as a cheap effort to minimize scars and elevate healing [208]. A nanocomposite was manufactures with titanium dioxide nanoparticles on carbon nanotubes and loaded further in cellulose acetate-collagen film. They showed a broad range of antimicrobial activity [209]. The increase in antibacterial properties were observed in nanofibers made of polycaprolactone and gelatin with copper oxide nanoparticle embedded on it [210]. Apart from the nanoparticle's incorporation in dressings, a lot of nanomaterials have emerged such as polymeric nanofibers, nanoceria and polymeric nano scaffolds [211]. Carbon based nanomaterials have a significant role in mediating wound healing owing to their unique properties of high surface area, targeted drug delivery and inherent antimicrobial properties [212]. Graphene induces ROS independent oxidative stress in bacteria thereby displaying antibacterial properties [213,214]. Graphene oxide nanofilm in epoxy coating accelerates wound healing [215]. Another attempt to disrupt the genetic material of bacterium to combat their proliferation is by combining graphene to the chitosan polyvinyl alcohol nanofiber [216]. Single walled carbon nanotube along with mercury was effective in treating infections in burns by increasing the expression of mRNA coding for wound healing factors [217]. Fullerene was found to immunomodulate and accelerate healing [218]. Some elucidated significant examples of organic nanoparticles in wound healing are: Liposomes as a drug delivery system for growth factors like glypican-1 induces angiogenesis to improve healing [219]. Polymeric nanoparticle combinations like PLGA, gelatine, chitosan, alginate is used in alleviating the challenges of wound healing [220]. Loading of EGF in poly-L-lactic acid enhances healing in gastric ulcers by acting as a nanocarrier [221]. Nitric oxide-controlled release from materials promotes skin homeostasis, induces keratinocyte proliferation and angiogenesis [222]. Nanofibers that mimic the physiology of skin display profound applications in dressings for wound healing. It can be used to develop novel scaffolds to support and carry bioactive materials [223]. Polycaprolactone can be an alternative to treat burns by facilitating adhesion and fibroblast proliferation [224]. Other natural polymer-based nanofibers that can be used to treat wounds are collagen, silk fibroin, alginate polysaccharides, chitosan and hyaluronic acid, dextran, xantham. These give an insight into the development of customized wound bandages in order to accelerate wound healing. Though the manufacturing cost is less the structural defects should be considered while manufacturing in large scale [225].

11. Limitations and future perspectives of using nanoparticles in wound healing:

Though the use of nanoparticles in wound healing has proven to be significant, it is important to be mindful of their biological safety before using [226]. There are significant reports of nanoparticles causing skin irritation, noxiousness, allergies, hypersensitivity on the skin, inflammation, irritation when applied on the wound site [227]. Hence, it stresses the need to optimize the dose, shape and stability to decrease their harmful effects on the skin [228]. It is also reported that exposure to nanoparticles leads to DNA damage, reduces the gene methylation and prompts towards tumor formation [229]. Some nanoparticles are known to cause hemolysis like Ag NPs and ZnO NPs through establishing a direct contact with the blood. In order to overcome the limitations of using them in the biological system, it is important to manipulate their physicochemical properties or to wrap the nanoparticles with bioactive substance like polysaccharides/phospholipids [230]. Accumulation of nanoparticles in the organs of the body can induce multisystem defects like organ damage or tumors [231]. To have a better understanding on their toxicity and its fate upon reaching the biological system, a detailed investigation is necessary to resolve the complications.

The multifunctional dressings should accelerate healing through controlling bacterial overburden, provide Ideal physio-chemical environment, regulate the wound healing physiology, improve angiogenesis and alleviate inflammation. In order to address the above challenges, nanotechnology could be an advantage. The major limitation with the existing studies is the usage of rodent models for *in vivo* studies where they differ structurally and functionally from human skin. Rodent skin heals through contraction whereas human skin heals by epithelization. Pig's skin is more analogous but the practical difficulties working with bigger animals is a big limitation. Another hurdle in using nanoparticles towards wound healing lies with its formulation costing. Improving the formulation, drug release, dosage optimization would significantly reduce the production/manufacturing expenses.

Though there are significant advances in designing multipurpose dressings, there is need of research in the following areas. Efforts can be made towards developing biomarkers through nanomedicine in understanding the pathophysiology of wound healing at the molecular level. Specific drugs should be developed for each stage of healing. Significant research has been carried out to develop smart delivery systems to deliver therapeutics. In improving translational prospects, clinical trials should be done extensively to devise products for patient's usage. Further, significant research is being done to investigate the role of nanoparticles to promote regeneration of hair follicles, evading scar development and in developing electronic skin. The research in this would pave the way towards developing smart strategies for wound healing.

Conclusion

Application of Nanoparticles in regenerative medicines had revolutions in the healing and inflammation processes towards speedy recovery from chronic inflammatory disease. Extensive research had emphasized the potential of Nanoparticles to stimulate and alter the molecular and cellular mechanism resulting in change in microenvironment that promotes the entire process of wound healing. Nanoparticles of different species are employed as anti-inflammatory, anti-microbial, and as efficient drug delivery system. It performs multiple tasks at various levels of healing like regulation of cytokines production, reducing oxidative stress, induce macrophage polarization, promote neo-angiogenesis. The efforts towards identification of biosafe, compatible nanoparticles to formulate better wound dressing is a major focus for navigating the nanomaterials to clinical science.

References

[1] D.G. Armstrong, J. Wrobel, J.M. Robbins, Guest Editorial: are diabetes-related wounds and amputations worse than cancer?, Int Wound J 4 (2007) 286–287. https://doi.org/10.1111/j.1742-481X.2007.00392.x.

[2] A. Clinton, T. Carter, Chronic Wound Biofilms: Pathogenesis and Potential Therapies, Lab Med 46 (2015) 277–284. https://doi.org/10.1309/LMBNSWKUI4JPN7SO.

[3] R.K. Thapa, K.L. Kiick, M.O. Sullivan, Encapsulation of collagen mimetic peptide-tethered vancomycin liposomes in collagen-based scaffolds for infection control in wounds, Acta Biomater 103 (2020) 115–128. https://doi.org/10.1016/j.actbio.2019.12.014.

[4] N. Monteiro, M. Martins, A. Martins, N.A. Fonseca, J.N. Moreira, R.L. Reis, N.M. Neves, Antibacterial activity of chitosan nanofiber meshes with liposomes immobilized releasing gentamicin, Acta Biomater 18 (2015) 196–205. https://doi.org/10.1016/j.actbio.2015.02.018.

[5] S. Ellis, E.J. Lin, D. Tartar, Immunology of Wound Healing, Curr Dermatol Rep 7 (2018) 350–358. https://doi.org/10.1007/s13671-018-0234-9.

[6] G. Hosgood, Stages of Wound Healing and Their Clinical Relevance, Veterinary Clinics of North America: Small Animal Practice 36 (2006) 667–685. https://doi.org/10.1016/j.cvsm.2006.02.006.

[7] G. Gethin, Understanding the inflammatory process in wound healing, Br J Community Nurs 17 (2012) S17–S22. https://doi.org/10.12968/bjcn.2012.17.Sup3.S17.

[8] J. Li, J. Chen, R. Kirsner, Pathophysiology of acute wound healing, Clin Dermatol 25 (2007) 9–18. https://doi.org/10.1016/j.clindermatol.2006.09.007.

[9] A.C. de O. Gonzalez, T.F. Costa, Z. de A. Andrade, A.R.A.P. Medrado, Wound healing - A literature review, An Bras Dermatol 91 (2016) 614–620. https://doi.org/10.1590/abd1806-4841.20164741.

[10] T.N. Demidova-Rice, M.R. Hamblin, I.M. Herman, Acute and Impaired Wound Healing, Adv Skin Wound Care 25 (2012) 304–314. https://doi.org/10.1097/01.ASW.0000416006.55218.d0.

[11] L.F. Rose, R.K. Chan, The Burn Wound Microenvironment, Adv Wound Care (New Rochelle) 5 (2016) 106–118. https://doi.org/10.1089/wound.2014.0536.

[12] S. Barrientos, H. Brem, O. Stojadinovic, M. Tomic-Canic, Clinical application of growth factors and cytokines in wound healing, Wound Repair and Regeneration 22 (2014) 569–578. https://doi.org/10.1111/wrr.12205.

[13] P. Olczyk, Ł. Mencner, K. Komosinska-Vassev, Diverse Roles of Heparan Sulfate and Heparin in Wound Repair, Biomed Res Int 2015 (2015) 1–7. https://doi.org/10.1155/2015/549417.

[14] S. Barrientos, O. Stojadinovic, M.S. Golinko, H. Brem, M. Tomic-Canic, PERSPECTIVE ARTICLE: Growth factors and cytokines in wound healing, Wound Repair and Regeneration 16 (2008) 585–601. https://doi.org/10.1111/j.1524-475X.2008.00410.x.

[15]   R.D. Burgoyne, A. Morgan, Secretory Granule Exocytosis, Physiol Rev 83 (2003) 581–632. https://doi.org/10.1152/physrev.00031.2002.

[16]   S. Muzammil, J. Neves Cruz, R. Mumtaz, I. Rasul, S. Hayat, M.A. Khan, A.M. Khan, M.U. Ijaz, R.R. Lima, M. Zubair, Effects of Drying Temperature and Solvents on In Vitro Diabetic Wound Healing Potential of Moringa oleifera Leaf Extracts, Molecules 28 (2023). https://doi.org/10.3390/molecules28020710.

[17]   G. Gainza, S. Villullas, J.L. Pedraz, R.M. Hernandez, M. Igartua, Advances in drug delivery systems (DDSs) to release growth factors for wound healing and skin regeneration, Nanomedicine 11 (2015) 1551–1573. https://doi.org/10.1016/j.nano.2015.03.002.

[18]   N.B. Menke, K.R. Ward, T.M. Witten, D.G. Bonchev, R.F. Diegelmann, Impaired wound healing, Clin Dermatol 25 (2007) 19–25. https://doi.org/10.1016/j.clindermatol.2006.12.005.

[19]   S. Barrientos, O. Stojadinovic, M.S. Golinko, H. Brem, M. Tomic-Canic, PERSPECTIVE ARTICLE: Growth factors and cytokines in wound healing, Wound Repair and Regeneration 16 (2008) 585–601. https://doi.org/10.1111/j.1524-475X.2008.00410.x.

[20]   P. Wee, Z. Wang, Epidermal Growth Factor Receptor Cell Proliferation Signaling Pathways, Cancers (Basel) 9 (2017) 52. https://doi.org/10.3390/cancers9050052.

[21]   Y. Zhao, A.A. Adjei, Targeting Angiogenesis in Cancer Therapy: Moving Beyond Vascular Endothelial Growth Factor, Oncologist 20 (2015) 660–673. https://doi.org/10.1634/theoncologist.2014-0465.

[22]   K. Hyldig, S. Riis, C. Pennisi, V. Zachar, T. Fink, Implications of Extracellular Matrix Production by Adipose Tissue-Derived Stem Cells for Development of Wound Healing Therapies, Int J Mol Sci 18 (2017) 1167. https://doi.org/10.3390/ijms18061167.

[23]   M.B. Dreifke, A.A. Jayasuriya, A.C. Jayasuriya, Current wound healing procedures and potential care, Materials Science and Engineering: C 48 (2015) 651–662. https://doi.org/10.1016/j.msec.2014.12.068.

[24]   J. Larouche, S. Sheoran, K. Maruyama, M.M. Martino, Immune Regulation of Skin Wound Healing: Mechanisms and Novel Therapeutic Targets, Adv Wound Care (New Rochelle) 7 (2018) 209–231. https://doi.org/10.1089/wound.2017.0761.

[25]   J. Domínguez-Bendala, L. Inverardi, C. Ricordi, Regeneration of pancreatic beta-cell mass for the treatment of diabetes, Expert Opin Biol Ther 12 (2012) 731–741. https://doi.org/10.1517/14712598.2012.679654.

[26]   G.S. Schultz, A. Wysocki, Interactions between extracellular matrix and growth factors in wound healing, Wound Repair and Regeneration 17 (2009) 153–162. https://doi.org/10.1111/j.1524-475X.2009.00466.x.

[27]   M. Xue, C.J. Jackson, Extracellular Matrix Reorganization During Wound Healing and Its Impact on Abnormal Scarring, Adv Wound Care (New Rochelle) 4 (2015) 119–136. https://doi.org/10.1089/wound.2013.0485.

[28]   M.F. Brizzi, G. Tarone, P. Defilippi, Extracellular matrix, integrins, and growth factors as tailors of the stem cell niche, Curr Opin Cell Biol 24 (2012) 645–651. https://doi.org/10.1016/j.ceb.2012.07.001.

[29]   A. Opneja, S. Kapoor, E.X. Stavrou, Contribution of platelets, the coagulation and fibrinolytic systems to cutaneous wound healing, Thromb Res 179 (2019) 56–63. https://doi.org/10.1016/j.thromres.2019.05.001.

[30]   C.M. Murphy, M.G. Haugh, F.J. O'Brien, The effect of mean pore size on cell attachment, proliferation and migration in collagen–glycosaminoglycan scaffolds for bone tissue engineering, Biomaterials 31 (2010) 461–466. https://doi.org/10.1016/j.biomaterials.2009.09.063.

[31]   P. Rousselle, F. Braye, G. Dayan, Re-epithelialization of adult skin wounds: Cellular mechanisms and therapeutic strategies, Adv Drug Deliv Rev 146 (2019) 344–365. https://doi.org/10.1016/j.addr.2018.06.019.

[32]   R.B. Diller, A.J. Tabor, The Role of the Extracellular Matrix (ECM) in Wound Healing: A Review, Biomimetics 7 (2022) 87. https://doi.org/10.3390/biomimetics7030087.

[33]   C.S. Oliver Cassell, O.P. Stefan Hofer, W.A. Morrison, K.R. Knight, Vascularisation of tissue-engineered grafts: the regulation of angiogenesis in reconstructive surgery and in disease states, Br J Plast Surg 55 (2002) 603–610. https://doi.org/10.1054/bjps.2002.3950.

[34]   A.D. Theocharis, D. Manou, N.K. Karamanos, The extracellular matrix as a multitasking player in disease, FEBS J 286 (2019) 2830–2869. https://doi.org/10.1111/febs.14818.

[35]   C. Walker, E. Mojares, A. del Río Hernández, Role of Extracellular Matrix in Development and Cancer Progression, Int J Mol Sci 19 (2018) 3028. https://doi.org/10.3390/ijms19103028.

[36]   J.D. Humphrey, E.R. Dufresne, M.A. Schwartz, Mechanotransduction and extracellular matrix homeostasis, Nat Rev Mol Cell Biol 15 (2014) 802–812. https://doi.org/10.1038/nrm3896.

[37]   G.S. Schultz, A. Wysocki, Interactions between extracellular matrix and growth factors in wound healing, Wound Repair and Regeneration 17 (2009) 153–162. https://doi.org/10.1111/j.1524-475X.2009.00466.x.

[38]   R.B. Diller, A.J. Tabor, The Role of the Extracellular Matrix (ECM) in Wound Healing: A Review, Biomimetics 7 (2022) 87. https://doi.org/10.3390/biomimetics7030087.

[39]   M. KJÆR, Role of Extracellular Matrix in Adaptation of Tendon and Skeletal Muscle to Mechanical Loading, Physiol Rev 84 (2004) 649–698. https://doi.org/10.1152/physrev.00031.2003.

[40]   A.L. Berrier, K.M. Yamada, Cell–matrix adhesion, J Cell Physiol 213 (2007) 565–573. https://doi.org/10.1002/jcp.21237.

[41]   A. Baroni, E. Buommino, V. De Gregorio, E. Ruocco, V. Ruocco, R. Wolf, Structure and function of the epidermis related to barrier properties, Clin Dermatol 30 (2012) 257–262. https://doi.org/10.1016/j.clindermatol.2011.08.007.

[42]   J.F. Almine, S.G. Wise, A.S. Weiss, Elastin signaling in wound repair, Birth Defects Res C Embryo Today 96 (2012) 248–257. https://doi.org/10.1002/bdrc.21016.

[43]   S. Barrientos, O. Stojadinovic, M.S. Golinko, H. Brem, M. Tomic-Canic, PERSPECTIVE ARTICLE: Growth factors and cytokines in wound healing, Wound Repair and Regeneration 16 (2008) 585–601. https://doi.org/10.1111/j.1524-475X.2008.00410.x.

[44]   B.M. Delavary, W.M. van der Veer, M. van Egmond, F.B. Niessen, R.H.J. Beelen, Macrophages in skin injury and repair, Immunobiology 216 (2011) 753–762. https://doi.org/10.1016/j.imbio.2011.01.001.

[45]   M. Fioranelli, M.G. Roccia, D. Flavin, L. Cota, Regulation of Inflammatory Reaction in Health and Disease, Int J Mol Sci 22 (2021) 5277. https://doi.org/10.3390/ijms22105277.

[46]   A. Yang, Y. Wu, G. Yu, H. Wang, Role of specialized pro-resolving lipid mediators in pulmonary inflammation diseases: mechanisms and development, Respir Res 22 (2021) 204. https://doi.org/10.1186/s12931-021-01792-y.

[47]   G. Fredman, N. Kamaly, S. Spolitu, J. Milton, D. Ghorpade, R. Chiasson, G. Kuriakose, M. Perretti, O. Farokhzad, I. Tabas, Targeted nanoparticles containing the proresolving peptide Ac2-26 protect against advanced atherosclerosis in hypercholesterolemic mice, Sci Transl Med 7 (2015). https://doi.org/10.1126/scitranslmed.aaa1065.

[48]   E.H. Steen, X. Wang, S. Balaji, M.J. Butte, P.L. Bollyky, S.G. Keswani, The Role of the Anti-Inflammatory Cytokine Interleukin-10 in Tissue Fibrosis, Adv Wound Care (New Rochelle) 9 (2020) 184–198. https://doi.org/10.1089/wound.2019.1032.

[49]   K. Zhang, V.C.H. Lui, Y. Chen, C.N. Lok, K.K.Y. Wong, Delayed application of silver nanoparticles reveals the role of early inflammation in burn wound healing, Sci Rep 10 (2020) 6338. https://doi.org/10.1038/s41598-020-63464-z.

[50]   S.M. Hirst, A.S. Karakoti, R.D. Tyler, N. Sriranganathan, S. Seal, C.M. Reilly, Anti-inflammatory Properties of Cerium Oxide Nanoparticles, Small 5 (2009) 2848–2856. https://doi.org/10.1002/smll.200901048.

[51]   Y. Wu, Y. Yang, W. Zhao, Z.P. Xu, P.J. Little, A.K. Whittaker, R. Zhang, H.T. Ta, Novel iron oxide–cerium oxide core–shell nanoparticles as a potential theranostic material for ROS related inflammatory diseases, J Mater Chem B 6 (2018) 4937–4951. https://doi.org/10.1039/C8TB00022K.

[52]   Z. Zhai, W. Ouyang, Y. Yao, Y. Zhang, H. Zhang, F. Xu, C. Gao, Dexamethasone-loaded ROS-responsive poly(thioketal) nanoparticles suppress inflammation and oxidative stress of acute lung injury, Bioact Mater 14 (2022) 430–442. https://doi.org/10.1016/j.bioactmat.2022.01.047.

[53]   T. Yu, S. Gan, Q. Zhu, D. Dai, N. Li, H. Wang, X. Chen, D. Hou, Y. Wang, Q. Pan, J. Xu, X. Zhang, J. Liu, S. Pei, C. Peng, P. Wu, S. Romano, C. Mao, M. Huang, X. Zhu, K. Shen, J. Qin, Y. Xiao, Modulation of M2 macrophage polarization by the crosstalk between Stat6 and Trim24, Nat Commun 10 (2019) 4353. https://doi.org/10.1038/s41467-019-12384-2.

[54]   H. Yen, S. Hsu, C. Tsai, Cytotoxicity and Immunological Response of Gold and Silver Nanoparticles of Different Sizes, Small 5 (2009) 1553–1561. https://doi.org/10.1002/smll.200900126.

[55]   S. Zanganeh, G. Hutter, R. Spitler, O. Lenkov, M. Mahmoudi, A. Shaw, J.S. Pajarinen, H. Nejadnik, S. Goodman, M. Moseley, L.M. Coussens, H.E. Daldrup-Link, Iron oxide nanoparticles inhibit tumour growth by inducing pro-inflammatory macrophage polarization in tumour tissues, Nat Nanotechnol 11 (2016) 986–994. https://doi.org/10.1038/nnano.2016.168.

[56]   Y. Yang, L. Guo, Z. Wang, P. Liu, X. Liu, J. Ding, W. Zhou, Targeted silver nanoparticles for rheumatoid arthritis therapy via macrophage apoptosis and Re-polarization, Biomaterials 264 (2021) 120390. https://doi.org/10.1016/j.biomaterials.2020.120390.

[57]   M. Sun, Z. Deng, F. Shi, Z. Zhou, C. Jiang, Z. Xu, X. Cui, W. Li, Y. Jing, B. Han, W. Zhang, S. Xia, Rebamipide-loaded chitosan nanoparticles accelerate prostatic wound healing by inhibiting M1 macrophage-mediated inflammation via the NF-κB signaling pathway, Biomater Sci 8 (2020) 912–925. https://doi.org/10.1039/C9BM01512D.

[58]   D. Kwon, B.G. Cha, Y. Cho, J. Min, E.-B. Park, S.-J. Kang, J. Kim, Extra-Large Pore Mesoporous Silica Nanoparticles for Directing in Vivo M2 Macrophage Polarization by Delivering IL-4, Nano Lett 17 (2017) 2747–2756. https://doi.org/10.1021/acs.nanolett.6b04130.

[59]   L. Wang, H. Zhang, L. Sun, W. Gao, Y. Xiong, A. Ma, X. Liu, L. Shen, Q. Li, H. Yang, Manipulation of macrophage polarization by peptide-coated gold nanoparticles and its protective effects on acute lung injury, J Nanobiotechnology 18 (2020) 38. https://doi.org/10.1186/s12951-020-00593-7.

[60]   L. Kou, H. Huang, Y. Tang, M. Sun, Y. Li, J. Wu, S. Zheng, X. Zhao, D. Chen, Z. Luo, X. Zhang, Q. Yao, R. Chen, Opsonized nanoparticles target and regulate macrophage polarization for osteoarthritis therapy: A trapping strategy, Journal of Controlled Release 347 (2022) 237–255. https://doi.org/10.1016/j.jconrel.2022.04.037.

[61]   M. Mittal, M.R. Siddiqui, K. Tran, S.P. Reddy, A.B. Malik, Reactive Oxygen Species in Inflammation and Tissue Injury, Antioxid Redox Signal 20 (2014) 1126–1167. https://doi.org/10.1089/ars.2012.5149.

[62]   D. Martinvalet, M. Walch, Editorial: The Role of Reactive Oxygen Species in Protective Immunity, Front Immunol 12 (2022). https://doi.org/10.3389/fimmu.2021.832946.

[63]   Y. Ranneh, F. Ali, A.M. Akim, H.Abd. Hamid, H. Khazaai, A. Fadel, Crosstalk between reactive oxygen species and pro-inflammatory markers in developing various chronic diseases: a review, Appl Biol Chem 60 (2017) 327–338. https://doi.org/10.1007/s13765-017-0285-9.

[64]   M. Mittal, M.R. Siddiqui, K. Tran, S.P. Reddy, A.B. Malik, Reactive oxygen species in inflammation and tissue injury., Antioxid Redox Signal 20 (2014) 1126–67. https://doi.org/10.1089/ars.2012.5149.

[65]   S. Watanabe, M. Alexander, A. V. Misharin, G.R.S. Budinger, The role of macrophages in the resolution of inflammation, Journal of Clinical Investigation 129 (2019) 2619–2628. https://doi.org/10.1172/JCI124615.

[66]   M.V.D.Z. Park, A.M. Neigh, J.P. Vermeulen, L.J.J. de la Fonteyne, H.W. Verharen, J.J. Briedé, H. van Loveren, W.H. de Jong, The effect of particle size on the cytotoxicity,

inflammation, developmental toxicity and genotoxicity of silver nanoparticles, Biomaterials 32 (2011) 9810–9817. https://doi.org/10.1016/j.biomaterials.2011.08.085.

[67]   E.-J. Park, K. Park, Oxidative stress and pro-inflammatory responses induced by silica nanoparticles in vivo and in vitro, Toxicol Lett 184 (2009) 18–25. https://doi.org/10.1016/j.toxlet.2008.10.012.

[68]   R.P. Nishanth, R.G. Jyotsna, J.J. Schlager, S.M. Hussain, P. Reddanna, Inflammatory responses of RAW 264.7 macrophages upon exposure to nanoparticles: Role of ROS-NFκB signaling pathway, Nanotoxicology 5 (2011) 502–516. https://doi.org/10.3109/17435390.2010.541604.

[69]   L. Sui, J. Wang, Z. Xiao, Y. Yang, Z. Yang, K. Ai, ROS-Scavenging Nanomaterials to Treat Periodontitis, Front Chem 8 (2020). https://doi.org/10.3389/fchem.2020.595530.

[70]   Q. Li, Y. Liu, X. Dai, W. Jiang, H. Zhao, Nanozymes Regulate Redox Homeostasis in ROS-Related Inflammation, Front Chem 9 (2021). https://doi.org/10.3389/fchem.2021.740607.

[71]   S. Zhao, Y. Li, Q. Liu, S. Li, Y. Cheng, C. Cheng, Z. Sun, Y. Du, C.J. Butch, H. Wei, An Orally Administered CeO_2 @Montmorillonite Nanozyme Targets Inflammation for Inflammatory Bowel Disease Therapy, Adv Funct Mater 30 (2020) 2004692. https://doi.org/10.1002/adfm.202004692.

[72]   R. Li, X. Hou, L. Li, J. Guo, W. Jiang, W. Shang, Application of Metal-Based Nanozymes in Inflammatory Disease: A Review., Front Bioeng Biotechnol 10 (2022) 920213. https://doi.org/10.3389/fbioe.2022.920213.

[73]   N. Singh, M.A. Savanur, S. Srivastava, P. D'Silva, G. Mugesh, A Redox Modulatory Mn_3O_4 Nanozyme with Multi-Enzyme Activity Provides Efficient Cytoprotection to Human Cells in a Parkinson's Disease Model, Angewandte Chemie International Edition 56 (2017) 14267–14271. https://doi.org/10.1002/anie.201708573.

[74]   J. Yao, Y. Cheng, M. Zhou, S. Zhao, S. Lin, X. Wang, J. Wu, S. Li, H. Wei, ROS scavenging Mn_3O_4 nanozymes for *in vivo* anti-inflammation, Chem Sci 9 (2018) 2927–2933. https://doi.org/10.1039/C7SC05476A.

[75]   C. Xu, O.U. Akakuru, X. Ma, J. Zheng, J. Zheng, A. Wu, Nanoparticle-Based Wound Dressing: Recent Progress in the Detection and Therapy of Bacterial Infections, Bioconjug Chem 31 (2020) 1708–1723. https://doi.org/10.1021/acs.bioconjchem.0c00297.

[76]   M. Ahamed, M.S. AlSalhi, M.K.J. Siddiqui, Silver nanoparticle applications and human health, Clinica Chimica Acta 411 (2010) 1841–1848. https://doi.org/10.1016/j.cca.2010.08.016.

[77]   J.N. Cruz, S. Muzammil, A. Ashraf, M.U. Ijaz, M.H. Siddique, R. Abbas, M. Sadia, Saba, S. Hayat, R.R. Lima, A review on mycogenic metallic nanoparticles and their potential role as antioxidant, antibiofilm and quorum quenching agents, Heliyon 10 (2024). https://doi.org/10.1016/j.heliyon.2024.e29500.

[78]   X.-F. Zhang, Z.-G. Liu, W. Shen, S. Gurunathan, Silver Nanoparticles: Synthesis, Characterization, Properties, Applications, and Therapeutic Approaches, Int J Mol Sci 17 (2016) 1534. https://doi.org/10.3390/ijms17091534.

Advances in Healthcare and Nanoparticle Toxicology　　　　　　Materials Research Forum LLC
Materials Research Foundations 171 (2024) 93-134　　　　　https://doi.org/10.21741/9781644903339-4

[79]　A. Kędziora, M. Speruda, E. Krzyżewska, J. Rybka, A. Łukowiak, G. Bugla-Płoskońska, Similarities and Differences between Silver Ions and Silver in Nanoforms as Antibacterial Agents, Int J Mol Sci 19 (2018) 444. https://doi.org/10.3390/ijms19020444.

[80]　M. Konop, J. Czuwara, E. Kłodzińska, A.K. Laskowska, D. Sulejczak, T. Damps, U. Zielenkiewicz, I. Brzozowska, A. Sureda, T. Kowalkowski, R.A. Schwartz, L. Rudnicka, Evaluation of keratin biomaterial containing silver nanoparticles as a potential wound dressing in full-thickness skin wound model in diabetic mice, J Tissue Eng Regen Med 14 (2020) 334–346. https://doi.org/10.1002/term.2998.

[81]　M. Ruiz-Serrano, J.C. Menéndez, Multicomponent reactions for the synthesis of natural products and natural product-like libraries, in: J.N. Cruz (Ed.), Drug Discovery and Design Using Natural Products, Springer Nature Switzerland, Cham, 2023: pp. 273–322. https://doi.org/10.1007/978-3-031-35205-8_10.

[82]　G.A. Govindasamy, R.B. S. M. N. Mydin, W.N.F.W.E. Effendy, S. Sreekantan, Novel dual-ionic ZnO/CuO embedded in porous chitosan biopolymer for wound dressing application: Physicochemical, bactericidal, cytocompatibility and wound healing profiles, Mater Today Commun 33 (2022) 104545. https://doi.org/10.1016/j.mtcomm.2022.104545.

[83]　M.I.N. Ahamed, S. Sankar, P.M. Kashif, S.K.H. Basha, T.P. Sastry, Evaluation of biomaterial containing regenerated cellulose and chitosan incorporated with silver nanoparticles, Int J Biol Macromol 72 (2015) 680–686. https://doi.org/10.1016/j.ijbiomac.2014.08.055.

[84]　Z. Qian, Y. Bai, J. Zhou, L. Li, J. Na, Y. Fan, X. Guo, H. Liu, A moisturizing chitosan-silk fibroin dressing with silver nanoparticles-adsorbed exosomes for repairing infected wounds, J Mater Chem B 8 (2020) 7197–7212. https://doi.org/10.1039/D0TB01100B.

[85]　I.N. de F. Ramos, M.F. da Silva, J.M.S. Lopes, J.N. Cruz, F.S. Alves, J. de A.R. do Rego, M.L. da Costa, P.P. de Assumpção, D. do S. Barros Brasil, A.S. Khayat, Extraction, Characterization, and Evaluation of the Cytotoxic Activity of Piperine in Its Isolatèd form and in Combination with Chemotherapeutics against Gastric Cancer, Molecules 28 (2023). https://doi.org/10.3390/molecules28145587.

[86]　N. Tra Thanh, M. Ho Hieu, N. Tran Minh Phuong, T. Do Bui Thuan, H. Nguyen Thi Thu, V.P. Thai, T. Do Minh, H. Nguyen Dai, V.T. Vo, H. Nguyen Thi, Optimization and characterization of electrospun polycaprolactone coated with gelatin-silver nanoparticles for wound healing application, Materials Science and Engineering: C 91 (2018) 318–329. https://doi.org/10.1016/j.msec.2018.05.039.

[87]　M. Konop, E. Kłodzińska, J. Borowiec, A.K. Laskowska, J. Czuwara, P. Konieczka, B. Cieślik, E. Waraksa, L. Rudnicka, Application of micellar electrokinetic chromatography for detection of silver nanoparticles released from wound dressing, Electrophoresis 40 (2019) 1565–1572. https://doi.org/10.1002/elps.201900020.

[88]　R. Singla, S. Soni, V. Patial, P.M. Kulurkar, A. Kumari, M. S., Y.S. Padwad, S.K. Yadav, In vivo diabetic wound healing potential of nanobiocomposites containing bamboo cellulose nanocrystals impregnated with silver nanoparticles, Int J Biol Macromol 105 (2017) 45–55. https://doi.org/10.1016/j.ijbiomac.2017.06.109.

Materials Research Forum LLC
https://doi.org/10.21741/9781644903339-4

[89]   A.K. Chatterjee, R. Chakraborty, T. Basu, Mechanism of antibacterial activity of copper nanoparticles, Nanotechnology 25 (2014) 135101. https://doi.org/10.1088/0957-4484/25/13/135101.

[90]   S. Alizadeh, B. Seyedalipour, S. Shafieyan, A. Kheime, P. Mohammadi, N. Aghdami, Copper nanoparticles promote rapid wound healing in acute full thickness defect via acceleration of skin cell migration, proliferation, and neovascularization, Biochem Biophys Res Commun 517 (2019) 684–690. https://doi.org/10.1016/j.bbrc.2019.07.110.

[91]   S.M. Bauer, R.J. Bauer, O.C. Velazquez, Angiogenesis, Vasculogenesis, and Induction of Healing in Chronic Wounds, Vasc Endovascular Surg 39 (2005) 293–306. https://doi.org/10.1177/153857440503900401.

[92]   Q. Li, F. Lu, G. Zhou, K. Yu, B. Lu, Y. Xiao, F. Dai, D. Wu, G. Lan, Silver Inlaid with Gold Nanoparticle/Chitosan Wound Dressing Enhances Antibacterial Activity and Porosity, and Promotes Wound Healing, Biomacromolecules 18 (2017) 3766–3775. https://doi.org/10.1021/acs.biomac.7b01180.

[93]   M.H. Sarfraz, M. Zubair, B. Aslam, A. Ashraf, M.H. Siddique, S. Hayat, J.N. Cruz, S. Muzammil, M. Khurshid, M.F. Sarfraz, A. Hashem, T.M. Dawoud, G.D. Avila-Quezada, E.F. Abd_Allah, Comparative analysis of phyto-fabricated chitosan, copper oxide, and chitosan-based CuO nanoparticles: antibacterial potential against Acinetobacter baumannii isolates and anticancer activity against HepG2 cell lines, Front Microbiol 14 (2023). https://doi.org/10.3389/fmicb.2023.1188743.

[94]   M.G. Arafa, R.F. El-Kased, M.M. Elmazar, Thermoresponsive gels containing gold nanoparticles as smart antibacterial and wound healing agents, Sci Rep 8 (2018) 13674. https://doi.org/10.1038/s41598-018-31895-4.

[95]   P. Victor, D. Sarada, K.M. Ramkumar, Pharmacological activation of Nrf2 promotes wound healing, Eur J Pharmacol 886 (2020) 173395. https://doi.org/10.1016/j.ejphar.2020.173395.

[96]   H. Kawaguchi, S. Jingushi, T. Izumi, M. Fukunaga, T. Matsushita, T. Nakamura, K. Mizuno, T. Nakamura, K. Nakamura, Local application of recombinant human fibroblast growth factor-2 on bone repair: A dose–escalation prospective trial on patients with osteotomy, Journal of Orthopaedic Research 25 (2007) 480–487. https://doi.org/10.1002/jor.20315.

[97]   K.P. Hoversten, L.J. Kiemele, A.M. Stolp, P.Y. Takahashi, B.P. Verdoorn, Prevention, Diagnosis, and Management of Chronic Wounds in Older Adults, Mayo Clin Proc 95 (2020) 2021–2034. https://doi.org/10.1016/j.mayocp.2019.10.014.

[98]   F.S. Alves, J.N. Cruz, I.N. de Farias Ramos, D.L. do Nascimento Brandão, R.N. Queiroz, G.V. da Silva, G.V. da Silva, M.F. Dolabela, M.L. da Costa, A.S. Khayat, J. de Arimatéia Rodrigues do Rego, D. do Socorro Barros Brasil, Evaluation of Antimicrobial Activity and Cytotoxicity Effects of Extracts of Piper nigrum L. and Piperine, Separations 10 (2023). https://doi.org/10.3390/separations10010021.

[99]   P.D. Marcato, L.B. De Paula, P.S. Melo, I.R. Ferreira, A.B.A. Almeida, A.S. Torsoni, O.L. Alves, In Vivo Evaluation of Complex Biogenic Silver Nanoparticle and Enoxaparin in Wound Healing, J Nanomater 2015 (2015) 1–10. https://doi.org/10.1155/2015/439820.

[100] M.I.N. Ahamed, S. Sankar, P.M. Kashif, S.K.H. Basha, T.P. Sastry, Evaluation of biomaterial containing regenerated cellulose and chitosan incorporated with silver nanoparticles, Int J Biol Macromol 72 (2015) 680–686. https://doi.org/10.1016/j.ijbiomac.2014.08.055.

[101] X. Dai, Q. Guo, Y. Zhao, P. Zhang, T. Zhang, X. Zhang, C. Li, Functional Silver Nanoparticle as a Benign Antimicrobial Agent That Eradicates Antibiotic-Resistant Bacteria and Promotes Wound Healing, ACS Appl Mater Interfaces 8 (2016) 25798–25807. https://doi.org/10.1021/acsami.6b09267.

[102] M. Adibhesami, M. Ahmadi, A.A. Farshid, F. Sarrafzadeh-Rezaei, B. Dalir-Naghadeh, Effects of silver nanoparticles on Staphylococcus aureus contaminated open wounds healing in mice: An experimental study., Vet Res Forum 8 (2017) 23–28.

[103] C. You, Q. Li, X. Wang, P. Wu, J.K. Ho, R. Jin, L. Zhang, H. Shao, C. Han, Silver nanoparticle loaded collagen/chitosan scaffolds promote wound healing via regulating fibroblast migration and macrophage activation, Sci Rep 7 (2017) 10489. https://doi.org/10.1038/s41598-017-10481-0.

[104] H. Ye, J. Cheng, K. Yu, In situ reduction of silver nanoparticles by gelatin to obtain porous silver nanoparticle/chitosan composites with enhanced antimicrobial and wound-healing activity, Int J Biol Macromol 121 (2019) 633–642. https://doi.org/10.1016/j.ijbiomac.2018.10.056.

[105] F. Kong, C. Fan, Y. Yang, B.H. Lee, K. Wei, 5-hydroxymethylfurfural-embedded poly (vinyl alcohol)/sodium alginate hybrid hydrogels accelerate wound healing, Int J Biol Macromol 138 (2019) 933–949. https://doi.org/10.1016/j.ijbiomac.2019.07.152.

[106] V. Nagarjuna Reddy, S. Nyamathulla, K. Abdul Kadir Pahirulzaman, S.I. Mokhtar, N. Giribabu, V.R. Pasupuleti, Gallocatechin-silver nanoparticles embedded in cotton gauze patches accelerated wound healing in diabetic rats by promoting proliferation and inhibiting apoptosis through the Wnt/β-catenin signaling pathway, PLoS One 17 (2022) e0268505. https://doi.org/10.1371/journal.pone.0268505.

[107] W. Klinkajon, P. Supaphol, Novel copper (II) alginate hydrogels and their potential for use as anti-bacterial wound dressings, Biomedical Materials 9 (2014) 045008. https://doi.org/10.1088/1748-6041/9/4/045008.

[108] B. Tao, C. Lin, Y. Deng, Z. Yuan, X. Shen, M. Chen, Y. He, Z. Peng, Y. Hu, K. Cai, Copper-nanoparticle-embedded hydrogel for killing bacteria and promoting wound healing with photothermal therapy, J Mater Chem B 7 (2019) 2534–2548. https://doi.org/10.1039/C8TB03272F.

[109] Q. Xu, M. Chang, Y. Zhang, E. Wang, M. Xing, L. Gao, Z. Huan, F. Guo, J. Chang, PDA/Cu Bioactive Hydrogel with "Hot Ions Effect" for Inhibition of Drug-Resistant Bacteria and Enhancement of Infectious Skin Wound Healing, ACS Appl Mater Interfaces 12 (2020) 31255–31269. https://doi.org/10.1021/acsami.0c08890.

[110] Y. Yang, Z. Dong, M. Li, L. Liu, H. Luo, P. Wang, D. Zhang, X. Yang, K. Zhou, S. Lei, <p>Graphene Oxide/Copper Nanoderivatives-Modified Chitosan/Hyaluronic Acid Dressings for Facilitating Wound Healing in Infected Full-Thickness Skin Defects</p>, Int J Nanomedicine Volume 15 (2020) 8231–8247. https://doi.org/10.2147/IJN.S278631.

[111] E. Ghasemian Lemraski, H. Jahangirian, M. Dashti, E. Khajehali, Mis.S. Sharafinia, R. Rafiee-Moghaddam, T.J. Webster, Antimicrobial Double-Layer Wound Dressing Based on Chitosan/Polyvinyl Alcohol/Copper: In vitro and in vivo Assessment, Int J Nanomedicine Volume 16 (2021) 223–235. https://doi.org/10.2147/IJN.S266692.

[112] X. Sun, M. Dong, Z. Guo, H. Zhang, J. Wang, P. Jia, T. Bu, Y. Liu, L. Li, L. Wang, Multifunctional chitosan-copper-gallic acid based antibacterial nanocomposite wound dressing, Int J Biol Macromol 167 (2021) 10–22. https://doi.org/10.1016/j.ijbiomac.2020.11.153.

[113] B. Tao, C. Lin, A. Guo, Y. Yu, X. Qin, K. Li, H. Tian, W. Yi, D. Lei, L. Chen, Fabrication of copper ions-substituted hydroxyapatite/polydopamine nanocomposites with high antibacterial and angiogenesis effects for promoting infected wound healing, Journal of Industrial and Engineering Chemistry 104 (2021) 345–355. https://doi.org/10.1016/j.jiec.2021.08.035.

[114] M. Wang, A. Xia, S. Wu, J. Shen, Facile Synthesis of the Cu, N-CDs@GO-CS Hydrogel with Enhanced Antibacterial Activity for Effective Treatment of Wound Infection, Langmuir 37 (2021) 7928–7935. https://doi.org/10.1021/acs.langmuir.1c00529.

[115] H. Gu, P.L. Ho, E. Tong, L. Wang, B. Xu, Presenting Vancomycin on Nanoparticles to Enhance Antimicrobial Activities, Nano Lett 3 (2003) 1261–1263. https://doi.org/10.1021/nl034396z.

[116] S. Hsu, Y.-B. Chang, C.-L. Tsai, K.-Y. Fu, S.-H. Wang, H.-J. Tseng, Characterization and biocompatibility of chitosan nanocomposites, Colloids Surf B Biointerfaces 85 (2011) 198–206. https://doi.org/10.1016/j.colsurfb.2011.02.029.

[117] M.J. Cozad, S.L. Bachman, S.A. Grant, Assessment of decellularized porcine diaphragm conjugated with gold nanomaterials as a tissue scaffold for wound healing, J Biomed Mater Res A 99A (2011) 426–434. https://doi.org/10.1002/jbm.a.33182.

[118] Mohd.A. Sherwani, S. Tufail, A.A. Khan, M. Owais, Gold Nanoparticle-Photosensitizer Conjugate Based Photodynamic Inactivation of Biofilm Producing Cells: Potential for Treatment of C. albicans Infection in BALB/c Mice, PLoS One 10 (2015) e0131684. https://doi.org/10.1371/journal.pone.0131684.

[119] N. Volkova, M. Yukhta, O. Pavlovich, A. Goltsev, Application of Cryopreserved Fibroblast Culture with Au Nanoparticles to Treat Burns, Nanoscale Res Lett 11 (2016) 22. https://doi.org/10.1186/s11671-016-1242-y.

[120] O. Akturk, K. Kismet, A.C. Yasti, S. Kuru, M.E. Duymus, F. Kaya, M. Caydere, S. Hucumenoglu, D. Keskin, Collagen/gold nanoparticle nanocomposites: A potential skin wound healing biomaterial, J Biomater Appl 31 (2016) 283–301. https://doi.org/10.1177/0885328216644536.

[121] O. Akturk, K. Kismet, A.C. Yasti, S. Kuru, M.E. Duymus, F. Kaya, M. Caydere, S. Hucumenoglu, D. Keskin, Collagen/gold nanoparticle nanocomposites: A potential skin wound healing biomaterial, J Biomater Appl 31 (2016) 283–301. https://doi.org/10.1177/0885328216644536.

[122] Q. Li, F. Lu, G. Zhou, K. Yu, B. Lu, Y. Xiao, F. Dai, D. Wu, G. Lan, Silver Inlaid with Gold Nanoparticle/Chitosan Wound Dressing Enhances Antibacterial Activity and Porosity, and Promotes Wound Healing, Biomacromolecules 18 (2017) 3766–3775. https://doi.org/10.1021/acs.biomac.7b01180.

[123] A. Pan, M. Zhong, H. Wu, Y. Peng, H. Xia, Q. Tang, Q. Huang, L. Wei, L. Xiao, C. Peng, Topical Application of Keratinocyte Growth Factor Conjugated Gold Nanoparticles Accelerate Wound Healing, Nanomedicine 14 (2018) 1619–1628. https://doi.org/10.1016/j.nano.2018.04.007.

[124] N.N. Mahmoud, S. Hikmat, D. Abu Ghith, M. Hajeer, L. Hamadneh, D. Qattan, E.A. Khalil, Gold nanoparticles loaded into polymeric hydrogel for wound healing in rats: Effect of nanoparticles' shape and surface modification, Int J Pharm 565 (2019) 174–186. https://doi.org/10.1016/j.ijpharm.2019.04.079.

[125] Y. Zou, R. Xie, E. Hu, P. Qian, B. Lu, G. Lan, F. Lu, Protein-reduced gold nanoparticles mixed with gentamicin sulfate and loaded into konjac/gelatin sponge heal wounds and kill drug-resistant bacteria, Int J Biol Macromol 148 (2020) 921–931. https://doi.org/10.1016/j.ijbiomac.2020.01.190.

[126] K. Wang, Z. Qi, S. Pan, S. Zheng, H. Wang, Y. Chang, H. Li, P. Xue, X. Yang, C. Fu, Preparation, characterization and evaluation of a new film based on chitosan, arginine and gold nanoparticle derivatives for wound-healing efficacy, RSC Adv 10 (2020) 20886–20899. https://doi.org/10.1039/D0RA03704D.

[127] P. Chen, L. Bian, X. Hu, Synergic Fabrication of Gold Nanoparticles Embedded Dextran/ Silk Sericin Nanomaterials for the Treatment and Care of Wound Healing, J Clust Sci 33 (2022) 2147–2156. https://doi.org/10.1007/s10876-021-02131-3.

[128] C. Xia, B. Ren, N. Liu, Y. Zheng, A Feasible Strategy of Fabricating of Gold-Encapsulated Dextran/Polyvinyl Alcohol Nanoparticles for the Treatment and Care of Wound Healing, J Clust Sci 33 (2022) 2179–2187. https://doi.org/10.1007/s10876-021-02132-2.

[129] S. Sharifi, M.J. Hajipour, L. Gould, M. Mahmoudi, Nanomedicine in Healing Chronic Wounds: Opportunities and Challenges, Mol Pharm 18 (2021) 550–575. https://doi.org/10.1021/acs.molpharmaceut.0c00346.

[130] L. Tang, J. Cheng, Nonporous silica nanoparticles for nanomedicine application, Nano Today 8 (2013) 290–312. https://doi.org/10.1016/j.nantod.2013.04.007.

[131] L. Chen, X. Zhou, C. He, Mesoporous silica nanoparticles for tissue-engineering applications, WIREs Nanomedicine and Nanobiotechnology 11 (2019). https://doi.org/10.1002/wnan.1573.

[132] C.S. Yah, G.S. Simate, Nanoparticles as potential new generation broad spectrum antimicrobial agents, DARU Journal of Pharmaceutical Sciences 23 (2015) 43. https://doi.org/10.1186/s40199-015-0125-6.

[133] F. Menaa, A. Abdelghani, B. Menaa, Graphene nanomaterials as biocompatible and conductive scaffolds for stem cells: impact for tissue engineering and regenerative medicine, J Tissue Eng Regen Med 9 (2015) 1321–1338. https://doi.org/10.1002/term.1910.

[134] J. Zhou, X. Qi, Multi-walled carbon nanotubes/epilson-polylysine nanocomposite with enhanced antibacterial activity, Lett Appl Microbiol 52 (2011) 76–83. https://doi.org/10.1111/j.1472-765X.2010.02969.x.

[135] G.M. Neelgund, A. Oki, Z. Luo, Antimicrobial activity of CdS and Ag2S quantum dots immobilized on poly(amidoamine) grafted carbon nanotubes, Colloids Surf B Biointerfaces 100 (2012) 215–221. https://doi.org/10.1016/j.colsurfb.2012.05.012.

[136] F. Cui, J. Sun, J. Ji, X. Yang, K. Wei, H. Xu, Q. Gu, Y. Zhang, X. Sun, Carbon dots-releasing hydrogels with antibacterial activity, high biocompatibility, and fluorescence performance as candidate materials for wound healing, J Hazard Mater 406 (2021) 124330. https://doi.org/10.1016/j.jhazmat.2020.124330.

[137] S. Aslan, M. Deneufchatel, S. Hashmi, N. Li, L.D. Pfefferle, M. Elimelech, E. Pauthe, P.R. Van Tassel, Carbon nanotube-based antimicrobial biomaterials formed via layer-by-layer assembly with polypeptides, J Colloid Interface Sci 388 (2012) 268–273. https://doi.org/10.1016/j.jcis.2012.08.025.

[138] K.K. Patel, D.B. Surekha, M. Tripathi, Md.M. Anjum, M.S. Muthu, R. Tilak, A.K. Agrawal, S. Singh, Antibiofilm Potential of Silver Sulfadiazine-Loaded Nanoparticle Formulations: A Study on the Effect of DNase-I on Microbial Biofilm and Wound Healing Activity, Mol Pharm 16 (2019) 3916–3925. https://doi.org/10.1021/acs.molpharmaceut.9b00527.

[139] F. Saporito, G. Sandri, M.C. Bonferoni, S. Rossi, C. Boselli, A. Icaro Cornaglia, B. Mannucci, P. Grisoli, B. Vigani, F. Ferrari, Essential oil-loaded lipid nanoparticles for wound healing, Int J Nanomedicine Volume 13 (2017) 175–186. https://doi.org/10.2147/IJN.S152529.

[140] R.A.-B. Sanad, H.M. Abdel-Bar, Chitosan–hyaluronic acid composite sponge scaffold enriched with Andrographolide-loaded lipid nanoparticles for enhanced wound healing, Carbohydr Polym 173 (2017) 441–450. https://doi.org/10.1016/j.carbpol.2017.05.098.

[141] M. Fumakia, E.A. Ho, Nanoparticles Encapsulated with LL37 and Serpin A1 Promotes Wound Healing and Synergistically Enhances Antibacterial Activity, Mol Pharm 13 (2016) 2318–2331. https://doi.org/10.1021/acs.molpharmaceut.6b00099.

[142] A.B. Scriboni, V.M. Couto, L.N. de M. Ribeiro, I.A. Freires, F.C. Groppo, E. de Paula, M. Franz-Montan, K. Cogo-Müller, Fusogenic Liposomes Increase the Antimicrobial Activity of Vancomycin Against Staphylococcus aureus Biofilm, Front Pharmacol 10 (2019). https://doi.org/10.3389/fphar.2019.01401.

[143] Z. Rukavina, M. Šegvić Klarić, J. Filipović-Grčić, J. Lovrić, Ž. Vanić, Azithromycin-loaded liposomes for enhanced topical treatment of methicillin-resistant Staphyloccocus aureus (MRSA) infections, Int J Pharm 553 (2018) 109–119. https://doi.org/10.1016/j.ijpharm.2018.10.024.

[144] S. Wang, C. Yan, X. Zhang, D. Shi, L. Chi, G. Luo, J. Deng, Antimicrobial peptide modification enhances the gene delivery and bactericidal efficiency of gold nanoparticles for accelerating diabetic wound healing, Biomater Sci 6 (2018) 2757–2772. https://doi.org/10.1039/C8BM00807H.

Materials Research Forum LLC
https://doi.org/10.21741/9781644903339-4

[145] N. Monteiro, M. Martins, A. Martins, N.A. Fonseca, J.N. Moreira, R.L. Reis, N.M. Neves, Antibacterial activity of chitosan nanofiber meshes with liposomes immobilized releasing gentamicin, Acta Biomater 18 (2015) 196–205. https://doi.org/10.1016/j.actbio.2015.02.018.

[146] R.K. Thapa, K.L. Kiick, M.O. Sullivan, Encapsulation of collagen mimetic peptide-tethered vancomycin liposomes in collagen-based scaffolds for infection control in wounds, Acta Biomater 103 (2020) 115–128. https://doi.org/10.1016/j.actbio.2019.12.014.

[147] A. Kushwaha, L. Goswami, B.S. Kim, Nanomaterial-Based Therapy for Wound Healing, Nanomaterials 12 (2022) 618. https://doi.org/10.3390/nano12040618.

[148] M. Ramasamy, J. Lee, Recent Nanotechnology Approaches for Prevention and Treatment of Biofilm-Associated Infections on Medical Devices, Biomed Res Int 2016 (2016) 1–17. https://doi.org/10.1155/2016/1851242.

[149] Y.H. Ngo, D. Li, G.P. Simon, G. Garnier, Paper surfaces functionalized by nanoparticles, Adv Colloid Interface Sci 163 (2011) 23–38. https://doi.org/10.1016/j.cis.2011.01.004.

[150] T. Wang, Y. Zheng, Y. Shen, Y. Shi, F. Li, C. Su, L. Zhao, Chitosan nanoparticles loaded hydrogels promote skin wound healing through the modulation of reactive oxygen species, Artif Cells Nanomed Biotechnol 46 (2018) 138–149. https://doi.org/10.1080/21691401.2017.1415212.

[151] A. Ibrahim, H. Fatima, M.M. Babar, Targeted delivery of natural products, in: J.N. Cruz (Ed.), Drug Discovery and Design Using Natural Products, Springer Nature Switzerland, Cham, 2023: pp. 377–393. https://doi.org/10.1007/978-3-031-35205-8_12.

[152] ' H. Liu, C. Wang, C. Li, Y. Qin, Z. Wang, F. Yang, Z. Li, J. Wang, A functional chitosan-based hydrogel as a wound dressing and drug delivery system in the treatment of wound healing, RSC Adv 8 (2018) 7533–7549. https://doi.org/10.1039/C7RA13510F.

[153] M.J. Mitchell, M.M. Billingsley, R.M. Haley, M.E. Wechsler, N.A. Peppas, R. Langer, Engineering precision nanoparticles for drug delivery, Nat Rev Drug Discov 20 (2021) 101–124. https://doi.org/10.1038/s41573-020-0090-8.

[154] R. Li, K. Liu, X. Huang, D. Li, J. Ding, B. Liu, X. Chen, Bioactive Materials Promote Wound Healing through Modulation of Cell Behaviors, Advanced Science 9 (2022). https://doi.org/10.1002/advs.202105152.

[155] S. Pina, J.M. Oliveira, R.L. Reis, Natural-Based Nanocomposites for Bone Tissue Engineering and Regenerative Medicine: A Review, Advanced Materials 27 (2015) 1143–1169. https://doi.org/10.1002/adma.201403354.

[156] F. Paladini, M. Pollini, Antimicrobial Silver Nanoparticles for Wound Healing Application: Progress and Future Trends, Materials 12 (2019) 2540. https://doi.org/10.3390/ma12162540.

[157] M. Ovais, I. Ahmad, A.T. Khalil, S. Mukherjee, R. Javed, M. Ayaz, A. Raza, Z.K. Shinwari, Wound healing applications of biogenic colloidal silver and gold nanoparticles: recent trends and future prospects, Appl Microbiol Biotechnol 102 (2018) 4305–4318. https://doi.org/10.1007/s00253-018-8939-z.

[158] G. Rath, T. Hussain, G. Chauhan, T. Garg, A.K. Goyal, Development and characterization of cefazolin loaded zinc oxide nanoparticles composite gelatin nanofiber mats for postoperative surgical wounds, Materials Science and Engineering: C 58 (2016) 242–253. https://doi.org/10.1016/j.msec.2015.08.050.

[159] F. Rezkita, K.G.P. Wibawa, A.P. Nugraha, Curcumin loaded Chitosan Nanoparticle for Accelerating the Post Extraction Wound Healing in Diabetes Mellitus Patient: A Review, Res J Pharm Technol 13 (2020) 1039. https://doi.org/10.5958/0974-360X.2020.00191.2.

[160] A. V Samrot, M.S. Sree, D. Rajalakshmi, L.N.R. Prakash, P. Prakash, Natural biopolymers as scaffold, in: J.N. Cruz (Ed.), Drug Discovery and Design Using Natural Products, Springer Nature Switzerland, Cham, 2023: pp. 23–36. https://doi.org/10.1007/978-3-031-35205-8_20.

[161] P. Dam, M. Celik, M. Ustun, S. Saha, C. Saha, E.A. Kacar, S. Kugu, E.N. Karagulle, S. Tasoglu, F. Buyukserin, R. Mondal, P. Roy, M.L.R. Macedo, O.L. Franco, M.H. Cardoso, S. Altuntas, A.K. Mandal, Wound healing strategies based on nanoparticles incorporated in hydrogel wound patches, RSC Adv 13 (2023) 21345–21364. https://doi.org/10.1039/D3RA03477A.

[162] S. Ahn, C.O. Chantre, A.R. Gannon, J.U. Lind, P.H. Campbell, T. Grevesse, B.B. O'Connor, K.K. Parker, Soy Protein/Cellulose Nanofiber Scaffolds Mimicking Skin Extracellular Matrix for Enhanced Wound Healing, Adv Healthc Mater 7 (2018) 1701175. https://doi.org/10.1002/adhm.201701175.

[163] R. Jangde, D. Singh, Preparation and optimization of quercetin-loaded liposomes for wound healing, using response surface methodology, Artif Cells Nanomed Biotechnol 44 (2016) 635–641. https://doi.org/10.3109/21691401.2014.975238.

[164] N. Hasan, J. Cao, J. Lee, S.P. Hlaing, M.A. Oshi, M. Naeem, M.-H. Ki, B.L. Lee, Y. Jung, J.-W. Yoo, Bacteria-Targeted Clindamycin Loaded Polymeric Nanoparticles: Effect of Surface Charge on Nanoparticle Adhesion to MRSA, Antibacterial Activity, and Wound Healing, Pharmaceutics 11 (2019) 236. https://doi.org/10.3390/pharmaceutics11050236.

[165] P. Dam, M. Celik, M. Ustun, S. Saha, C. Saha, E.A. Kacar, S. Kugu, E.N. Karagulle, S. Tasoglu, F. Buyukserin, R. Mondal, P. Roy, M.L.R. Macedo, O.L. Franco, M.H. Cardoso, S. Altuntas, A.K. Mandal, Wound healing strategies based on nanoparticles incorporated in hydrogel wound patches, RSC Adv 13 (2023) 21345–21364. https://doi.org/10.1039/D3RA03477A.

[166] J. Wu, M.D. Weir, M.A.S. Melo, H.H.K. Xu, Development of novel self-healing and antibacterial dental composite containing calcium phosphate nanoparticles, J Dent 43 (2015) 317–326. https://doi.org/10.1016/j.jdent.2015.01.009.

[167] Y. Liang, X. Zhao, T. Hu, Y. Han, B. Guo, Mussel-inspired, antibacterial, conductive, antioxidant, injectable composite hydrogel wound dressing to promote the regeneration of infected skin, J Colloid Interface Sci 556 (2019) 514–528. https://doi.org/10.1016/j.jcis.2019.08.083.

[168] J. Anjana, V.K. Rajan, R. Biswas, R. Jayakumar, Controlled Delivery of Bioactive Molecules for the Treatment of Chronic Wounds, Curr Pharm Des 23 (2017). https://doi.org/10.2174/1381612823666170503145528.

[169] K. Rahim, S. Saleha, X. Zhu, L. Huo, A. Basit, O.L. Franco, Bacterial Contribution in Chronicity of Wounds, Microb Ecol 73 (2017) 710–721. https://doi.org/10.1007/s00248-016-0867-9.

[170] M.A. Pérez-Díaz, L. Boegli, G. James, C. Velasquillo, R. Sánchez-Sánchez, R.-E. Martínez-Martínez, G.A. Martínez-Castañón, F. Martinez-Gutierrez, Silver nanoparticles with antimicrobial activities against Streptococcus mutans and their cytotoxic effect, Materials Science and Engineering: C 55 (2015) 360–366. https://doi.org/10.1016/j.msec.2015.05.036.

[171] A.J. Kora, R.B. Sashidhar, Antibacterial activity of biogenic silver nanoparticles synthesized with gum ghatti and gum olibanum: a comparative study, J Antibiot (Tokyo) 68 (2015) 88–97. https://doi.org/10.1038/ja.2014.114.

[172] M. Stevanović, Kovačević, Petković, Filipič, Uskoković, Effect of poly-α, γ, L-glutamic acid as a capping agent on morphology and oxidative stress-dependent toxicity of silver nanoparticles, Int J Nanomedicine (2011) 2837. https://doi.org/10.2147/IJN.S24889.

[173] N.J. Amruthraj, J.P. Preetam Raj, A. Lebel, Capsaicin-capped silver nanoparticles: its kinetics, characterization and biocompatibility assay, Appl Nanosci 5 (2015) 403–409. https://doi.org/10.1007/s13204-014-0330-5.

[174] S.K. Nethi, A.K. Barui, S. Mukherjee, C.R. Patra, Engineered Nanoparticles for Effective Redox Signaling During Angiogenic and Antiangiogenic Therapy, Antioxid Redox Signal 30 (2019) 786–809. https://doi.org/10.1089/ars.2017.7383.

[175] K.P. Claffey, L.F. Brown, L.F. del Aguila, K. Tognazzi, K.T. Yeo, E.J. Manseau, H.F. Dvorak, Expression of vascular permeability factor/vascular endothelial growth factor by melanoma cells increases tumor growth, angiogenesis, and experimental metastasis., Cancer Res 56 (1996) 172–81.

[176] H. Xu, F. Lv, Y. Zhang, Z. Yi, Q. Ke, C. Wu, M. Liu, J. Chang, Hierarchically micro-patterned nanofibrous scaffolds with a nanosized bio-glass surface for accelerating wound healing, Nanoscale 7 (2015) 18446–18452. https://doi.org/10.1039/C5NR04802H.

[177] Y. Zhang, M. Chang, F. Bao, M. Xing, E. Wang, Q. Xu, Z. Huan, F. Guo, J. Chang, Multifunctional Zn doped hollow mesoporous silica/polycaprolactone electrospun membranes with enhanced hair follicle regeneration and antibacterial activity for wound healing, Nanoscale 11 (2019) 6315–6333. https://doi.org/10.1039/C8NR09818B.

[178] S.-C. Tao, B.-Y. Rui, Q.-Y. Wang, D. Zhou, Y. Zhang, S.-C. Guo, Extracellular vesicle-mimetic nanovesicles transport LncRNA-H19 as competing endogenous RNA for the treatment of diabetic wounds, Drug Deliv 25 (2018) 241–255. https://doi.org/10.1080/10717544.2018.1425774.

[179] S. Chigurupati, M.R. Mughal, E. Okun, S. Das, A. Kumar, M. McCaffery, S. Seal, M.P. Mattson, Effects of cerium oxide nanoparticles on the growth of keratinocytes, fibroblasts and vascular endothelial cells in cutaneous wound healing, Biomaterials 34 (2013) 2194–2201. https://doi.org/10.1016/j.biomaterials.2012.11.061.

[180] J.E. Kim, J. Lee, M. Jang, M.H. Kwak, J. Go, E.K. Kho, S.H. Song, J.E. Sung, J. Lee, D.Y. Hwang, Accelerated healing of cutaneous wounds using phytochemically stabilized gold

nanoparticle deposited hydrocolloid membranes, Biomater Sci 3 (2015) 509–519.
https://doi.org/10.1039/C4BM00390J.

[181] P. Lau, N. Bidin, S. Islam, W.N.B.W.M. Shukri, N. Zakaria, N. Musa, G. Krishnan, Influence of gold nanoparticles on wound healing treatment in rat model: Photobiomodulation therapy, Lasers Surg Med 49 (2017) 380–386. https://doi.org/10.1002/lsm.22614.

[182] R.H. Adams, A. Eichmann, Axon Guidance Molecules in Vascular Patterning, Cold Spring Harb Perspect Biol 2 (2010) a001875–a001875. https://doi.org/10.1101/cshperspect.a001875.

[183] J. Yang, The role of reactive oxygen species in angiogenesis and preventing tissue injury after brain ischemia, Microvasc Res 123 (2019) 62–67. https://doi.org/10.1016/j.mvr.2018.12.005.

[184] S.Y. Wang, H. Kim, G. Kwak, H.Y. Yoon, S.D. Jo, J.E. Lee, D. Cho, I.C. Kwon, S.H. Kim, Development of Biocompatible HA Hydrogels Embedded with a New Synthetic Peptide Promoting Cellular Migration for Advanced Wound Care Management, Advanced Science 5 (2018) 1800852. https://doi.org/10.1002/advs.201800852.

[185] D. Zhao, M. Liu, Q. Li, X. Zhang, C. Xue, Y. Lin, X. Cai, Tetrahedral DNA Nanostructure Promotes Endothelial Cell Proliferation, Migration, and Angiogenesis via Notch Signaling Pathway, ACS Appl Mater Interfaces 10 (2018) 37911–37918. https://doi.org/10.1021/acsami.8b16518.

[186] S. Schoors, U. Bruning, R. Missiaen, K.C.S. Queiroz, G. Borgers, I. Elia, A. Zecchin, A.R. Cantelmo, S. Christen, J. Goveia, W. Heggermont, L. Goddé, S. Vinckier, P.P. Van Veldhoven, G. Eelen, L. Schoonjans, H. Gerhardt, M. Dewerchin, M. Baes, K. De Bock, B. Ghesquière, S.Y. Lunt, S.-M. Fendt, P. Carmeliet, Fatty acid carbon is essential for dNTP synthesis in endothelial cells, Nature 520 (2015) 192–197. https://doi.org/10.1038/nature14362.

[187] J.E. Kim, J. Lee, M. Jang, M.H. Kwak, J. Go, E.K. Kho, S.H. Song, J.E. Sung, J. Lee, D.Y. Hwang, Accelerated healing of cutaneous wounds using phytochemically stabilized gold nanoparticle deposited hydrocolloid membranes, Biomater Sci 3 (2015) 509–519. https://doi.org/10.1039/C4BM00390J.

[188] P. Lau, N. Bidin, S. Islam, W.N.B.W.M. Shukri, N. Zakaria, N. Musa, G. Krishnan, Influence of gold nanoparticles on wound healing treatment in rat model: Photobiomodulation therapy, Lasers Surg Med 49 (2017) 380–386. https://doi.org/10.1002/lsm.22614.

[189] W.-Y. Chen, H.-Y. Chang, J.-K. Lu, Y.-C. Huang, S.G. Harroun, Y.-T. Tseng, Y.-J. Li, C.-C. Huang, H.-T. Chang, Self-Assembly of Antimicrobial Peptides on Gold Nanodots: Against Multidrug-Resistant Bacteria and Wound-Healing Application, Adv Funct Mater 25 (2015) 7189–7199. https://doi.org/10.1002/adfm.201503248.

[190] F. Öri, R. Dietrich, C. Ganz, M. Dau, D. Wolter, A. Kasten, T. Gerber, B. Frerich, Silicon-dioxide–polyvinylpyrrolidone as a wound dressing for skin defects in a murine model, Journal of Cranio-Maxillofacial Surgery 45 (2017) 99–107. https://doi.org/10.1016/j.jcms.2016.10.002.

[191] X. Wang, F. Lv, T. Li, Y. Han, Z. Yi, M. Liu, J. Chang, C. Wu, Electrospun Micropatterned Nanocomposites Incorporated with Cu$_2$S Nanoflowers for Skin Tumor Therapy and Wound Healing, ACS Nano 11 (2017) 11337–11349. https://doi.org/10.1021/acsnano.7b05858.

[192] Y. Xiao, D. Xu, H. Song, F. Shu, P. Wei, X. Yang, C. Zhong, X. Wang, W.E. Müller, Y. Zheng, S. Xiao, Z. Xia, <p>Cuprous oxide nanoparticles reduces hypertrophic scarring by inducing fibroblast apoptosis</p>, Int J Nanomedicine Volume 14 (2019) 5989–6000. https://doi.org/10.2147/IJN.S196794.

[193] S.K. Nethi, A.K. Barui, V.S. Bollu, B.R. Rao, C.R. Patra, Pro-angiogenic Properties of Terbium Hydroxide Nanorods: Molecular Mechanisms and Therapeutic Applications in Wound Healing, ACS Biomater Sci Eng 3 (2017) 3635–3645. https://doi.org/10.1021/acsbiomaterials.7b00457.

[194] M. Shi, L. Xia, Z. Chen, F. Lv, H. Zhu, F. Wei, S. Han, J. Chang, Y. Xiao, C. Wu, Europium-doped mesoporous silica nanosphere as an immune-modulating osteogenesis/angiogenesis agent, Biomaterials 144 (2017) 176–187. https://doi.org/10.1016/j.biomaterials.2017.08.027.

[195] A.K. Barui, S.K. Nethi, S. Haque, P. Basuthakur, C.R. Patra, Recent Development of Metal Nanoparticles for Angiogenesis Study and Their Therapeutic Applications, ACS Appl Bio Mater 2 (2019) 5492–5511. https://doi.org/10.1021/acsabm.9b00587.

[196] D. Hao, G. Zhang, Y. Gong, Z. Ma, Development and biological evaluation of cerium oxide loaded polycaprolactone dressing on cutaneous wound healing in nursing care, Mater Lett 265 (2020) 127401. https://doi.org/10.1016/j.matlet.2020.127401.

[197] S.M. Carvalho, C.D.F. Moreira, A.C.X. Oliveira, A.A.R. Oliveira, E.M.F. Lemos, M.M. Pereira, Bioactive glass nanoparticles for periodontal regeneration and applications in dentistry, in: Nanobiomaterials in Clinical Dentistry, Elsevier, 2019: pp. 351–383. https://doi.org/10.1016/B978-0-12-815886-9.00015-2.

[198] K. Varaprasad, Co-assembled ZnO (shell) – CuO (core) nano-oxide materials for microbial protection, Phosphorus Sulfur Silicon Relat Elem 193 (2018) 74–80. https://doi.org/10.1080/10426507.2017.1417301.

[199] S. Varshney, A. Nigam, S.J. Pawar, N. Mishra, An overview on biomedical applications of versatile silica nanoparticles, synthesized via several chemical and biological routes: A review, Phosphorus Sulfur Silicon Relat Elem 197 (2022) 72–88. https://doi.org/10.1080/10426507.2021.2017434.

[200] M. Sharifiaghdam, E. Shaabani, F. Asghari, R. Faridi-Majidi, Chitosan coated metallic nanoparticles with stability, antioxidant, and antibacterial properties: Potential for wound healing application, J Appl Polym Sci 139 (2022). https://doi.org/10.1002/app.51766.

[201] P. Deng, X. Liang, F. Chen, Y. Chen, J. Zhou, Novel multifunctional dual-dynamic-bonds crosslinked hydrogels for multi-strategy therapy of MRSA-infected wounds, Appl Mater Today 26 (2022) 101362. https://doi.org/10.1016/j.apmt.2022.101362.

[202] K. Zhang, G. Zhao, An Effective Wound Healing Material Based on Gold Incorporation into a Heparin-Polyvinyl Alcohol Nanocomposite: Enhanced In Vitro and In Vivo Care of

Perioperative Period, J Clust Sci 33 (2022) 1655–1665. https://doi.org/10.1007/s10876-021-02078-5.

[203] E. Ghasemian Lemraski, H. Jahangirian, M. Dashti, E. Khajehali, Mis.S. Sharafinia, R. Rafiee-Moghaddam, T.J. Webster, Antimicrobial Double-Layer Wound Dressing Based on Chitosan/Polyvinyl Alcohol/Copper: In vitro and in vivo Assessment, Int J Nanomedicine Volume 16 (2021) 223–235. https://doi.org/10.2147/IJN.S266692.

[204] M. Bandeira, B.S. Chee, R. Frassini, M. Nugent, M. Giovanela, M. Roesch-Ely, J. da S. Crespo, D.M. Devine, Antimicrobial PAA/PAH Electrospun Fiber Containing Green Synthesized Zinc Oxide Nanoparticles for Wound Healing, Materials 14 (2021) 2889. https://doi.org/10.3390/ma14112889.

[205] V.A.T. Le, T.X. Trinh, P.N. Chien, N.N. Giang, X.-R. Zhang, S.-Y. Nam, C.-Y. Heo, Evaluation of the Performance of a ZnO-Nanoparticle-Coated Hydrocolloid Patch in Wound Healing, Polymers (Basel) 14 (2022) 919. https://doi.org/10.3390/polym14050919.

[206] A. Sathiyaseelan, K. Saravanakumar, A.V.A. Mariadoss, M.-H. Wang, Antimicrobial and Wound Healing Properties of FeO Fabricated Chitosan/PVA Nanocomposite Sponge, Antibiotics 10 (2021) 524. https://doi.org/10.3390/antibiotics10050524.

[207] K. Zamani, N. Allah-Bakhshi, F. Akhavan, M. Yousefi, R. Golmoradi, M. Ramezani, H. Bach, S. Razavi, G.-R. Irajian, M. Gerami, A. Pakdin-Parizi, M. Tafrihi, F. Ramezani, Antibacterial effect of cerium oxide nanoparticle against Pseudomonas aeruginosa, BMC Biotechnol 21 (2021) 68. https://doi.org/10.1186/s12896-021-00727-1.

[208] Y. Lv, Y. Xu, X. Sang, C. Li, Y. Liu, Q. Guo, S. Ramakrishna, C. Wang, P. Hu, H.S. Nanda, PLLA–gelatin composite fiber membranes incorporated with functionalized CeNPs as a sustainable wound dressing substitute promoting skin regeneration and scar remodeling, J Mater Chem B 10 (2022) 1116–1127. https://doi.org/10.1039/D1TB02677A.

[209] M.E. David, R.M. Ion, R.M. Grigorescu, L. Iancu, A.M. Holban, F. Iordache, A.I. Nicoara, E. Alexandrescu, R. Somoghi, S. Teodorescu, A.I. Gheboianu, Biocompatible and Antimicrobial Cellulose Acetate-Collagen Films Containing MWCNTs Decorated with TiO2 Nanoparticles for Potential Biomedical Applications, Nanomaterials 12 (2022) 239. https://doi.org/10.3390/nano12020239.

[210] S.K. Karuppannan, R. Ramalingam, S.B. Mohamed Khalith, S.A. Musthafa, M.J.H. Dowlath, G. Munuswamy-Ramanujam, K.D. Arunachalam, Copper oxide nanoparticles infused electrospun polycaprolactone/gelatin scaffold as an antibacterial wound dressing, Mater Lett 294 (2021) 129787. https://doi.org/10.1016/j.matlet.2021.129787.

[211] I. Kalashnikova, S. Das, S. Seal, Nanomaterials for wound healing: scope and advancement, Nanomedicine 10 (2015) 2593–2612. https://doi.org/10.2217/nnm.15.82.

[212] P. Sundaram, H. Abrahamse, Phototherapy Combined with Carbon Nanomaterials (1D and 2D) and Their Applications in Cancer Therapy, Materials 13 (2020) 4830. https://doi.org/10.3390/ma13214830.

[213] S. Gurunathan, J. Woong Han, A. Abdal Daye, V. Eppakayala, J. Kim, Oxidative stress-mediated antibacterial activity of graphene oxide and reduced graphene oxide in

Pseudomonas aeruginosa, Int J Nanomedicine (2012) 5901.
https://doi.org/10.2147/IJN.S37397.

[214] H. Fatima, A. Ibrahim, S.D.A. Hamdani, T.A. Rajput, A. Noor, A. Gul, M.M. Babar, Natural product formulations to overcome poor ADMET properties, in: J.N. Cruz (Ed.), Drug Discovery and Design Using Natural Products, Springer Nature Switzerland, Cham, 2023: pp. 435–452. https://doi.org/10.1007/978-3-031-35205-8_15.

[215] Y. Zheng, S. Li, D. Han, L. Kong, J. Wang, M. Zhao, W. Cheng, H. Ju, Z. Yang, S. Ding, Eco-Friendly Preparation of Epoxy-Rich Graphene Oxide for Wound Healing, ACS Biomater Sci Eng 7 (2021) 752–763. https://doi.org/10.1021/acsbiomaterials.0c01598.

[216] B. Lu, T. Li, H. Zhao, X. Li, C. Gao, S. Zhang, E. Xie, Graphene-based composite materials beneficial to wound healing, Nanoscale 4 (2012) 2978. https://doi.org/10.1039/c2nr11958g.

[217] K. Banihashemi, N. Amirmozafari, I. Mehregan, R. Bakhtiari, B. Sobouti, Antibacterial effect of carbon nanotube containing chemical compounds on drug-resistant isolates of Acinetobacter baumannii, Iran J Microbiol (2021). https://doi.org/10.18502/ijm.v13i1.5501.

[218] B. Lu, T. Li, H. Zhao, X. Li, C. Gao, S. Zhang, E. Xie, Graphene-based composite materials beneficial to wound healing, Nanoscale 4 (2012) 2978. https://doi.org/10.1039/c2nr11958g.

[219] A.J. Monteforte, B. Lam, S. Das, S. Mukhopadhyay, C.S. Wright, P.E. Martin, A.K. Dunn, A.B. Baker, Glypican-1 nanoliposomes for potentiating growth factor activity in therapeutic angiogenesis, Biomaterials 94 (2016) 45–56. https://doi.org/10.1016/j.biomaterials.2016.03.048.

[220] K.K. Chereddy, C.-H. Her, M. Comune, C. Moia, A. Lopes, P.E. Porporato, J. Vanacker, M.C. Lam, L. Steinstraesser, P. Sonveaux, H. Zhu, L.S. Ferreira, G. Vandermeulen, V. Préat, PLGA nanoparticles loaded with host defense peptide LL37 promote wound healing, Journal of Controlled Release 194 (2014) 138–147. https://doi.org/10.1016/j.jconrel.2014.08.016.

[221] K. Han, K.-D. Lee, Z.-G. Gao, J.-S. Park, Preparation and evaluation of poly(l-lactic acid) microspheres containing rhEGF for chronic gastric ulcer healing, Journal of Controlled Release 75 (2001) 259–269. https://doi.org/10.1016/S0168-3659(01)00400-X.

[222] A. Schwentker, Y. Vodovotz, R. Weller, T.R. Billiar, Nitric oxide and wound repair: role of cytokines?, Nitric Oxide 7 (2002) 1–10. https://doi.org/10.1016/S1089-8603(02)00002-2.

[223] S. Chen, B. Liu, M.A. Carlson, A.F. Gombart, D.A. Reilly, J. Xie, Recent advances in electrospun nanofibers for wound healing, Nanomedicine 12 (2017) 1335–1352. https://doi.org/10.2217/nnm-2017-0017.

[224] W. Cheng, R. Xu, D. Li, C. Bortolini, J. He, M. Dong, F. Besenbacher, Y. Huang, M. Chen, Artificial extracellular matrix delivers TGFb1 regulating myofibroblast differentiation, RSC Adv 6 (2016) 21922–21928. https://doi.org/10.1039/C5RA26164C.

[225] J.E. Martín-Alfonso, A.A. Cuadri, M. Berta, M. Stading, Relation between concentration and shear-extensional rheology properties of xanthan and guar gum solutions, Carbohydr Polym 181 (2018) 63–70. https://doi.org/10.1016/j.carbpol.2017.10.057.

Materials Research Forum LLC
https://doi.org/10.21741/9781644903339-4

[226] L.A. De Louise, Morphology-Dependent Titanium Dioxide Nanoparticle-Induced Keratinocyte Toxicity And Exacerbation Of Allergic Contact Dermatitis, Toxicology: Current Research 4 (2020) 1–7. https://doi.org/10.24966/TCR-3735/100019.

[227] X. Lai, M. Wang, Y. Zhu, X. Feng, H. Liang, J. Wu, L. Nie, L. Li, L. Shao, ZnO NPs delay the recovery of psoriasis-like skin lesions through promoting nuclear translocation of p-NFκB p65 and cysteine deficiency in keratinocytes, J Hazard Mater 410 (2021) 124566. https://doi.org/10.1016/j.jhazmat.2020.124566.

[228] S. Hashempour, S. Ghanbarzadeh, H.I. Maibach, M. Ghorbani, H. Hamishehkar, Skin toxicity of topically applied nanoparticles, Ther Deliv 10 (2019) 383–396. https://doi.org/10.4155/tde-2018-0060.

[229] K. Sooklert, S. Nilyai, R. Rojanathanes, D. Jindatip, N. Sae-liang, N. Kitkumthorn, A. Mutirangura, A. Sereemaspun, <p>N-acetylcysteine reverses the decrease of DNA methylation status caused by engineered gold, silicon, and chitosan nanoparticles</p>, Int J Nanomedicine Volume 14 (2019) 4573–4587. https://doi.org/10.2147/IJN.S204372.

[230] M.S. Bakshi, Nanotoxicity in Systemic Circulation and Wound Healing, Chem Res Toxicol 30 (2017) 1253–1274. https://doi.org/10.1021/acs.chemrestox.7b00068.

[231] N. Hadrup, A.K. Sharma, K. Loeschner, Toxicity of silver ions, metallic silver, and silver nanoparticle materials after in vivo dermal and mucosal surface exposure: A review, Regulatory Toxicology and Pharmacology 98 (2018) 257–267. https://doi.org/10.1016/j.yrtph.2018.08.007.

Advances in Healthcare and Nanoparticle Toxicology
Materials Research Foundations 171 (2024) 135-153

Materials Research Forum LLC
https://doi.org/10.21741/9781644903339-5

Chapter 5

Interaction of Nanoparticles with Nucleic Acids

Sankara Rao Miditana[1*], Saivenkatesh Korlam[2]

[1]Department of Chemistry, Government. Degree College, Puttur, Tirupati, A.P-517583, India

[2]Department of Botany, SVA Government College, Srikalahasti, A.P-517644, India

* sraom90@gmail.com

Abstract

The interaction of nanoparticles with nucleic acids has gained significant attention in recent years due to its potential applications in various fields, including gene therapy, biosensing, immunotherapy and drug delivery. The mechanisms of interaction between nanoparticles and nucleic acids are complex and involve multiple factors, including surface chemistry, size, and shape of nanoparticles. Once nanoparticles bind to nucleic acids, they can form complexes that can alter the physical and chemical properties of nucleic acids, leading to changes in their enzymatic activity and stability. The interaction between nanoparticles and nucleic acids also provides a promising avenue for targeted delivery of nucleic acids to specific cells or tissues.

Keywords

Drug Delivery, DNA, Gene Delivery, mRNA, Nucleic Acid, Nanoparticles

Contents

1.   Introduction

The word "nanotechnology" is used in the scientific community when a nanostructure, measured on a scale of 1 to 100 nm, allows a particular material to exhibit its special properties. Nanotechnology has attracted a lot of attention over the past decade and is making substantial advancements in a wide range of science and technological fields [1]. The development of nanomaterials and nanoparticles (NPs) for application in diverse fields such as catalysis, electrochemistry, biomedicine, pharmaceuticals, sensors, food technology, cosmetics, etc., is the focus of a new branch of study called nanotechnology [2-4].

In recent years, the rapid progress in nanotechnology has opened exciting possibilities for innovative biomedical applications. Among the myriads of nanoscale structures, nucleic acid nanoparticles (NANPs) have emerged as a promising and versatile class of nanocarriers, garnering significant attention in the fields of research and therapeutics. Although nucleic acids are best known as the molecules that carry genetic information, they are also a flexible material to produce nanometer-scale structures because their sequences may be designed so that the strands fold into unique secondary structures. In 1982, Seeman first proposed using branched DNA building blocks to construct ordered arrays [5]. NANPs, composed primarily of DNA or RNA, possess unique properties that make them highly programmable, biocompatible, and amenable to precise functionalization. These attributes have positioned NANPs as a transformative platform for drug delivery, gene therapy, and diagnostic applications, addressing critical challenges in modern medicine. In addition, NANPs offer a stronger molecular system compared to other NPs derived from small molecules, peptides, and polymers due to their high-fidelity hydrogen bonding and structural programmability (Table.1).

The interaction of nanoparticles with nucleic acids, including DNA and RNA, has become an area of significant research interest due to the potential applications in various fields, such as gene therapy, biosensing, and drug delivery. The interaction between nanoparticles and nucleic acids can occur through several mechanisms, including electrostatic interactions, hydrogen bonding, and hydrophobic interactions. Once the nanoparticles come into contact with nucleic acids, they can bind to them and form complexes that can alter the physical and chemical properties of the nucleic acids [7-10]. NANPs are nanostructures formed by self-assembly of nucleic acids into complex architectures. These nanoparticles can be designed to possess specific physicochemical properties, stability, and functionalities for various applications. They can be classified into several categories based on their composition, size, and structure.

Materials Research Forum LLC
https://doi.org/10.21741/9781644903339-5

Table. 1. Comparison of NPs designed from Nucleic Acids and Other Molecules [6].

		Procs	Cons
Nucleic acid NPs	DNA	High stability, programmable, biocompatible, non-toxic, ease of synthesis and modification, and well-established pH-dependent secondary conformations.	Limited stability toward extreme pH and temperature.
	RNA	Diverse secondary and tertiary conformations.	Highly unstable, expensive, and sensitive to RNases.
Other nanoparticles	Small molecules	Bioavailability, metabolic stability, small size, ease of synthesis and broad tunability.	Unique properties and pH dependency of small molecules cannot be generalized.
	Peptides/ Proteins	High potency and selectivity, high chemical and biological diversity, ease of synthesis and structural tunability, and pH-responsive nature is reasonably established.	Poor metabolic stability, weak membrane permeability, low oral bioavailability and predictable behavior toward pH.
	Polymer	Alternative for nucleic acids, small molecule and protein-based Nanomaterials.	Limited tunability, complex procedure for synthesis and processing, high cost, and pH response not well-defined.

Nucleic acid nanoparticles represent a rapidly evolving field at the intersection of nanotechnology and molecular biology. These nanoparticles are constructed from various types of nucleic acids, such as DNA, RNA, or their analogs, and have emerged as versatile and promising nanoscale materials with diverse applications [11]. The unique properties of nucleic acids, including their programmability, biocompatibility, and ability to carry genetic information, make them attractive building blocks for designing nanoscale structures with precise control over their properties and functions [12]. NANPs can be engineered through rational design or self-assembly processes, allowing the creation of complex and functional nanoarchitectures. They can be modified to carry specific cargos, such as therapeutic agents, genes, or imaging probes, enabling targeted delivery and controlled release at the desired site. Additionally, nucleic acids' base-pairing interactions allow for the incorporation of functional elements, such as targeting ligands or stimuli-responsive components, further expanding their capabilities in biomedical applications [6].

When nanoparticles interact with nucleic acids, several physical and chemical changes are observed. The binding of nucleic acids to nanoparticles can lead to alterations in the size and shape of the NPs, with DNA wrapping around the nanoparticle core being a prominent example. This interaction can also modify the surface charge of nanoparticles, influencing their colloidal stability and cellular uptake. The presence of nucleic acids can induce aggregation or disaggregation of nanoparticles depending on the system and environmental conditions [13]. Furthermore, the interaction can lead to surface functionalization of nanoparticles with biomolecules like DNA or RNA, enabling specific targeting and improved biocompatibility. Changes in the conformation of nucleic acids can occur upon binding to nanoparticles, potentially affecting their biological activity and binding affinity to target molecules. Additionally, the interaction with nanoparticles provides stability and protection to nucleic acids, shielding them from enzymatic degradation and enhancing their potential as therapeutic agents. These observed physical and chemical changes are crucial for optimizing the design and application of nanoparticles in nucleic acid-based therapies and diagnostics [14-18].

The use of nanotechnology in medicine for prevention, diagnosis, and treatment is known as nanomedicine. It takes advantage of the distinctive chemical, physical, and biological characteristics of materials at the nanoscale [19]. Using nucleic acids—DNA, RNA, and their numerous modifications to design and create nanostructures for therapeutic applications, or nucleic acid nanotechnology, is one of the emerging subfields of nanomedicine [20]. Single-stranded DNA or RNA molecules are coherently built into modular nucleic acid nanoparticles (NANPs), which are easily customized into supramolecular three-dimensional structures composed only of nucleic acids, in addition to the programmability and inherent activities of nucleic acids. Rings, fibers, and polygonal nanostructures can be assembled from canonical and noncanonical base pairs formed by RNA and DNA to generate a variety of higher-order structures [21-23]. The selection of nucleic acid components benefits NANPs by offering tunability for their physicochemical characteristics, biological activities, and multifunctionality [24].

Numerous studies in biotechnology and biomedicine recommend employing NANPs as bioactive drug transporters, molecular tools for imaging and biosensing, scaffolds for biochemical activities, or multifunctional nanoparticles combining the aforementioned uses into a single complex [25-27]. Multiple NANP synthesis methods had been developed, their in vitro and in vivo characterization methods had been created, and proof-of-concept data for employing NANPs in diverse therapeutic applications had been obtained by the rapidly growing discipline of nucleic acid nanotechnology [28-30].

The interaction between nanoparticles and nucleic acids can also be used to deliver nucleic acids to specific cells or tissues [31]. This is because nanoparticles can protect nucleic acids from degradation in the bloodstream and allow them to be delivered to their target site. Additionally, the size and surface chemistry of nanoparticles can be engineered to improve their targeting efficiency and reduce their toxicity. The present study primarily centers upon the attributes inherent to NANPs, delineating alterations manifest within nanoparticles following interactions with nucleic acids. This discourse encompasses an exploration of diverse types of NANPs, an examination of their manifold applications, and an elucidation of the latest strides achieved within the realm of nanoparticle-nucleic acid interactions.

2. Types of nucleic acid nanoparticles

Nucleic acid nanoparticles encompass various types of nanostructures constructed from nucleic acids, such as DNA, RNA, or their analogs. Each type of NANP possesses unique characteristics and functionalities, making them suitable for different applications. Some common types of nucleic acid nanoparticles are reported below.

2.1 DNA nanoparticles

While many other ligands have been created to control how nanoparticles are put together, DNA has garnered the most attention [32]. Compared to other ligands, DNA has a number of advantages, including the following: (i) a DNA strand binds selectively and specifically to its complementary sequence; (ii) a variety of DNA structures can be created using straightforward techniques; (iii) DNA can interact specifically with other functional molecules, such as enzymes; and (iv) the strands of DNA can be chemically altered to include functional chemical structures.

DNA nanoparticles are constructed using the principles of DNA nanotechnology, where single-stranded DNA molecules are designed to self-assemble into complex nanostructures through complementary base-pairing interactions. These nanostructures can be programmed to have precise shapes, sizes, and functionalities. DNA nanoparticles have shown promise in drug delivery, biosensing, and nanoelectronics due to their programmability and biocompatibility. Some key features of DNA nanoparticles include:

a. DNA Origami: This technique involves folding a long single-stranded DNA scaffold using shorter "staple" strands to create two- and three-dimensional shapes with nanometer-scale precision [33].

b. DNA Tiles: DNA tiles are short DNA molecules that form programmable two-dimensional patterns on a surface through specific hybridization [34].

c. DNA Nanotubes and Nanowires: These structures are formed by rolling or folding DNA molecules to create tubular or wire-like nanostructures [35].

d. DNA Nanocages: These are three-dimensional, hollow structures with the potential to encapsulate small molecules or functional entities [36].

2.2 RNA nanoparticles

RNA nanoparticles are primarily composed of RNA molecules, such as small interfering RNA (siRNA), messenger RNA (mRNA), or ribozymes. These nanoparticles are designed to perform specific functions, such as gene silencing, gene editing, or gene expression modulation [37]. Key features of RNA nanoparticles include:

a. siRNA Nanoparticles: siRNA molecules are incorporated into nanoparticles to silence specific target genes, thereby offering potential therapeutic applications in treating genetic diseases and cancer [38].

b. mRNA Nanoparticles: mRNA nanoparticles are designed to deliver exogenous mRNA into cells to express therapeutic proteins, making them promising candidates for vaccines and gene therapies [39].

c. RNA-Based Immunotherapies: RNA nanoparticles can stimulate the immune system, making them useful in cancer immunotherapy [40].

2.3 Hybrid nucleic acid nanoparticles

Hybrid NANPs are composed of both DNA and RNA components, combining the strengths of both nucleic acids to achieve specific functionalities. These nanoparticles are designed to leverage the unique properties of DNA and RNA and offer enhanced stability, targeting capabilities, and

Advances in Healthcare and Nanoparticle Toxicology   Materials Research Forum LLC
Materials Research Foundations 171 (2024) 135-153   https://doi.org/10.21741/9781644903339-5

gene regulation functionalities. Hybrid nucleic acid nanoparticles are an exciting area of research with potential applications in gene therapy, drug delivery, and biomedical imaging [41].

a. DNA-RNA Hybrid Nanoparticles: These nanoparticles combine the programmability and structural diversity of DNA with the gene regulation capabilities of RNA [42].

b. DNA-RNA Chimeras: Chimeric nanoparticles are formed by incorporating both DNA and RNA segments, allowing for multifunctionality and improved therapeutic effects [43].

3.   Advantages of interaction of nucleic acids with NPs

The interaction of nanoparticles with nucleic acids offers several major advantages compared to using nucleic acids without such interaction. Some of the key advantages are discussed below.

a. Enhanced Stability and Protection: Nanoparticles can protect nucleic acids, such as RNA and DNA, from enzymatic degradation and nuclease attack, leading to increased stability and preservation of their integrity during storage and delivery [44]. This protection is crucial for maintaining the therapeutic efficacy of nucleic acid-based therapies.

b. Improved Cellular Uptake and Intracellular Delivery: Nanoparticles facilitate the efficient delivery of nucleic acids into target cells, overcoming cellular barriers that often limit the entry of naked nucleic acids [45]. This improved cellular uptake enhances the bioavailability and efficacy of nucleic acid-based therapeutics.

c. Targeted Delivery: Functionalization of nanoparticles enables specific targeting of nucleic acids to particular cell types or tissues, reducing off-target effects and enhancing the accumulation of therapeutic nucleic acids at the intended site of action [46].

d. Controlled Release: Nanoparticles can be engineered to release nucleic acids in a controlled and sustained manner, allowing for prolonged therapeutic effects and potentially reducing the frequency of administration [47].

e. Co-Delivery of Multiple Nucleic Acids: Nanoparticles provide a platform for co-delivery of multiple nucleic acids, such as siRNAs and mRNAs, allowing for combination therapies and synergistic effects in treating complex diseases [48].

f. Non-Viral Delivery: Nanoparticles serve as non-viral delivery systems for nucleic acids, avoiding potential risks associated with viral vectors, such as immunogenicity and insertional mutagenesis [49].

g. Personalized Medicine: Nanoparticles can be tailored for individual patients, allowing personalized medicine approaches by targeting specific genetic mutations or disease-associated nucleic acids [50]. This individualized treatment can improve therapeutic outcomes.

4.   Characteristics of NANPs

Nucleic acid nanoparticles possess several key characteristics that make them unique and versatile nanomaterials for various applications. These characteristics arise from the properties of nucleic acids, such as DNA and RNA, and their ability to self-assemble into complex nanostructures. Below are some of the detailed characteristics of NANPs are discussed.

Advances in Healthcare and Nanoparticle Toxicology   Materials Research Forum LLC
Materials Research Foundations 171 (2024) 135-153   https://doi.org/10.21741/9781644903339-5

a. Programmability: One of the most significant advantages of nucleic acid nanoparticles is their programmability. DNA and RNA molecules follow well-defined base-pairing rules, allowing researchers to design and control the precise structure and function of NANPs [51]. This programmability enables the creation of diverse nanoarchitectures with tailored properties and functionalities.

b. Biocompatibility: Nucleic acids are natural biomolecules present in all living organisms. As such, nucleic acid nanoparticles exhibit excellent biocompatibility, reducing the risk of adverse reactions or immune responses when used in biological applications [49]. Furthermore, they can be modified to enhance their stability and biostability in various physiological conditions.

c. Size and Shape Control: With the ability to design the sequence and arrangement of nucleic acids, NANPs can be precisely engineered to achieve specific sizes and shapes [52]. Researchers can create nanoparticles with nanometer-scale dimensions, making them suitable for targeted drug delivery, cellular uptake, and interactions at the molecular level [53].

d. Targeting Capabilities: Nucleic acid nanoparticles can be functionalized with ligands, antibodies, or other targeting moieties to achieve specific targeting to particular cells or tissues [54]. This targeting ability enhances the selectivity and efficiency of drug delivery and gene therapy approaches, minimizing off-target effects and improving therapeutic outcomes.

e. Cargo Loading: NANPs can encapsulate various cargos, such as therapeutic drugs, siRNA, mRNA, imaging agents, or other functional entities. This cargo-loading capacity allows for the efficient delivery of therapeutic payloads to specific cellular targets or tissues [55].

f. Stability and Protection: By incorporating modified nucleotides or employing advanced nanotechnology approaches, researchers can enhance the stability of nucleic acid nanoparticles, protecting them from enzymatic degradation and harsh environmental conditions during delivery and circulation [56].

g. Triggered Release: Nucleic acid nanoparticles can be engineered to respond to specific environmental cues, such as changes in pH, temperature, or the presence of specific biomolecules. This property allows for triggered release of payloads at the target site, enabling controlled drug release or gene expression regulation [57].

5.   Applications of nucleic acid nanoparticles

Nucleic acid nanoparticles have shown great potential in various applications, particularly in the field of nanomedicine and therapeutics. These nanoparticles are composed of nucleic acids, such as DNA or RNA, and can be designed to carry out specific functions, including targeted drug delivery, gene therapy, and immunotherapy. Here are some of the key applications of nucleic acid nanoparticles are reported.

5.1   Nanoparticles used for nucleic acid drug delivery

Nanoparticles have emerged as versatile carriers for nucleic acid drug delivery, revolutionizing the field of gene therapy and personalized medicine. These nanoparticles can protect and efficiently deliver nucleic acids, such as DNA, siRNA, and mRNA, to target cells or tissues, offering promising therapeutic applications for various genetic and acquired diseases. Here, we will delve into the diverse types of nanoparticles used for nucleic acid drug delivery and explore their advantages, challenges, and potential.

a. Lipid-Based Nanoparticles:

Lipid-based nanoparticles, such as liposomes and lipid nanoparticles (LNPs), are among the most extensively studied carriers for nucleic acid delivery [58]. They consist of lipid bilayers that can encapsulate nucleic acids within their aqueous core or associate with the lipids through electrostatic interactions. Lipid-based nanoparticles offer biocompatibility, easy modification, and the ability to efficiently deliver nucleic acids to the cytoplasm [59]. The success of COVID-19 mRNA vaccines, which employ LNPs to deliver the mRNA encoding the spike protein, has highlighted the potential of lipid-based nanoparticles in nucleic acid drug delivery.

b. Polymeric Nanoparticles:

Polymeric nanoparticles, made from biodegradable and biocompatible polymers, have garnered considerable interest as nucleic acid carriers. These nanoparticles can be finely tuned for size, surface charge, and stability, enabling controlled release and targeted delivery [60]. Polymeric nanoparticles can encapsulate a variety of nucleic acids and protect them from enzymatic degradation, ensuring their safe and effective delivery to specific cells [61]. Additionally, they can be functionalized with ligands to enhance targeting and internalization into specific cell types.

c. Inorganic Nanoparticles:

Inorganic nanoparticles, such as gold nanoparticles, iron oxide nanoparticles, and silica nanoparticles, have also been explored for nucleic acid delivery. They offer unique properties, including tunable size, shape, and surface chemistry, making them versatile carriers for therapeutic nucleic acids. Inorganic nanoparticles can be coated with polymers or functionalized with targeting ligands to enhance their biocompatibility and specificity for target cells [62].

d. Dendrimers:

Dendrimers are highly branched, synthetic polymers that can encapsulate nucleic acids within their cavities. They offer precise control over size and surface functionality, enabling efficient cellular uptake and intracellular delivery of nucleic acids. However, dendrimers can be relatively complex to synthesize and may present some toxicity concerns, necessitating careful design and optimization for safe and effective nucleic acid delivery [63].

e. Hybrid Nanoparticles:

Hybrid nanoparticles combine the benefits of different materials to create multifunctional carriers. For instance, combining lipids and polymers can result in lipopolyplexes, which combine the advantages of both lipid-based and polymeric nanoparticles [64]. These hybrid systems can offer improved stability, transfection efficiency, and targeting capabilities.

5.2   Gene delivery for gene therapy

Nanoparticles have been extensively explored as carriers for gene delivery in gene therapy applications [65]. Lipid-based nanoparticles, such as liposomes and lipid nanoparticles (LNPs), have shown remarkable success in delivering therapeutic DNA to correct or replace defective genes. Additionally, polymeric nanoparticles and dendrimers offer versatile platforms for delivering gene-editing tools like CRISPR-Cas9, enabling precise gene modifications [66]. These nanoparticles protect the nucleic acids from degradation, facilitate their cellular uptake, and promote successful gene expression.

a. RNA Interference (RNAi):

Nanoparticles have been extensively used for delivering siRNA to silence specific genes via RNA interference. Lipid-based nanoparticles and polymeric nanoparticles efficiently encapsulate and deliver siRNA to target cells, leading to post-transcriptional gene silencing [67]. This approach shows great promise for treating genetic diseases, viral infections, and certain types of cancer.

b. mRNA-Based Vaccines:

The development of mRNA-based vaccines against infectious diseases has been a breakthrough application of nanoparticles interacting with nucleic acids. Lipid-based nanoparticles, specifically LNPs, have been utilized to deliver synthetic mRNA encoding viral antigens, such as the spike protein in SARS-CoV-2, for inducing an immune response [68]. These nanoparticles protect the mRNA from degradation and promote its translation into antigenic proteins, leading to the production of protective antibodies.

c. Cancer Therapeutics:

Nanoparticles that interact with nucleic acids have shown promise in cancer therapeutics. RNA interference through siRNA delivery can specifically target oncogenes, while the delivery of anticancer drugs via nucleic acid-based nanoparticles enhances drug efficacy and reduces off-target effects [69].

Nanoparticles that interact with nucleic acids have shown great promise in cancer research and therapy. These innovative approaches leverage the unique properties of nanoparticles to deliver therapeutic nucleic acids, such as small interfering RNA (siRNA), microRNA (miRNA), and plasmid DNA, for cancer treatment. Detailed account of applications of nanoparticles in cancer therapy are discussed below.

i. siRNA Delivery for Gene Silencing:

Nanoparticles serve as efficient carriers for delivering siRNA to silence specific oncogenes or cancer-promoting genes [70]. By targeting the aberrant genes responsible for tumor growth and progression, siRNA-loaded nanoparticles can suppress cancer cell proliferation, induce apoptosis, and inhibit angiogenesis. Lipid-based nanoparticles, polymeric nanoparticles, and inorganic nanoparticles have been widely explored as carriers for siRNA delivery in cancer therapy.

ii. mRNA-Based Cancer Vaccines:

Nanoparticles, particularly lipid-based nanoparticles, have facilitated the development of mRNA-based cancer vaccines. These vaccines use mRNA to encode tumor-specific antigens, which, when delivered into dendritic cells or cancer cells, elicit a strong immune response against cancer cells expressing those antigens [71]. mRNA-loaded nanoparticles have demonstrated promising results in preclinical and clinical studies for cancer immunotherapy.

iii. Plasmid DNA Delivery for Gene Therapy:

Nanoparticles have been investigated as carriers for plasmid DNA delivery in cancer gene therapy. These nanoparticles can deliver therapeutic genes encoding tumor suppressors, apoptosis-inducing proteins, or immune-stimulating cytokines directly to tumor cells. This approach aims to correct genetic defects or modulate cellular pathways, leading to the inhibition of tumor growth and enhanced anticancer immune responses.

iv. Photothermal Therapy (PTT):

Inorganic nanoparticles, such as gold nanorods or nanoshells, can absorb near-infrared light and convert it into heat, enabling targeted photothermal therapy (PTT) for cancer treatment. When nanoparticles are localized to the tumor site and exposed to laser irradiation, they generate localized hyperthermia, leading to tumor cell destruction and enhancing the therapeutic effect.

5.3   Immunotherapy

Nucleic acid nanoparticles can be used to deliver immunostimulatory nucleic acids, such as Toll-like receptor (TLR) agonists or cytokines, to enhance the immune response against cancer or infectious diseases [72]. Nucleic acid nanoparticles have emerged as a versatile and promising tool in immunotherapy, offering innovative approaches to combat diseases. These nanoparticles, often comprising DNA or RNA molecules, can be precisely engineered to carry genetic information or act as immune modulators. In cancer immunotherapy, they facilitate the development of personalized vaccines by presenting tumor-specific antigens to stimulate targeted immune responses against cancer cells [73]. Nucleic acid nanoparticles also excel in delivering therapeutic RNAs, like messenger RNA (mRNA) vaccines, contributing to the rapid response against infectious diseases [74]. Furthermore, their ability to modulate immune reactions and regulate gene expression opens avenues for treating autoimmune disorders and enhancing immunity against pathogens. As research advances, nucleic acid nanoparticles are poised to play a pivotal role in revolutionizing the field of immunotherapy, potentially offering safer and more effective treatments for a range of diseases.

5.4   Diagnosis and imaging

Nucleic acid nanoparticles have been explored as diagnostic agents due to their ability to bind to specific targets, making them useful for detecting disease biomarkers [75]. Nucleic acid nanoparticles have emerged as a promising tool in molecular imaging, offering a unique and adaptable approach to visualizing biological processes at the molecular level. By engineering these nanoparticles with specific targeting ligands and imaging agents, such as fluorescent dyes or nanoparticles, researchers can create highly sensitive probes for detecting and monitoring cellular and molecular events. In cancer diagnostics, nucleic acid nanoparticles can be designed to target tumor-specific markers, enabling non-invasive imaging of tumor growth, metastasis, and response to treatment [76]. Moreover, these nanoparticles can serve as contrast agents in various imaging modalities, including magnetic resonance imaging (MRI), positron emission tomography (PET), and optical imaging. Their ability to combine targeted specificity with diverse imaging capabilities positions nucleic acid nanoparticles as a valuable asset in advancing molecular imaging techniques, facilitating earlier disease detection and personalized medicine approaches.

6.   Applications of metal nucleic acid nanoparticles

Metal nucleic acid nanoparticles (MNPs) represent a fascinating and versatile class of nanomaterials that combine the unique properties of both metals and nucleic acids. These nanoparticles have found in various therapeutic applications due to their tunable properties, biocompatibility, and potential for targeted interactions. Here, we reported some of the MNPs.

Advances in Healthcare and Nanoparticle Toxicology
Materials Research Foundations 171 (2024) 135-153

Materials Research Forum LLC
https://doi.org/10.21741/9781644903339-5

a. Gold Nanoparticles (AuNPs):

Gold nanoparticles have been extensively studied due to their biocompatibility and ease of functionalization. They are used in various applications, including gene delivery and sensing. AuNPs can interact with DNA through electrostatic interactions and hydrogen bonding. Surface functionalization with cationic ligands can enable AuNPs to bind to DNA and facilitate gene delivery processes [77].

b. Silver Nanoparticles (AgNPs):

Silver nanoparticles have antimicrobial properties and are used in various applications, including wound dressings and antibacterial coatings. However, AgNPs can interact with DNA and RNA, potentially leading to genotoxic effects. There is ongoing research to understand the mechanisms underlying these interactions and to evaluate the potential risks [78].

c. Copper Nanoparticles (CuNPs):

Copper nanoparticles with nucleic acids is potentially enable for gene delivery [79]. It should be considered that the impact of nanoparticles on nucleic acids can vary based on factors such as nanoparticle properties (size, surface chemistry, shape), concentration, exposure time, and cell type. Some of the MNPs and their applications are shown in the Table.no.2.

Table.2: Metal nucleic acid nanoparticles applications.

S.No.	Metal NP used	Nucleic Acid used	Application	Ref. No.
1	Au	siRNA, miRNA	Gene Silencing/ Cancer Treatment	80
2	Au	RNAs	Gene Silencing	81
3	Ag	DNA	Biosensing	78
4	Ag	DNA	Gene, ion, or small-molecule sensors	82
5	Cu	DNA	Gene Delivery	83

7.   Recent developments of NANPs

In recent years, nucleic acid nanomaterials have seen significant advancements and breakthroughs, driven by the integration of nanotechnology, molecular biology, and materials science. These developments have expanded the applications of NANPs in various fields, such as drug delivery, gene therapy, diagnostics, and biotechnology. Detailed accounts of some of the recent developments in this area are reported.

a. RNA Nanoparticles for Vaccines and Gene Therapies:

Recent years have witnessed substantial progress in the use of RNA nanoparticles for vaccine development and gene therapies. Messenger RNA (mRNA) vaccines have emerged as a groundbreaking technology, exemplified by the rapid development and global deployment of mRNA-based COVID-19 vaccines [84]. The design and delivery of stabilized mRNA nanoparticles have been optimized for improved efficacy, safety, and immune response. Moreover, advances in lipid-based nanoparticle formulations have significantly enhanced the delivery and translation efficiency of mRNA, making these platforms highly promising for infectious disease and cancer vaccines.

b. DNA Origami Nanoparticles for Drug Delivery and Imaging:

DNA origami, a powerful bottom-up nanofabrication method, has emerged as a promising tool for designing drug delivery vehicles with precise control over shape and cargo loading. By functionalizing DNA origami with targeting ligands and stimuli-responsive elements, researchers have achieved enhanced drug delivery specificity and controlled release. Additionally, DNA origami nanostructures have been employed in molecular imaging and biosensing applications, providing valuable insights into cellular processes and disease detection [85].

c. Hybrid Nucleic Acid Nanomaterials for Therapeutics:

Hybrid nucleic acid nanoparticles, combining DNA and RNA segments, have been explored for their synergistic effects and multifunctional properties. These hybrid nanomaterials have demonstrated enhanced stability, target specificity, and therapeutic effects in gene regulation, cancer therapy, and immunotherapy applications [86,62]. By incorporating both DNA and RNA components exploit the strengths of both nucleic acids to achieve novel therapeutic outcomes.

d. Nucleic Acid Nanoparticles for Theragnostic:

Recent developments have integrated therapeutic and diagnostic functions into a single nucleic acid nanoparticle, known as theranostic nanoparticles. These nanomaterials allow real-time monitoring of therapeutic efficacy and can be used for personalized medicine and patient-specific treatment regimens [75].

Thus, recent developments in nucleic acid nanomaterials have propelled the field forward, expanding their applications in therapeutics, diagnostics, and nanotechnology. The integration of DNA origami, RNA nanoparticles, and hybrid nucleic acid structures has unlocked novel functionalities, leading to potential breakthroughs in personalized medicine and targeted therapies.

Conclusion

In summary, Nanoparticles have garnered significant attention in various fields due to their unique properties and potential applications. NANPs present a transformative approach to address current challenges in biomedical research and therapeutics. Their programmability, biocompatibility, and capacity to carry diverse payloads make them a promising tool for targeted, efficient, and precise delivery of therapeutic and diagnostic agents. As the field continues to evolve, nucleic acid nanoparticles hold the potential to revolutionize modern medicine, offering new treatment modalities for a wide array of diseases. NANPs represent a diverse and promising class of nanomaterials with various applications in biotechnology, medicine, and nanotechnology.

References

[1]A.S. Abdelsattar, A. Dawoud, M.A. Helal, Interaction of nanoparticles with biological macromolecules: a review of molecular docking studies, Nanotoxicology. 15 (2021) 66–95. https://doi.org/10.1080/17435390.2020.1842537.

[2]C. Vanlalveni, S. Lallianrawna, A. Biswas, M. Selvaraj, B. Changmai, S.L. Rokhum, Green synthesis of silver nanoparticles using plant extracts and their antimicrobial activities: a

Materials Research Foundations 171 (2024) 135-153   https://doi.org/10.21741/9781644903339-5

review of recent literature, RSC Adv. 11 (2021) 2804–2837. https://doi.org/10.1039/d0ra09941d.

[3]A. Bera, H. Belhaj, Application of nanotechnology by means of nanoparticles and nanodispersions in oil recovery - A comprehensive review, J. Nat. Gas Sci. Eng. 34 (2016) 1284–1309. https://doi.org/10.1016/j.jngse.2016.08.023.

[4]L.J. Frewer, N. Gupta, S. George, A.R.H. Fischer, E.L. Giles, D. Coles, Consumer attitudes towards nanotechnologies applied to food production, Trends Food Sci. Technol. 40 (2014) 211–225. https://doi.org/10.1016/j.tifs.2014.06.005.

[5]B.J. Casey, L.H. Somerville, I.H. Gotlib, O. Ayduk, N.T. Franklin, M.K. Askren, J. Jonides, M.G. Berman, N.L. Wilson, T. Teslovich, G. Glover, V. Zayas, W. Mischel, Y. Shoda, Behavioral and neural correlates of delay of gratification 40 years later, Proc. Natl. Acad. Sci. U. S. A. 108 (2011) 14998–15003. https://doi.org/10.1073/pnas.1108561108.

[6]M.N. Mattath, D. Ghosh, S. Pratihar, S. Shi, T. Govindaraju, Nucleic Acid Architectonics for pH-Responsive DNA Systems and Devices, ACS Omega. 7 (2022) 3167–3176. https://doi.org/10.1021/acsomega.1c06464.

[7]D. Peer, J.M. Karp, S. Hong, O.C. Farokhzad, R. Margalit, R. Langer, Nanocarriers as an emerging platform for cancer therapy, Nat. Nanotechnol. 2 (2007) 751–760. https://doi.org/10.1038/nnano.2007.387.

[8]S. Muzammil, J. Neves Cruz, R. Mumtaz, I. Rasul, S. Hayat, M.A. Khan, A.M. Khan, M.U. Ijaz, R.R. Lima, M. Zubair, Effects of Drying Temperature and Solvents on In Vitro Diabetic Wound Healing Potential of Moringa oleifera Leaf Extracts, Molecules. 28 (2023). https://doi.org/10.3390/molecules28020710.

[9]N.D. Sonawane, F.C. Szoka, A.S. Verkman, Chloride Accumulation and Swelling in Endosomes Enhances DNA Transfer by Polyamine-DNA Polyplexes, J. Biol. Chem. 278 (2003) 44826–44831. https://doi.org/10.1074/jbc.M308643200.

[10] L. Jiang, P. Vader, R.M. Schiffelers, Extracellular vesicles for nucleic acid delivery: Progress and prospects for safe RNA-based gene therapy, Gene Ther. 24 (2017) 157–166. https://doi.org/10.1038/gt.2017.8.

[11] A. Nel, T. Xia, L. Mädler, N. Li, Toxic potential of materials at the nanolevel, Science (80-.). 311 (2006) 622–627. https://doi.org/10.1126/science.1114397.

[12] I. Khan, K. Saeed, I. Khan, Nanoparticles: Properties, applications and toxicities, Arab. J. Chem. 12 (2019) 908–931. https://doi.org/10.1016/j.arabjc.2017.05.011.

[13] G. Tao, Y. Chen, R. Lin, J. Zhou, X. Pei, F. Liu, N. Li, How G-quadruplex topology and loop sequences affect optical properties of DNA-templated silver nanoclusters, Nano Res. 11 (2018) 2237–2247. https://doi.org/10.1007/s12274-017-1844-4.

[14] B. Purohit, Community Based Health Insurance in India: Prospects and Challenges, Health (Irvine. Calif). 06 (2014) 1237–1245. https://doi.org/10.4236/health.2014.611152.

[15] M. Beld, C. Sol, J. Goudsmit, R. Boom, Fractionation of nucleic acids into single-stranded and double-stranded forms, Nucleic Acids Res. 24 (1996) 2618–2619. https://doi.org/10.1093/nar/24.13.2618.

[16] X. Cai, S. Conley, M. Naash, Nanoparticle applications in ocular gene therapy, Vision Res. 48 (2008) 319–324. https://doi.org/10.1016/j.visres.2007.07.012.

[17] S. Mukherjee, R.N. Ghosh, F.R. Maxfield, Endocytosis, Physiol. Rev. 77 (1997) 759–803. https://doi.org/10.1152/physrev.1997.77.3.759.

[18] G.B. Pinto, A. dos Reis Corrêa, G.N.C. da Silva, J.S. da Costa, P.L.B. Figueiredo, Drug development from essential oils: New discoveries and perspectives, in: J.N. Cruz (Ed.), Drug Discov. Des. Using Nat. Prod., Springer Nature Switzerland, Cham, 2023: pp. 79–101. https://doi.org/10.1007/978-3-031-35205-8_4.

[19] D. Bila, Y. Radwan, M.A. Dobrovolskaia, M. Panigaj, K.A. Afonin, The recognition of and reactions to nucleic acid nanoparticles by human immune cells, Molecules. 26 (2021) 4231. https://doi.org/10.3390/molecules26144231.

[20] K.A. Afonin, M. Viard, A.Y. Koyfman, A.N. Martins, W.K. Kasprzak, M. Panigaj, R. Desai, A. Santhanam, W.W. Grabow, L. Jaeger, E. Heldman, J. Reiser, W. Chiu, E.O. Freed, B.A. Shapiro, Multifunctional RNA nanoparticles, Nano Lett. 14 (2014) 5662–5671. https://doi.org/10.1021/nl502385k.

[21] E.F. Khisamutdinov, H. Li, D.L. Jasinski, J. Chen, J. Fu, P. Guo, Enhancing immunomodulation on innate immunity by shape transition among RNA triangle, square and pentagon nanovehicles, Nucleic Acids Res. 42 (2014) 9996–10004. https://doi.org/10.1093/nar/gku516.

[22] L. Rackley, J.M. Stewart, J. Salotti, A. Krokhotin, A. Shah, J.R. Halman, R. Juneja, J. Smollett, L. Lee, K. Roark, M. Viard, M. Tarannum, J. Vivero-Escoto, P.F. Johnson, M.A. Dobrovolskaia, N. V. Dokholyan, E. Franco, K.A. Afonin, RNA Fibers as Optimized Nanoscaffolds for siRNA Coordination and Reduced Immunological Recognition, Adv. Funct. Mater. 28 (2018). https://doi.org/10.1002/adfm.201805959.

[23] F.S. Alves, J.N. Cruz, I.N. de Farias Ramos, D.L. do Nascimento Brandão, R.N. Queiroz, G.V. da Silva, G.V. da Silva, M.F. Dolabela, M.L. da Costa, A.S. Khayat, J. de Arimatéia Rodrigues do Rego, D. do Socorro Barros Brasil, Evaluation of Antimicrobial Activity and Cytotoxicity Effects of Extracts of Piper nigrum L. and Piperine, Separations. 10 (2023). https://doi.org/10.3390/separations10010021.

[24] N.B. Leontis, J. Stombaugh, E. Westhof, The non-Watson-Crick base pairs and their associated isostericity matrices, Nucleic Acids Res. 30 (2002) 3497–3531. https://doi.org/10.1093/nar/gkf481.

[25] C.J. Delebecque, A.B. Lindner, P.A. Silver, F.A. Aldaye, Organization of intracellular reactions with rationally designed RNA assemblies, Science (80-.). 333 (2011) 470–474. https://doi.org/10.1126/science.1206938.

[26] M. Panigaj, M.B. Johnson, W. Ke, J. McMillan, E.A. Goncharova, M. Chandler, K.A. Afonin, Aptamers as Modular Components of Therapeutic Nucleic Acid Nanotechnology, ACS Nano. 13 (2019) 12301–12321. https://doi.org/10.1021/acsnano.9b06522.

[27] L. Gong, Z. Zhao, Y.F. Lv, S.Y. Huan, T. Fu, X.B. Zhang, G.L. Shen, R.Q. Yu, DNAzyme-based biosensors and nanodevices, Chem. Commun. 51 (2015) 979–995. https://doi.org/10.1039/c4cc06855f.

[28]   P. Guo, The emerging field of RNA nanotechnology, Nat. Nanotechnol. 5 (2010) 833–842. https://doi.org/10.1038/nnano.2010.231.

[29]   J. Kim, E. Franco, RNA nanotechnology in synthetic biology, Curr. Opin. Biotechnol. 63 (2020) 135–141. https://doi.org/10.1016/j.copbio.2019.12.016.

[30]   M. Panigaj, M.A. Dobrovolskaia, K.A. Afonin, 2021: An immunotherapy odyssey and the rise of nucleic acid nanotechnology, Nanomedicine. 16 (2021) 1635–1640. https://doi.org/10.2217/nnm-2021-0097.

[31]   D.C. Luther, R. Huang, T. Jeon, X. Zhang, Y.W. Lee, H. Nagaraj, V.M. Rotello, Delivery of drugs, proteins, and nucleic acids using inorganic nanoparticles, Adv. Drug Deliv. Rev. 156 (2020) 188–213. https://doi.org/10.1016/j.addr.2020.06.020.

[32]   A.F. De Fazio, D. Misatziou, Y.R. Baker, O.L. Muskens, T. Brown, A.G. Kanaras, Chemically modified nucleic acids and DNA intercalators as tools for nanoparticle assembly, Chem. Soc. Rev. 50 (2021) 13410–13440. https://doi.org/10.1039/d1cs00632k.

[33]   F. Hong, F. Zhang, Y. Liu, H. Yan, DNA Origami: Scaffolds for Creating Higher Order Structures, Chem. Rev. 117 (2017) 12584–12640. https://doi.org/10.1021/acs.chemrev.6b00825.

[34]   G.B. Pinto, A. dos Reis Corrêa, G.N.C. da Silva, J.S. da Costa, P.L.B. Figueiredo, Drug development from essential oils: New discoveries and perspectives, in: J.N. Cruz (Ed.), Drug Discov. Des. Using Nat. Prod., Springer Nature Switzerland, Cham, 2023: pp. 79–101. https://doi.org/10.1007/978-3-031-35205-8_4.

[35]   S. Agarwal, M.A. Klocke, P.E. Pungchai, E. Franco, Dynamic self-assembly of compartmentalized DNA nanotubes, Nat. Commun. 12 (2021). https://doi.org/10.1038/s41467-021-23850-1.

[36]   M. Scherf, F. Scheffler, C. Maffeo, U. Kemper, J. Ye, A. Aksimentiev, R. Seidel, U. Reibetanz, Trapping of protein cargo molecules inside DNA origami nanocages, Nanoscale. 14 (2022) 18041–18050. https://doi.org/10.1039/d2nr05356j.

[37]   J.M. Sasso, B.J.B. Ambrose, R. Tenchov, R.S. Datta, M.T. Basel, R.K. Delong, Q.A. Zhou, The Progress and Promise of RNA Medicine-An Arsenal of Targeted Treatments, J. Med. Chem. 65 (2022) 6975–7015. https://doi.org/10.1021/acs.jmedchem.2c00024.

[38]   A. Babu, R. Muralidharan, N. Amreddy, M. Mehta, A. Munshi, R. Ramesh, Nanoparticles for siRNA-based gene silencing in tumor therapy, IEEE Trans. Nanobioscience. 15 (2016) 849–863. https://doi.org/10.1109/TNB.2016.2621730.

[39]   X. Hou, T. Zaks, R. Langer, Y. Dong, Lipid nanoparticles for mRNA delivery, Nat. Rev. Mater. 6 (2021) 1078–1094. https://doi.org/10.1038/s41578-021-00358-0.

[40]   C. Liu, Q. Shi, X. Huang, S. Koo, N. Kong, W. Tao, mRNA-based cancer therapeutics, Nat. Rev. Cancer. 23 (2023) 526–543. https://doi.org/10.1038/s41568-023-00586-2.

[41]   N. Stephanopoulos, Hybrid Nanostructures from the Self-Assembly of Proteins and DNA, Chem. 6 (2020) 364–405. https://doi.org/10.1016/j.chempr.2020.01.012.

[42]   H.B. Gamper, H. Parekh, M.C. Rice, M. Bruner, H. Youkey, E.B. Kmiec, The DNA strand of chimeric RNA/DNA oligonucleotides can direct gene repair/conversion activity in

mammalian and plant cell-free extracts, Nucleic Acids Res. 28 (2000) 4332–4339. https://doi.org/10.1093/nar/28.21.4332.

[43]   W. Park, H. Shin, B. Choi, W.K. Rhim, K. Na, D. Keun Han, Advanced hybrid nanomaterials for biomedical applications, Prog. Mater. Sci. 114 (2020) 100686. https://doi.org/10.1016/j.pmatsci.2020.100686.

[44]   W. Ho, M. Gao, F. Li, Z. Li, X.Q. Zhang, X. Xu, Next-Generation Vaccines: Nanoparticle-Mediated DNA and mRNA Delivery, Adv. Healthc. Mater. 10 (2021). https://doi.org/10.1002/adhm.202001812.

[45]   S.A. Dilliard, D.J. Siegwart, Passive, active and endogenous organ-targeted lipid and polymer nanoparticles for delivery of genetic drugs, Nat. Rev. Mater. 8 (2023) 282–300. https://doi.org/10.1038/s41578-022-00529-7.

[46]   A. Friedman, S. Claypool, R. Liu, The Smart Targeting of Nanoparticles, Curr. Pharm. Des. 19 (2013) 6315–6329. https://doi.org/10.2174/13816128113199990375.

[47]   S.A.A. Rizvi, A.M. Saleh, Applications of nanoparticle systems in drug delivery technology, Saudi Pharm. J. 26 (2018) 64–70. https://doi.org/10.1016/j.jsps.2017.10.012.

[48]   I. Pontón, A.M. del Rio, M.G. Gómez, D. Sánchez-García, Preparation and applications of organo-silica hybrid mesoporous silica nanoparticles for the co-delivery of drugs and nucleic acids, Nanomaterials. 10 (2020) 1–37. https://doi.org/10.3390/nano10122466.

[49]   M.H. Sarfraz, M. Zubair, B. Aslam, A. Ashraf, M.H. Siddique, S. Hayat, J.N. Cruz, S. Muzammil, M. Khurshid, M.F. Sarfraz, A. Hashem, T.M. Dawoud, G.D. Avila-Quezada, E.F. Abd_Allah, Comparative analysis of phyto-fabricated chitosan, copper oxide, and chitosan-based CuO nanoparticles: antibacterial potential against Acinetobacter baumannii isolates and anticancer activity against HepG2 cell lines, Front. Microbiol. 14 (2023). https://doi.org/10.3389/fmicb.2023.1188743.

[50]   M.A. Alghamdi, A.N. Fallica, N. Virzì, P. Kesharwani, V. Pittalà, K. Greish, The Promise of Nanotechnology in Personalized Medicine, J. Pers. Med. 12 (2022) 673. https://doi.org/10.3390/jpm12050673.

[51]   Z. Lu, H.Y. Chang, The RNA base-pairing problem and base-pairing solutions, Cold Spring Harb. Perspect. Biol. 10 (2018) a034926. https://doi.org/10.1101/cshperspect.a034926.

[52]   A. Sharma, K. Vaghasiya, R.K. Verma, A.B. Yadav, DNA nanostructures: Chemistry, self-assembly, and applications, Emerg. Appl. Nanoparticles Archit. Nanostructures Curr. Prospect. Futur. Trends. (2018) 71–94. https://doi.org/10.1016/B978-0-323-51254-1.00003-8.

[53]   M.J. Mitchell, M.M. Billingsley, R.M. Haley, M.E. Wechsler, N.A. Peppas, R. Langer, Engineering precision nanoparticles for drug delivery, Nat. Rev. Drug Discov. 20 (2021) 101–124. https://doi.org/10.1038/s41573-020-0090-8.

[54]   M. Windelspecht, Changes To The Genetic Material, Genet. 101. (2024) 101–124. https://doi.org/10.5040/9798400656194.ch-005.

[55]   M.B. Johnson, M. Chandler, K.A. Afonin, Nucleic acid nanoparticles (NANPs) as molecular tools to direct desirable and avoid undesirable immunological effects, Adv. Drug Deliv. Rev. 173 (2021) 427–438. https://doi.org/10.1016/j.addr.2021.04.011.

Materials Research Forum LLC
https://doi.org/10.21741/9781644903339-5

[56]   S. Nordmeier, W. Ke, K.A. Afonin, V. Portnoy, Exosome mediated delivery of functional nucleic acid nanoparticles (NANPs), Nanomedicine Nanotechnology, Biol. Med. 30 (2020) 102285. https://doi.org/10.1016/j.nano.2020.102285.

[57]   K. Duffy, S. Arangundy-Franklin, P. Holliger, Modified nucleic acids: Replication, evolution, and next-generation therapeutics, BMC Biol. 18 (2020). https://doi.org/10.1186/s12915-020-00803-6.

[58]   A. Ata, Bioactive natural products from medicinal plants, in: J.N. Cruz (Ed.), Drug Discov. Des. Using Nat. Prod., Springer Nature Switzerland, Cham, 2023: pp. 417–434. https://doi.org/10.1007/978-3-031-35205-8_14.

[59]   B. García-Pinel, C. Porras-Alcalá, A. Ortega-Rodríguez, F. Sarabia, J. Prados, C. Melguizo, J.M. López-Romero, Lipid-based nanoparticles: Application and recent advances in cancer treatment, Nanomaterials. 9 (2019) 638. https://doi.org/10.3390/nano9040638.

[60]   Y. Zhao, L. Huang, Lipid nanoparticles for gene delivery, Adv. Genet. 88 (2014) 13–36. https://doi.org/10.1016/B978-0-12-800148-6.00002-X.

[61]   A.G. Niculescu, A.M. Grumezescu, Polymer-based nanosystems-a versatile delivery approach, Materials (Basel). 14 (2021) 6812. https://doi.org/10.3390/ma14226812.

[62]   A. Kumar, A. Singam, G. Swaminathan, N. Killi, N.K. Tangudu, J. Jose, R. Gundloori Vn, L. Dinesh Kumar, Combinatorial therapy using RNAi and curcumin nano-architectures regresses tumors in breast and colon cancer models, Nanoscale. 14 (2022) 492–505. https://doi.org/10.1039/d1nr04411g.

[63]   R. Thiruppathi, S. Mishra, M. Ganapathy, P. Padmanabhan, B. Gulyás, Nanoparticle functionalization and its potentials for molecular imaging, Adv. Sci. 4 (2017). https://doi.org/10.1002/advs.201600279.

[64]   E. Abbasi, S.F. Aval, A. Akbarzadeh, M. Milani, H.T. Nasrabadi, S.W. Joo, Y. Hanifehpour, K. Nejati-Koshki, R. Pashaei-Asl, Dendrimers: Synthesis, applications, and properties, Nanoscale Res. Lett. 9 (2014) 1–10. https://doi.org/10.1186/1556-276X-9-247.

[65]   C.S. Morales, P.M. Valencia, A.B. Thakkar, E. Swanson, R. Langer, Recent developments in multifunctional hybrid nanoparticles: Opportunities and challenges in cancer therapy, Front. Biosci. - Elit. 4 E (2012) 529–545. https://doi.org/10.2741/398.

[66]   W. Ma, Y. Zhan, Y. Zhang, C. Mao, X. Xie, Y. Lin, The biological applications of DNA nanomaterials: current challenges and future directions, Signal Transduct. Target. Ther. 6 (2021). https://doi.org/10.1038/s41392-021-00727-9.

[67]   L. Duan, K. Ouyang, X. Xu, L. Xu, C. Wen, X. Zhou, Z. Qin, Z. Xu, W. Sun, Y. Liang, Nanoparticle Delivery of CRISPR/Cas9 for Genome Editing, Front. Genet. 12 (2021). https://doi.org/10.3389/fgene.2021.673286.

[68]   A.G. Niculescu, A.C. Bîrcă, A.M. Grumezescu, New applications of lipid and polymer-based nanoparticles for nucleic acids delivery, Pharmaceutics. 13 (2021) 2053. https://doi.org/10.3390/pharmaceutics13122053.

[69]   I.N. de F. Ramos, M.F. da Silva, J.M.S. Lopes, J.N. Cruz, F.S. Alves, J. de A.R. do Rego, M.L. da Costa, P.P. de Assumpção, D. do S. Barros Brasil, A.S. Khayat, Extraction,

Characterization, and Evaluation of the Cytotoxic Activity of Piperine in Its Isolated form and in Combination with Chemotherapeutics against Gastric Cancer, Molecules. 28 (2023). https://doi.org/10.3390/molecules28145587.

[70]   W. Alshaer, H. Zureigat, A. Al Karaki, A. Al-Kadash, L. Gharaibeh, M.M. Hatmal, A.A.A. Aljabali, A. Awidi, Corrigendum to "siRNA: Mechanism of action, challenges, and therapeutic approaches" [Eur. J. Pharmacol. 905 (2021) 174178] (European Journal of Pharmacology (2021) 905, (S0014299921003319), (10.1016/j.ejphar.2021.174178)), Eur. J. Pharmacol. 916 (2022) 174741. https://doi.org/10.1016/j.ejphar.2022.174741.

[71]   Y. Wang, Z. Li, Y. Han, L. Hwa Liang, A. Ji, Nanoparticle-Based Delivery System for Application of siRNA In Vivo, Curr. Drug Metab. 11 (2010) 182–196. https://doi.org/10.2174/138920010791110863.

[72]   C. Liu, Q. Shi, X. Huang, S. Koo, N. Kong, W. Tao, mRNA-based cancer therapeutics, Nat. Rev. Cancer. 23 (2023) 526–543. https://doi.org/10.1038/s41568-023-00586-2.

[73]   J.K. Durbin, D.K. Miller, J. Niekamp, E.F. Khisamutdinov, Modulating immune response with nucleic acid nanoparticles, Molecules. 24 (2019) 3740. https://doi.org/10.3390/molecules24203740.

[74]   M.J. Lin, J. Svensson-Arvelund, G.S. Lubitz, A. Marabelle, I. Melero, B.D. Brown, J.D. Brody, Cancer vaccines: the next immunotherapy frontier, Nat. Cancer. 3 (2022) 911–926. https://doi.org/10.1038/s43018-022-00418-6.

[75]   A. Gupta, J.L. Andresen, R.S. Manan, R. Langer, Nucleic acid delivery for therapeutic applications, Adv. Drug Deliv. Rev. 178 (2021) 113834. https://doi.org/10.1016/j.addr.2021.113834.

[76]   J.A. Kulkarni, D. Witzigmann, S.B. Thomson, S. Chen, B.R. Leavitt, P.R. Cullis, R. van der Meel, The current landscape of nucleic acid therapeutics, Nat. Nanotechnol. 16 (2021) 630–643. https://doi.org/10.1038/s41565-021-00898-0.

[77]   J.O. Jin, G. Kim, J. Hwang, K.H. Han, M. Kwak, P.C.W. Lee, Nucleic acid nanotechnology for cancer treatment, Biochim. Biophys. Acta - Rev. Cancer. 1874 (2020) 188377. https://doi.org/10.1016/j.bbcan.2020.188377.

[78]   M. Cordeiro, F.F. Carlos, P. Pedrosa, A. Lopez, P.V. Baptista, Gold nanoparticles for diagnostics: Advances towards points of care, Diagnostics. 6 (2016) 43. https://doi.org/10.3390/diagnostics6040043.

[79]   J. Jin, X. Ouyang, J. Li, J. Jiang, H. Wang, Y. Wang, R. Yang, Nucleic acid-modulated silver nanoparticles: A new electrochemical platform for sensing chloride ion, Analyst. 136 (2011) 3629–3634. https://doi.org/10.1039/c1an15283a.

[80]   J.N. Cruz, S. Muzammil, A. Ashraf, M.U. Ijaz, M.H. Siddique, R. Abbas, M. Sadia, Saba, S. Hayat, R.R. Lima, A review on mycogenic metallic nanoparticles and their potential role as antioxidant, antibiofilm and quorum quenching agents, Heliyon. 10 (2024). https://doi.org/10.1016/j.heliyon.2024.e29500.

[81]   J. Conde, J. Rosa, J.M. de la Fuente, P. V. Baptista, Gold-nanobeacons for simultaneous gene specific silencing and intracellular tracking of the silencing events, Biomaterials. 34 (2013) 2516–2523. https://doi.org/10.1016/j.biomaterials.2012.12.015.

[82]   W.X. Lei, Z.S. An, B.H. Zhang, Q. Wu, W.J. Gong, J.M. Li, W.L. Chen, Construction of gold-siRNANPR1nanoparticles for effective and quick silencing ofNPR1inArabidopsis thaliana, RSC Adv. 10 (2020) 19300–19308. https://doi.org/10.1039/d0ra02156c.

[83]   A. Latorre, Á. Somoza, DNA-Mediated Silver Nanoclusters: Synthesis, Properties and Applications, ChemBioChem. 13 (2012) 951–958. https://doi.org/10.1002/cbic.201200053.

[84]   A. Erxleben, Interactions of copper complexes with nucleic acids, Coord. Chem. Rev. 360 (2018) 92–121. https://doi.org/10.1016/j.ccr.2018.01.008.

[85]   N. Chaudhary, D. Weissman, K.A. Whitehead, mRNA vaccines for infectious diseases: principles, delivery and clinical translation, Nat. Rev. Drug Discov. 20 (2021) 817–838. https://doi.org/10.1038/s41573-021-00283-5.

[86]   J. Bush, S. Singh, M. Vargas, E. Oktay, C.H. Hu, R. Veneziano, Synthesis of DNA origami scaffolds: Current and emerging strategies, Molecules. 25 (2020) 3386. https://doi.org/10.3390/molecules25153386.

Advances in Healthcare and Nanoparticle Toxicology
Materials Research Foundations 171 (2024) 154-190

Materials Research Forum LLC
https://doi.org/10.21741/9781644903339-6

Chapter 6

Molecular Interactions between Nanoparticles and Biomolecules

Rashmi Niranjan[1] and Richa Priyadarshini[1]*

[1]Department of Life Sciences, School of Natural Sciences, Shiv Nadar Institution of Eminence, Gautam Buddha Nagar, Uttar Pradesh, 201314, India

*richa.priyadarshini@snu.edu.in

Abstract

Nanoparticles (NPs) and nanomaterials have applications in all sectors of present-day life like industrial, food-technology, medical, cosmetics, pharmaceutical, biomedical, etc. NPs interact with various biomolecules, including nucleic acids, proteins, lipids, and carbohydrates. NPs exhibit beneficial and adverse interactions with biomolecules depending on their physicochemical properties. NPs' mode of synthesis, duration of exposure, concentration, and charge influence their interactions with various organic molecules. This work discusses NPs' interaction with biomolecules and their role in various biological applications.

Keywords

Nanoparticles, Proteins, Lipids, Carbohydrates, Nucleic Acids, Biological Applications

Contents

1.   Introduction

Even though nanotechnology is a relatively new branch of science, it has undergone tremendous development and found its applications in various industrial and biomedical sectors. Nanoparticles (NP) are particles with at least one of three spatial dimensions less than 100 nm [1,2]. Nanomaterials behave ambiguously in biosystems, and investigating mechanisms that govern interactions between nanomaterials and biomolecules is important. Nanomaterials demonstrate both adverse and supportive effects on biomolecules, cells, and tissues [3,4]. The nanomaterials' properties, like surface properties, charge alterations, size, and shape, empower them to cooperate with the targets. Nanomaterials show an extensive range of reactivity given their physicochemical, biomolecular, biochemical, and biophysical features [5,6]. Scientists have fabricated NPs like iron oxide NPs, gold and silver NPs, quantum dots (QDs), polymeric NPs, lipid-based NPs, and carbon-based NPs with varied properties for biomedical applications. Most of these NPs are prepared either by a bottom-up or a top-down approach [7]. Synthesized NPs are immiscible in water and have extremely less water solvency, hindering their direct use in biological applications. Further surface refinements of NPs by covalent modification or non-covalent modification have to be performed to enhance biocompatibility. The covalent modification involves covalent linkage with biomolecules like carbohydrates, surfactants, proteins, DNA, RNA, and lipids, which eases the problem of immiscibility [8]. Metal and metal oxide, QDs, fullerene, and fibrous nanomaterials cause chromosomal fragmentation, DNA strand breaks, point mutation, and oxidative DNA damage [9]. While there are few reports available on the toxicity of NPs, a comprehensive study on NP's actions in biosystems is required. A better understanding of the mechanism employed by NPs at biological interfaces would aid in developing novel nanomaterials. This chapter describes the relationship between NPs and various biomolecules such as proteins, lipids, carbohydrates, and nucleic acids (NAs).

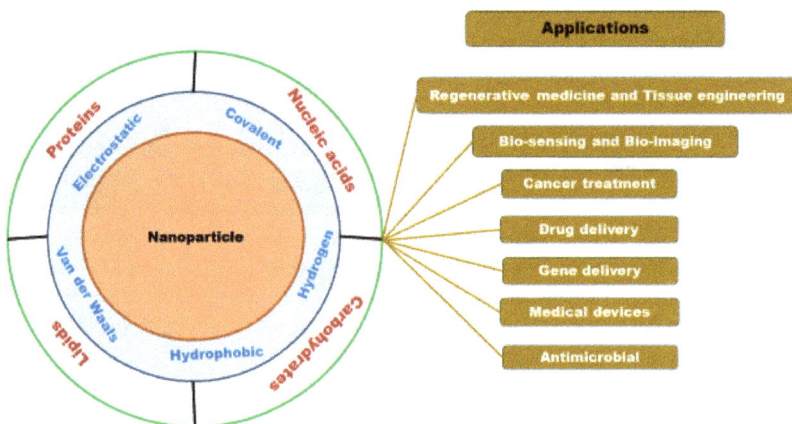

Figure 1: Interaction of NPs with biomolecules and their applications

2.   Proteins interactions with nanoparticles

Proteins are one of the most significant biomolecular constituents in a biosystem. Protein molecules consist of chains of amino acids linked by peptide bonds. Since proteins perform several important biological activities such as enzymes, antibodies, signaling, and transport molecules, studying the protein-NPs interactions is important to understand the role NPs may play in biological systems[10,11]. Interaction between NPs and proteins leads to the formation of a complex known as protein-corona (PC). Adsorption of protein on NP surface relies on hydrogen bonds, solvation forces, Van der Waals interactions, etc. Based on its composition, the protein corona can be divided into two broad categories: hard and soft [12].

Advances in Healthcare and Nanoparticle Toxicology   Materials Research Forum LLC
Materials Research Foundations 171 (2024) 154-190   https://doi.org/10.21741/9781644903339-6

In *hard corona*, proteins interact directly with NPs' surfaces and are termed "primary binders". Hard corona formation can vary from a few seconds to a few hours [13,14]. Lundqvist et al. studied hard corona formation involving different NPs and plasma and cytosolic fluid [15]. Lynch I and Dawson observed entropy-driven binding as one of the mechanisms of protein adsorption on NP surface forming hard corona [13]. Plasma proteins such as apolipoproteins, albumin, fibrinogen, and transferrin adhere to the NP surface quite efficiently. Protein corona is utilized to study NPs uptake and transport pathways in organisms [16].

The soft corona consists of molecules loosely bound to the hard corona, therefore, it is termed "secondary binders". The time taken for the formation of a soft corona varies from seconds to days. In contrast to the lifetime of hard corona, soft corona desorbs in a few minutes [17]. Proteins adsorbed to NPs can undergo desorption at any time allowing other proteins to adsorb to NPs thus always being in a dynamic state. This effect is called the Vroman effect and can be divided into "early" and "late" stages [18].

In the *"early stages"*, the proteins having higher association rates adsorb rapidly onto the NP surface. If the proteins adsorbed had short residence times, then during the second phase, the *"late stages"*, these proteins are replaced by proteins with slower association rates and longer residence times [19]. Goppert and Muller analyzed plasma protein adsorption and observed that during the early stage, the rapid adsorption of albumin, IgG, and fibrinogen occurs, being replaced by apolipoproteins and coagulation factors in the "late stages" due to the higher affinity of apolipoproteins for hydrophobic surfaces [20].

3. Factors affecting protein-corona

Adsorption of protein is dependent on properties like nature, size, shape, surface morphology, chemistry, surface charge, solubility, and modification of the nanomaterials. Other factors such as temperature, pH, dynamic shear stress, exposure time, and medium components also affect the adsorption of proteins [21,22]. How these physicochemical parameters affect NPs and protein interactions are discussed in the following subsections.

3.1 Size of nanoparticles

The size of the NP can have a significant impact on PC formation. Similar NPs differing in size vary in their PC composition. Surface curvature also significantly affects protein binding. NPs with highly curved surface areas show a decrease in protein interaction, and the proteins undergo fewer conformational changes in comparison to proteins adsorbed to low surface curvature [13,23]. A change of even 10 nm in NPs size can alter PC composition significantly. For example, the binding of approximately 37% of proteins gets affected by the size of SiO_2 NPs used. Clustsacaerin lipoproteins bind to smaller NPs, whereas, prothrombin and gelsolin adsorb readily on larger NPs [24]. Gold NPs of two different sizes when used, have more plasma proteins adsorbed to smaller-sized NPs. IgM-dependent complement activation is most efficient on dextran particles in the optimal size range, ~250 nm [25].

3.2 Surface charge of nanoparticles

Surface charge on NPs play a crucial role in PC formation and composition. Reports suggest that an increase in surface charge increases protein adsorption. Conformation of protein changes with the surface charge of the nanomaterial. Positively charged nanomaterials induce maximum

Advances in Healthcare and Nanoparticle Toxicology   Materials Research Forum LLC
Materials Research Foundations 171 (2024) 154-190   https://doi.org/10.21741/9781644903339-6

conformational variations whereas with negatively charged nanomaterials fewer variations are observed. Neutrally charged NPs results in minimal conformational variations [26]. Polystyrene NPs when positively charged, promote the adsorption of proteins with isoelectric points (pI) < 5.5, and negatively charged NPs adsorb proteins with pI > 5.5 [21] Increase in surface charge density also enhanced protein adsorption. Compared to neutral NPs, charged NPs have a higher rate of internalization, and positively charged NPs internalize rapidly compared to negatively charged NPs. Sometimes the surface charge could also have a detrimental effect on adsorbed proteins [27,28]. For example, proteins adsorbed on gold NPs with charged ligands undergo denaturation, whereas neutral ligands have no such effects [13].

3.3   Hydrophilicity/hydrophobicity

Hydrophilicity/Hydrophobicity is an important parameter that influences adsorption of proteins. Hydrophobic NPs adsorb more proteins in comparison to hydrophilic NPs as hydrophobic polymer chain clusters and form more protein-binding sites [29]. An increase in hydrophobicity and surface charge density enhances the affinity of NPs to proteins. Hydrophobicity affects not only the amount of protein adsorbed but also its composition [21]. For example, an increase in the negative charge density and hydrophobicity of polystyrene NPs improved plasma protein adsorption [14]. The study suggests that proteins preferentially adsorb onto hydrophobic NPs than hydrophilic NPs even though they have the same affinity for both. This enhanced adsorption increases the opsonization of hydrophobic NPs [21]. Fabricated poly [acrylonitrile-co- (N-vinyl pyrrolidone)] NPs are hydrophobic and enhance the amount of adsorbed protein in comparison to their hydrophilic counterparts. As reported by Moghimi and Patel, more proteins adsorb cholesterol-free liposomes than cholesterol-rich liposomes. Apolipoproteins also bind with a higher affinity to hydrophobic NPs [30].

3.4   Nanoparticle material

NP differentially interacts with various functional groups present on proteins; thus, PC composition varies with NP material. For example, different plasma proteins adsorbed on NPs ZnO, SiO_2, and TiO_2, though they had the same surface charge. TiO_2 adsorbed histidine-rich glycoprotein, kininogen-1, complement C9 and C1q, fetuin A, vitronectin, apolipoprotein A1, whereas SiO_2 adsorbed complement C8, and apolipoprotein A. ZnO showed affinity towards alpha-2-macroglobulin, transferrin, alpha-1-antichymotryspin. Similarly, serum albumin adsorbed most abundantly on single-walled carbon nanotubes (SWCNT) but not on silica NP (SiNPs) [31].

3.5   Temperature

Temperature is among the least studied factors affecting PC formation. Proteins are denatured at high temperatures and lose their three-dimensional conformation. Temperature not only changes the composition of the protein adsorbed but also the coverage of the surface area of NPs during PC formation. Temperature also regulates the pattern and rate of cellular uptake of NPs PC [26]. Dextran-coated superparamagnetic iron oxide NPs (SPIONs) incubated with fetal bovine serum demonstrated that PC composition varies with varying temperatures. Mahmoudi et al. reported the effects of local heat induction on PCs composition using cetyltrimethylammonium bromide-stabilized (CTAB-stabilized) gold nanorods [32].

3.6   pH

Changes in the pH of the environment induce alterations in protein binding affinity towards NP thus altering the composition of PC [33]. Nanomaterials and proteins have a specific isoionic point. A reduction in the stability of NPs is observed when the pH of the interacting medium is close to the isoelectric point. pH affects PC formation and surface charge on NPs. Positively charged NPs interact electrostatically with negatively charged prokaryotic and eukaryotic cells [34]. Humic acid and negatively charged NPs show a strong electrostatic attraction in acidic pH but are weakened in alkaline pH. Similarly, bacterial exopolysaccharides bind less onto silver NPs if pH is more than the isoelectric point [35].

3.7   Surface functionalization and coatings

Surface functionalization of NPs and pre-coating are utilized to modify protein corona composition. Pre-coating of SWCNT with Pluronic F127 results in decreased albumin adsorption. Similarly, Poloxamine 908 coated on polystyrene nanospheres yields reduced fibronectin adsorption [36]. This strategy can also be utilized for modification of NPs surface to allow it to escape the immune system successfully. For example, polyethylene glycol (PEG) coating on the NP surface makes it unrecognizable by the reticuloendothelial system (RES) [37]. Studies on NPs coated with materials such as   PEG, dextran, poloxamer, poloxamine, and polysorbate demonstrated improved protein-NP interactions and biodistribution [21].

3.8   Protein conformation, concentration, and exposure duration

The specific arrangement of amino acids in a polypeptide gives it characteristic primary, secondary, super secondary, tertiary, and quaternary structures. During protein and NP interactions, conformational changes take place. The secondary structure of BSA changes in the presence of cationic polystyrene NPs, but binds to the albumin receptors in the presence of anionic polystyrene NPs [38]. The adsorption of protein is influenced by its relative concentration. A Protein in high abundance may replace a protein present in low concentration even if it has a low binding affinity [39]. Exposure duration has a significant impact on the interaction between proteins and nanomaterials and the pattern of PC formed. With the increase in exposure time, an enhancement in the adsorption rate of proteins on NPs is observed [40,41].

4.   Applications of the protein corona

4.1   Protein fibrillation

Protein fibrillation is a process in which proteins form insoluble fibrils that aggregate and form plaques. One such group of proteins is amyloidogenic proteins. These aggregated fibrils are the cause of protein-misfolded neurodegenerative diseases like Alzheimer, Parkinson, and dialysis-related amyloidosis. Reports suggest that NPs have an impact on the protein fibrillation process. NPs can both enhance and inhibit fibrillation [13]. Carbon nanotube (CNT), cerium oxide, and PEG-coated QDs enhance the fibrillation of amyloidogenic protein β-2-microglobulin, whereas hydrated fullerene (C60) inhibits amyloid-beta 25–35 peptide fibrillation [42]. Hydrophobic polystyrene NPs display temperature-dependent 'dual' fibrillation kinetics [43]. Gold NPs inhibit protein fibrillation in pristine form but enhance fibrillation after coronation with (serum and plasma) proteins [44].

Advances in Healthcare and Nanoparticle Toxicology   Materials Research Forum LLC
Materials Research Foundations 171 (2024) 154-190   https://doi.org/10.21741/9781644903339-6

4.2   Cell targeting and drug delivery

Specific cell targeting approaches are employed for effective drug or gene delivery, imaging, and therapeutic delivery. PC interacts with the components of the cellular envelope, therefore, these interactions can be studied, manipulated, and modified for designing NPs with distinct cell-targeting properties [45,46]. However, the use of PCs is challenging in biomedical applications due to their inherent ability to modify native conformation, orientation, and display of the ligand [47]. Transferrin (Tf)-coated fluorescent SiNPs lose their targeting ability to both soluble Tf receptors and bound Tf receptors [48]. Similar observations were made by Mirshafiee et al. [49]. To eliminate this effect, coatings with zwitterionic materials or anti-fouling polymers (PEG) are done [50]. Specific cell targeting requires further designing the NP surface, and the following parameters need to be considered: i) protein that can deliver the NP to the desired location, ii) surfactants to increase the adhesion rate of the desired protein onto PC complex, iii) evolution of the protein corona concerning time [45,46].

"Protein corona effect" is used for targeting cancer cells [51]. Transferrin, insulin, folic acid, EGFP-EGF1, GRP, EGF, apoA-I, and apoE are surface ligands exploited for ligand-receptor targeting of cancer cells. The Blood-Brain barrier (BBB) prevents easy delivery of most of the therapeutics; however, covalently attached apolipoproteins (apoE, apoA-I, and apoB-100) enhance the NP transport across it. Hexadecylcyanoacrylate NPs facilitate apoE adsorption and passage through the BBB and target the brain efficiently [21]. Similarly, polysorbates stabilized lipid NPs enhance the crossing of these NPs through the BBB by preferential adsorption of apolipoprotein E. NP albumin-bound paclitaxel (nab-paclitaxel) exhibits improved tumor targeting and decreased toxicity [52].

PC is used as a carrier for charged payloads. Payloads holding capacity of the corona is ~ten times higher than covalent strategies. For example, the PC around CTAB-capped gold nanorod carries negatively charged DNA and positively charged doxorubicin (DOX) [53]. The composition and concentration of the corona, surfactant concentration on the nanomaterials, and ionic strength of the buffer directly affect the loading efficiency and release profile of the therapeutic payloads. Behzadi et al. observed that the PC significantly reduced payload release from SPIONs, whereas the thinner hard corona layer had no such effect [54].

4.3   Toxicity

The toxicity of some NPs hinders their usage in biomedical sectors. It has been suggested that PC formation on the surface of NPs helps to eliminate its toxic effect. Serum proteins in the PC complex reduce the inherent cytotoxicity of NPs due to reduced interaction between cells and NPs surfaces. Observations made by Alex et al. and Hu et al. support such an effect. Serum proteins such as Human Serum Albumin, IgG, and transferrin around the PC complex reduce the cytotoxicity of functionalized graphene nanoribbons along with improving cell viability. Protein adsorption on graphene oxide enhances its toxicity towards A549 cells [55,56]. Albumin adsorption on SWCNTs and silica NPs causes an anti-inflammatory response in the macrophages [13,57]. The factors affecting PC formation and applications of protein NP interaction are summarized in Table 1.

Table 1. Factors affecting PC formation and applications of protein NP interaction

Factors	Effects	Applications
Size	<ul><li>NP size alters the curvature of the NP surface, influencing PC composition.</li><li>Larger-sized NPs provide more coverage for proteins.</li><li>Thickness of the corona increases on the small-sized NPs.</li></ul>	<ul><li>**Protein fibrillation:** NPs both enhance and inhibit fibrillation.</li><li>**Cell targeting:** "Protein corona effect" is used for targeting specific cells such as cancer cells.</li><li>**Toxicity:** PC complex reduces the cytotoxicity of NPs towards cells.</li><li>**Drug Delivery:** PC is used as a carrier for charged payloads with a higher holding capacity.</li></ul>
Surface charge of NP	<ul><li>Positively charged promote adsorption of proteins with isoelectric points (pI) < 5.5</li><li>Negatively charged NPs adsorb proteins with pI > 5.5</li><li>Charged NPs internalize at a higher rate in comparison to neutral NPs.</li></ul>	
Hydrophilicity/Hydrophobicity	<ul><li>More proteins adsorb on hydrophobic NPs than hydrophilic NPs.</li><li>Protein adsorption onto NPs is directly proportional to the degree of hydrophobicity.</li></ul>	
Temperature	<ul><li>Regulates the pattern and rate of cellular uptake of NPs protein-corona.</li><li>Temperature changes the composition of the protein adsorbed on NPs.</li><li>Temperature influences the coverage of the surface area of NPs during PC formation.</li></ul>	
pH of medium	<ul><li>pH of the medium influences the interaction between NPs and proteins.</li><li>pH of medium lower than the isoelectric point induces a positive charge on the NP's surface.</li><li>In case of higher pH of the medium, the NPs have a negative charge on their surface</li></ul>	
Surface Functionalization and Coatings	Surface functionalization and pre-coating of NPs alters PC composition.	
Protein Conformation, Concentration, and Exposure Duration	<ul><li>Conformational changes occur during the NP and protein interactions, which provide functional variations in protein interaction.</li><li>Protein in high concentration replaces the protein in less amount even if it has a low binding affinity.</li><li>PC formation is affected quantitatively and qualitatively by the exposure time. With an increase in exposure time, the rate of adsorption of proteins increases.</li></ul>	

5.   Nucleic acids interactions with nanoparticles

NAs are polymeric molecules of nucleotides containing a phosphate group, a pentose, and a base (adenine, guanine, thymine, cytosine, or uracil). These NAs are responsible for encoding the genetic information of a cell [58]. The fundamental intermolecular forces involved in nanomaterial-genetic materials interactions are similar to nanomaterials-proteins. Studies involving such interactions have led to the development of new nanostructures finding applications in biomedical and therapeutics. Nanomaterials and NAs interactions form products desirable as the potential materials for DNA and RNA technologies. The interaction between some of the NPs and NAs is explained below.

5.1   Carbon nanoparticles

CNTs, graphene, fullerene, and nanodiamonds are the most common carbon-based nanomaterials explored for interaction with genetic materials [59]. CNTs are extensively being explored for drug and gene delivery because of their high-aspect ratio and flexible structure [60]. Graphene oxide (GO) is being exploited as nanocarriers as it has a large surface area, good biocompatibility, tunable surface chemistry, high water dispersibility, and low-cost scalable production [61,62]. CNTs are classified into two main categories: SWCNT (SWCNTs) and multi-walled carbon nanotubes (MWCNTs).

The interaction of CNTs with NAs (ssDNA) enhances the dispersion of CNTs in aqueous solutions. π-stacking is involved in single-stranded DNA (ssDNA) and SWCNTs complex [60]. DNA binding with a nanotube depends on the orientation of the four adjacent nucleotides concerning the long axis of a nanotube [63]. Studies on the binding affinity of ammonium-functionalized CNTs with NAs revealed that electrostatic and hydrogen-bonding (H-bonding) are dominant, followed by p-p stacking [64]. CNTs might cause a conformational transition in the DNA or RNA strand dependent on the sequence present in the nucleic acid. Along with conformational transition, SWCNTs lower the (Tm) melting time of DNA [65]. Adsorption of ssDNA around the surface of CNTs involves the van der Waal dispersive forces.

CNTs and ssDNA have stable binding configurations; however, double-stranded DNA (dsDNA) gets adsorbed partially with a longer simulation time. dsDNA undergoes structural deformations on interaction with CNTs but to a lesser extent than ssDNA. ssDNA wraps around SWCNTs in a helical manner because of electrostatic and torsion interaction within the sugar-phosphate backbone. One-half of the nucleobases of ssDNA reach the interior of CNTs within nanoseconds, and the other half remains outside and gets wrapped around the outer surface of CNTs. The dsDNA fails to reach the interior of CNTs because of the energy barrier. The binding energies of CNTs towards nucleobases vary with the hydroxylation and carboxylation of CNTs [66,67]. The binding of DNA strands to SWCNTs protects from enzymatic cleavage and interference from NA-binding proteins. SWCNT-DNA complex can target a specific mRNA inside cells [68].

Graphene is another carbon nanomaterial being used extensively and interacts with genetic material largely through van der Waals forces. The binding between graphene and nucleobases can be regulated by the addition or removal of electrons from the system during the interaction. Generally, the binding between GO and ssDNA or single-stranded RNA (ssRNA) involves π- π stacking, H-bonding, and van der Waals forces [69]. In the case of ssDNA, H-bonding between the oxygen atoms on the surface of graphene and the polar groups of the oligonucleotide strand, π- π stacking between the hexagonal aromatic rings of GO, and the ring structure of the

Advances in Healthcare and Nanoparticle Toxicology
Materials Research Foundations 171 (2024) 154-190

Materials Research Forum LLC
https://doi.org/10.21741/9781644903339-6

nucleobases is the stabilization force [70]. The interaction between GO and dsDNA or dsRNA is still not fully clear. As reported, the hydrophilic surface of dsDNA prevents the hydrophobic interactions of their bases with GO [71]. However, Lei et al. reported that dsDNA can bind to GO in the presence of a high salt concentration. This is possible because of electrostatic interactions [72]. On the contrary, works by Tang et al. demonstrated that electrostatic interactions have no such effects on binding [70]. In the presence of Cu^{2+}, GO and reduced graphene oxide (rGO) shows DNA cleavage properties because of π- π interaction [73]. GO easily recognizes DNA sequences during PCR amplification [74].

Fullerenes are a family of molecules of carbon atoms that form 3D shapes. Fullerenes 3D structures are hydrophobic and bind DNA through multiple electrostatic, and intercalative interactions. Charged ions bind electrostatically to the DNA, and some planar aromatic heterocyclic compounds show intercalative binding. The binding of fullerene with B-DNA does not cause any structural changes, whereas, on interaction with A-DNA, it breaks the H-bonds between the red base pairs [75]. If a defective site is present in dsDNA, fullerene binds and disrupts DNA functionality. The fullerenols cause changes in the structure and biochemical nature of bacterial DNA molecules. The interaction between fullerene and DNA varies with the weight of fullerene and forms of DNA. Small fullerenes < (C48) bind with the minor grooves of the B-DNA and of higher molecular weight to the major groves. In the case of A-DNA, fullerene binds to major groves only. Z-DNA, binding to fullerenes, shows random arrangement [76,77].

5.2   Quantum dots

Interaction of QDs and DNA results in a conjugate formation. This interaction depends on the nature and surface properties of QDs. For example, ZnO-QDs bind preferably to sites containing a lone pair, irrespective of its location in nucleic acid [78,79]. One of the most important parameters to study QDs-DNA interaction is the concentration of DNA. DNA concentrations with a range of 0.5 to 0.25 µg/ml are appropriate for the binding of DNA with QDs. Since QDs conjugate effectively with oligonucleotides, they can act as suitable probes to detect a specific nucleic acid [80,81].

5.3   Dendrimers

The dendrimers are systematically intensely branched, reactive nanomaterials. Dendrimers cause hemolysis and other hemocytotoxic impacts, dependent on the chemical core, surface end product, and cationic features of dendrimers. DNA, while interacting with amine-terminated ethylenediamine core polyamidoamine dendrimer, wraps around G-7 dendrimer but not around G-2 and G-4. The predominant binding force governing the interaction between the two is the electrostatic interactions which is in direct control of ionic strength and pH. Other parameters like the size of the dendrimer, length of DNA, and the ratio of charge between DNA and dendrimer affect the interaction between DNA and dendrimers too [80]. The binding between DNA and dendrimer reveals the presence of two distinct zones, namely tight bound DNA zone and linker-bound DNA zone [82]. In the case of a higher positive charge on the dendrimer, a very strong binding interaction is observed. The interaction of poly-amid-amine (PAMAM) dendrimer G4 generation and ssDNA involves the electrostatic attraction between the positive charge present on the dendrimer and the negative charge on the DNA backbone. Therefore, the charge ratio between the positively charged dendrimer and the negative charge on the DNA affects this interaction.

Advances in Healthcare and Nanoparticle Toxicology   Materials Research Forum LLC
Materials Research Foundations 171 (2024) 154-190   https://doi.org/10.21741/9781644903339-6

Lower generations of the dendrimers do not wrap around the DNA, whereas the higher generations of dendrimers do so in a manner similar to DNA wrapping around nucleosomes [67]. The interactive forces involved between NAs and NPs are summarized in Table 2.

Table 2. Interactive forces involved between nucleic acids and NPs

Nanoparticles	Nucleic acids	Interactive force
CNTs	ssDNA	π-stacking, adsorption involving van der Waal dispersive forces.
Ammonium-functionalized CNTs	DNA/RNA	Electrostatic, H-bonding, and p-p stacking.
SWCNTs	ssDNA	Electrostatic and torsion interaction
Graphene	DNA/RNA	Van der Waals forces
Graphene oxide	ssDNA/ssRNA	π- π stacking, H-bonding, and van der Waals forces
Graphene oxide	dsDNA	Electrostatic interactions (high salt concentration)
Fullerenes	DNA	Electrostatic, groove-binding, and intercalative interactions.
Dendrimers	DNA	Electrostatic interactions

6. Applications

6.1 Gene therapy

6.1.1 Plasmid DNA delivery

Gene therapy is being explored for treating gene-related disorders. A major challenge in nucleic acids based gene is crossing cellular barriers and reaching the targeted sites. In most cases, a vector is needed to protect the foreign genetic material from degradation, enable cellular uptake, and enhance transfection efficiency. Viruses are majorly employed as vectors but suffer from problems of toxicity and immunogenicity [83,84]. NPs have shown remarkable promise in this area. CNTs are being explored to overcome the problems caused by conventional vectors. Bianco et al. reported the use of CNTs for gene delivery. Plasmid DNA (pDNA) bound to positively charged SWCNT-NH^{3+}, enhanced DNA uptake and gene expression [85]. Amine functional group forms

Materials Research Forum LLC
https://doi.org/10.21741/9781644903339-6

a pDNA-CNTs complex to deliver DNA efficiently [86]. Polyethylenimine functionalized CNTs show high transfection efficiency and also protect DNA from degradation. Cationic polymers such as polyethyleneimine (PEI) or polyamidoamine (PAMAM) enhance the positive charge onto the surface of CNTs. Multiwalled carbon nanotubes (MWCNTs) with polymers like PEI, polyallylamine (PAA), or PEI/PAA transfer DNA without resulting in cytotoxicity [87,88]. Graphene is also used for gene delivery as it protects NAs from enzymatic degradation. The ability of GO to protect NAs is modulated by many factors, such as salt concentration, composition of the buffer solution, the incubation time, the concentration of the NAs, and the presence of nucleases. Graphene causes conformational changes in NAs, steric hindrance, and changes in local ion concentration. As a result, nucleases either cannot recognize the NAs or come in contact with NAs [70]. GO complexed with low-molecular-weight PEI, enhances cellular uptake of NAs with reduced cytotoxicity [89].

6.1.2 siRNA delivery (small interfering RNA)

Various reports suggest that grafting polycationic molecules onto nanomaterials reduces the toxicity of polycationic molecules toward cell membranes. Functionalization of multi-walled carbon nanotubes (MWCNTs) with PAMAM dendrons has proven effective for cellular delivery of siRNA. MWCNTs demonstrated a complexation efficacy of siRNA correlated to the number of branches on their dendritic structure. This high efficiency was due to an augmentation in positive charges on the conjugated CNT surfaces [90,91]. Kam et al. were the first to report SWCNTs coated with a lipophilic PEG that was covalently functionalized with siRNA or DNA by cleavable SS bonds. The disulfide reduction by glutathione initiates the release of DNA or siRNA following cell internalization. Only cellular internalization but no DNA nuclear translocation was observed in the absence of disulfide bonds [92]. The CNT conjugates exhibit a greater degree of gene silencing, as demonstrated by the comparison between siRNA-SWCNTs and siRNA complexed to lipofectamine. PEI functionalized SWCNTs, when used for complexation of hTERT-siRNA and with NGR peptide, successfully translocate siRNA into cells of PC3 tumor-bearing mice, and lead to tumor cell growth inhibition [93]. rGO functionalized with a phospholipid-based amphiphilic polymer (PL-PEG) and the cell-penetrating peptide octa-arginine (R8) possesses enhanced stability with no detrimental effect on cell viability. The positively charged surface on R8 helped the formation of complexes with siRNA and their subsequent uptake by the cells [94]. To enhance the cellular uptake and selectively deliver genes at specific tumor sites, different tumor-targeting moieties like antibodies, peptides, folic acid, etc. have been introduced onto the GO surface, which recognize the corresponding "molecular signature" on the cell surface [95,96]. For example, the co-delivery of DOX and siRNA using GO-PEI overcomes multidrug resistance of breast cancer cells [97].

6.1.3 Bioimaging

UpConversion NPs (UCNPs) upconvert near-infrared (NIR) radiation into shorter-wavelength photoluminescence. This property is useful for diagnostic and bioimaging. Li et al. functionalized phospholipids onto UCNP, which mimics lipids constituents of the outer cell membrane. The phospholipids behave in an amphiphilic manner driving the hydrophobic association of the tail of the lipids to the NP surface. UCNPs coated with DNA aptamers are used for targeted imaging of cancer cells. For example, AS1411-coated UCNPs, when incubated with MCF-7 cells exhibit a strong upconversion luminescent image of cells. Joining DNA onto UCNP surfaces involves three basic steps: change of UCNPs into water-soluble structures, surface alteration of functional groups, and conjugation of DNA [98]. Graphene nanoplatelets (GNPs) combined with complementary fluorescent materials can be used for multimodal bioimaging. It is possible because of the plasmonic resonance of GNPs and the fluorescent properties of UCNPs present in DNA–GNP-UCNPs conjugates. GNPs exhibit very unique absorption and emission properties, making them appropriate for multiple imaging modalities like photothermal imaging, two-photon luminescence imaging, and photoacoustic imaging. The structure and shape of GNPs vary with the nucleic acid. GNPs synthesized in the presence of 30-mer oligo-A or -C form flower-shaped NPs, whereas, in the presence of 30-mer Oligo-T, configure normal gold nanospheres. DNA-coupled NPs with different morphology are uptaken by cells with ease and visualized by light-scattering imaging. Higher brightness is observed with DNA–gold nanoflowers compared to nanospheres. A similar method of synthesis could be applied to other NPs with different morphologies, including gold, silver, and lead–gold core–shell NPs [99,100]. Another material used for bioimaging is QDs due to their resistance to photobleaching. Many different methods, like covalent interaction and electrostatic interaction, are being employed to conjugate DNA on the QDs surface. Biotemplated synthesis of QDs can be utilized to directly attach DNA on a nascent NP's surface [101].

6.1.4 Biosensing

Since DNA sequences themselves do not have any signal-generating functional groups, they are coupled to inorganic NPs through thiol−Au for use as sensing probes. Various DNA−NP-based colorimetric sensors have been developed in recent years, mostly following two methods- labeled and label-free methods. In the labeled method, DNA covalently conjugates to GNPs, followed by exposure and binding to analytes. As soon as the target binding triggers the assembly/ disassembly of GNPs, a color change in the solution is observed. Such colorimetric sensors are used for sensing metal ions and biomolecules. For a label-free method, noncovalent adsorption of NAs onto GNPs is used [102]. QDs, inorganic NPs with extremely bright fluorescence, are incorporated into GNP aggregates for photoluminescent detection of different targets in a single assay. Another example of NPs-NAs conjugate used for detection is superparamagnetic iron oxide NPs (SPIOs). Thrombin-detecting MRI sensor was constructed based on the thrombin-induced assembly of SPIOs, and aptamer-modified SPIOs are used for sensitive detection of cancer cells [103]. For one-step colorimetric detection of multiple targets, lateral flow devices (LFDs) are loaded with non-cross-linked DNAzyme−GNP conjugates/ aptamer-linked GNP aggregates [104].

6.1.5 Drug delivery

Drug delivery faces lots of challenges in the health sector. Since most drug carriers used to date have inherent toxicity, effective methods for targeted drug delivery are a necessity. DNA nanotechnology/DNA aptamer techniques can be used to develop drug delivery systems with a high affinity for the target tissue/cells based on the stimuli-responsive principle. Aptamers possess specific binding ability, and this property is being explored for drug delivery. The conformational switching of C-rich i-motif DNA was used successfully for different delivery purposes. A stem-loop DNA nanoswitch shows two conformations- a "Load" conformation having a DOX-intercalating domain and a "Release" conformation having a duplex portion recognizable by a specific transcription factor [105]. NPs such as mesoporous silica-based NPs (MSNs) are used as carriers of drug molecules for controlled release to the desired location. AS1411 aptamer coupled to the MSN surface recognizes nucleolin-overexpressing breast cancer cells specifically. Pores of

MSNs are loaded with DOX and capped with avidins for drug delivery. Because of the capping, the drug is not released during the delivery process; however, once inside the cancer cell, the cap opens in the presence of cytoplasmic vitamin H, releasing the drug [106]. Aptamer-modified DNA−lipid NPs in an aqueous medium are employed in ophthalmic drugs. Many DNA molecules for treatment have received the US Food and Drug Administration (FDA) approval too [107].

7. Lipids interaction with nanoparticles

Lipids comprise fatty acids, phospholipids, sterols, terpenes, etc. Fatty acids such as oleic acid and ricinoleic acid, when bound to iron oxide NPs (IONPs), show a bilayer formation around the IONPs surface, which provides high hydrophilicity forming stable colloidal NPs, thus preventing the colloids from being aggregated. A fatty acid oleate adsorbs on the nanomaterial surface in a small, carpet-like fashion covering the entire surface. Adsorption of phospholipids onto the NP surface is affected by the surface curvature of IONPs. DOTAP, a phospholipid, forms a monolayer around naked IONPs, which interacts with the adjacent layer, thus forming a bilayer. The interaction between phospholipids and NPs depends on the hydrophobicity of the surface [108].

7.1 Interaction of nanoparticles with cell membrane

The cell membrane structurally is a lipid bilayer with specific proteins. This lipid bilayer is the product of the self-assembly of amphipathic phospholipids. The cell membrane is selectively permeable because of the arrangement of the head and tail of amphipathic phospholipids. Some compounds like lipases and scramblase facilitate the translocation of some specific molecules through the lipid bilayer [109,110]. The interaction between nanomaterials and cell membranes is well studied. NPs and cell membrane dynamics are impacted by the size, shape, thickness, and surface features of nanomaterials. Moreover, the pH of the solvent, polarity, temperature, viscosity, and ionic strength of the environment also influences NPs' relationship with membranes. Membranes are fabricated either from lipids or polymers, forming liposomes or polymersomes [111]. Numerous studies have been performed to understand the biological activity of nanomaterials on both cells and artificial models. NPs are internalized by different mechanisms like phagocytosis, micropinocytosis, clathrin-dependent endocytosis, caveolae-dependent endocytosis, or by direct penetration (Figure 2). Opsonized NPs show ligand-receptor interaction, which activates a signaling cascade, enabling them to interact with the cell membrane. Mostly, NPs with a size of 500 nm or more interact with the cell membrane by phagocytosis. NPs' shape, opsonization, and surface properties determine the phagocytic uptake efficiency of NPs [112,113].

NPs are also internalized by caveolae-mediated endocytosis. The protein caveolin is responsible for flask-shaped structures in cells known as caveolae. Proteins involved in caveolae-mediated endocytosis are caveolin, caveolin-2, dynamin, synaptosomal-associated protein, and vesicle-associated membrane protein. Studies suggest that caveolae-mediated internalized NPs can evade degradation by lysosomes, making them useful in therapeutics. An example of such internalization is the uptake of albumin-bound NPs by cancer cells. NPs of around 120 nm size undergo caveolae-mediated endocytosis [112,113].

NPs demonstrate clathrin-mediated endocytosis to interact with cell membrane. Clathrin has a 3-legged structure assembly, occupying 2% of the cell surface. Clathrin forms vesicles in the membrane responsible for the nutrients' transport across the membrane. Macropinocytosis, the internalization process, involves the formation of 'ruffles' in the membrane. This uptake mechanism is independent of clathrin or caveolae. The invaginations formed in the membrane are of size 0.2 - 5µm. Macropinocytosis involves the uptake of NPs of size >1µm. For example, NPs modified with folic acid internalize by this mechanism. Macropinocytosis has the advantage of evading lysosomal enzymatic degradation, and the internalized contents are retained in the endocytic vesicles or directly released into the cytoplasm [112–114].

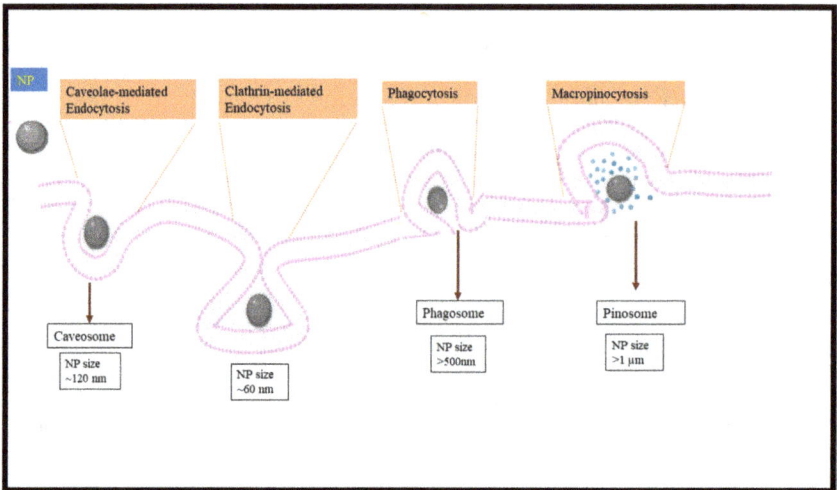

Figure 2: Different modes of internalization utilized by NPs during interaction with the cell membrane.

Some of the NPs interacting with cell membrane are discussed below:

7.2   Metal nanoparticles

To understand the interaction between metal NPs and cell membrane, gold NPs have been extensively studied. Gold NPs (small) are able to penetrate the hydrophobic region of the lipid bilayer. A large number of dodecanthiol-modified gold NPs were observed to be embedded in the L-α-phosphatidylcholine bilayer vesicles [115]. The incorporation of NPs into the bilayer creates voids around each individual NP attracting other NPs and leading to the formation of clusters, reducing the ΔG_{def} (energy cost for bilayer deformation) to a minimum. This forms Janus-type NP vesicle hybrids. The interparticle interaction ($\Delta\Omega$) [116] between two gold NPs during the clustering phenomenon involves capillary forces within the bilayer. This capillary force model is based on the three-phase contact angle between water, lipids, and embedded gold NPs. As per mathematical calculation, in the instance of the bilayer being in the gel phase, a repulsive force acts on gold NPs embedded in the bilayer, whereas attractive force acts on NPs when the bilayer is in the fluid phase. The embedded NPs undergo clustering and declustering during the bilayer's heating and cooling, respectively. In practice, strong van der Waals forces among the clustered

NPs prevent declustering during the cooling process. The effects of van der Waals interaction can be minimized in the presence of protecting agent like alkanethiol [116]. Interaction between NPs and lipid bilayer is influenced by charge present on NPs surface. NPs with a positive charge result in the destruction of the cell membrane, whereas, anionic NPs do not show such behavior [117].

7.3   Metal oxide nanoparticles

Roiter et al. studied the particle size-dependent interaction of silica NPs with liposomes. Silica NPs of size 1 to 140 nm in diameter interacted with the 1,2-dimyristoyl-sn-glycero-3-phosphocholine (DMPC) bilayer in two ways. The bilayer formed only on the substrate surface with the silica NPs of size 5 nm to 20 nm and on both the particle and substrate surfaces in the size range of 5 nm to 140 nm. The lipid bilayer covers the larger particles and is pierced by smaller particles. Similarly, for the particles with diameters smaller than the lipid bilayer, two different interactions were observed. This interactive behavior can be because of the hydrophilic surface of silica [118]. Another metal oxide NP studied is iron oxide NPs. Iron oxide NPs have electromagnetic properties. Superparamagnetic NPs convert electromagnetic energy into thermal energy enabling the heating of liposomes after being embedded into the membrane. One typical function for the superparamagnetic NPs is local heating by applying a radio frequency (RF) wave, which facilitates controlled drug release from liposomes at the target site. A study of the leakage activity of liposomes with Fe_2O_3 NPs demonstrated that the stability of leakage from liposomes improves under both no RF and RF irradiation [119]. Fe_2O_3 NPs modified with certain moieties, such as oleic acid, interacts with lipid molecules differently based on the chemical structure of the lipid. A reduced mean lipid molecular area was observed in the case of saturated lipid monolayers and an expanded mean molecular area in unsaturated monolayers [119]. Another example of metal oxide NP interacting is cerium oxide. The uptake of CeO_2 NPs by *Caenorhabditis elegans* is influenced by the charge present on NPs. The uptake of positively charged CeO_2 NPs is 20-25 times more as compared to negatively and neutral charged CeO_2 NPs [120].

7.4   Polyoxometalate clusters

NP clusters, such as surfactant-encapsulated clusters (SECs), show a strong interaction with lipids. The study of NP clusters is important to understand the interaction between NPs and lipids, as it is practically impossible to synthesize NPs of uniform diameter. SECs comprise surfactants and polyoxometalate (POM) clusters. POM clusters are anionic in composition, forming supramolecular structures with cationic surfactants with ease. Some categorical types of POM clusters are Keggin, Dawson, and Anderson, as per structural variations [121,122]. POMs bind onto zwitterionic lipid membrane (phosphatidylcholine or phosphatidylethanolamine) by electrostatic force and exhibit destructive activity against liposomes resulting in morphological changes in the membranes such as pore and multilayer formation. For example, SiW12O4042 Keggin POMs destroyed egg-PC liposomes. Dawson POMs exhibit similar destructive properties against liposomes. The amount of liposome destruction caused by negative POMs is greater than positive POMs. It is because of the negative charge present on POMs and the positive charge on the PC head group of liposomes. The mechanism proposed for POM-lipid interaction is as follows:
- POMs that are negatively charged are initially adsorbed onto the liposome surface and cause lipid rearrangement to correspond to the curvature of the POM surface. They desorb and create thermodynamically stable (POM-surfactant supramolecules) SEC-like structures because it is energetically undesirable for slanted and bulky structures to be embedded in the flat lipid layer

surface. It accelerates the change from an unstable defective liposome into a stable SEC-like structure when the quantity of lipid desorption approaches the threshold value because the liposomes are unable to maintain their vesicular shape [123]. It is believed that this structural change is where POMs first started acting destructively against the membrane model. Wheel-like $[(MnO_3)_{176}(H_2O)_{80}]^{32-}$ POMs and small POMs cause liposome damage by gelation, surface reconstruction, and pore formation on lipid membranes [124].

7.5 Polymeric nanoparticles

Polymeric NPs interact with lipid bilayers through adsorption. The charges present on polymeric NPs affect the interaction like other types of NPs. Adsorption of negatively charged polystyrene (PS) NPs onto the dilauroyl phosphatidylcholine (DLPC) bilayer leads to the phase transition of the bilayer from liquid to gel. Lipid gelation occurs only at the points of NPs interaction with the bilayer in a concentration-dependent manner. However, positively charged NPs cause gel-to-fluid phase transition. Exothermic and endothermic, both processes are involved in the interaction of negatively charged PS NPs with DLPC liposomes (in fluid phase), whereas only the endothermic process occurs during DPPC liposomes (in gel phase) interaction with positively charged PS NPs. Reports suggest that NP interactions with the cell membrane are affected by the surface state, having an impact on the uptake too [125]. The interaction of polymeric NPs with lipids bilayer is affected by particle concentration. At high NP concentrations, lipid diffusivity is suppressed, attributed to the formation of "pores". A significant transfer of lipid molecules from the bilayer to the NP surface occurs at higher concentrations. This transfer is driven by the hydrophobic interaction between the NP surface and the acyl chain of lipid molecules. Therefore, it is expected that a lipid monolayer with acyl chains interacting with the NP surface could be formed. Numerous studies have shown that polymeric NPs can serve as stable host materials for drug delivery systems [124]. It is crucial to understand the mechanism of each step involved in drug delivery to cells, including diffusion, adsorption, internalization, drug release, and unfavorable destructive activity. Therefore, further in vivo and in vitro research is required to explore the use of polymeric NPs for drug delivery systems.

7.6 Effect of nanoparticles on lipid membranes

When certain types of NPs, such as those made of iron oxide, CdSe-QDs, or gold NPs with a significant surface charge, come into contact with lipid membranes, they can cause significant damage to the membrane. This can lead to increased porosity of the cell membrane, which can have a range of negative consequences. There are two main types of disruption: nanoscale hole formation and membrane thinning [126]. The interaction between charged NPs and the lipid membrane can lead to deformation and changes in the distribution of the lipid head groups [127]. Electrostatic forces play a key role in this process, as positively charged NPs can influence the tilt angle of the lipid phosphate terminus, resulting in an enlargement of the head group area. A recent study revealed Recent molecular dynamics simulations have revealed that Gold NPs with positive charges can cause defects in negatively charged lipid vesicles by altering the bilayer's surface texture and inducing pore formation, which is dependent on the number of positive charges present on the NPs' surface [128]. These NPs can cause membrane depolarization by flipping membrane areas and particle inclusion. Positively charged NPs induce aggregation by binding to the plasma membrane. Another detrimental effect of NPs is membrane fusion [129].

Advances in Healthcare and Nanoparticle Toxicology　　　　　Materials Research Forum LLC
Materials Research Foundations 171 (2024) 154-190　　　https://doi.org/10.21741/9781644903339-6

Surface hydrophobicity affects NP interaction with lipid membranes. Numerous theoretical studies have demonstrated that the hydrophobicity or hydrophilicity of an NP surface exerts a significant influence on the morphology and the creation of holes in lipid membranes [130]. Embedding of functional hydrophobic NPs into lipid membranes has numerous applications, such as controlling bilayer permeabilization, biocompatibility, and liposomal release. For successful embedding, the NP must be small enough to fit within the bilayer dimension and possess a hydrophobic surface. Researchers have been able to create hybrid vesicles with hydrophobic CdSe NPs confined within a 4 nm thick lipid bilayer membrane, which were used for the release of small molecular weight compounds into living cells [131]. Hydrophobicity/ hydrophilicity moiety affects the location of the NPs while interacting with the membrane. For example, 2 nm hydrophobic Au-NPs are attracted to the hydrophobic lipid bilayer of a lipid vesicle membrane [132]. Hydrophobic decanethiol-covered Au-NPs cause an increase in bilayer fluidity of the gel phase by reduction of lipid ordering. Hydrophobic molecules attract and create patched lipid bilayers on large NPs, promoting low lipid areas on supported lipid bilayers [133].

The impact of NP size on the structure and morphology of NP-lipid assemblies and membrane disruption remains uncertain till now. Smaller dodecanethiol-capped gold NPs (5.7 nm) embed directly within the lipid bilayer; however, larger Au-NPs are dispersed in the aqueous phase by a lipid monolayer [133]. Smaller silica NPs form a hole in the lipid bilayer, while larger NPs are mostly covered with the lipid bilayers [118]. The bilayer disruption effect of aminated silica nanospheres was studied on 1,2-dioleoyl-sn-glycero-3-phosphocholine bilayers. The disruption effect was found to be more pronounced for the 500 nm-sized particles than for the 50 nm-sized particles. Calculations have demonstrated smaller, interacting NPs will cause a curving-away of the membrane from the NPs, whereas in the case of small, non-interacting NPs, the membrane will curve towards the NPs. Larger NPs cause inclusion effects and the formation of fission and budding structures in lipid membranes [134,135]. 2–8 nm sized NPs are able to embed within a lipid bilayer favorably as per computer simulations. Based on the bilayer phase behavior NPs with a diameter in proximity to or even exceeding the thickness of the bilayer are possible to embed in the lipid bilayer [130]. Dendrimers cause an increase in porosity in lipids. Amine-terminated dendrimers of generation 5 (G5) and generation 7 (G7) create a hole in the membrane, while uncharged (hydrophobic) dendrimers absorb onto the outer surface of the membrane. Additionally, these holes allow dendrimers to penetrate the cells resulting in the diffusion of cytosolic proteins outside the cells [136,137]. These discoveries could potentially open exciting new treatment and drug delivery options in the future.

8.　Carbohydrates interaction with nanoparticles

Carbohydrates are undoubtedly an essential component of biomolecules, accounting for more than 80% of biomass. One of their primary functions is to provide energy storage and support structural functions [138]. Monosaccharides are linked together by glycosidic bonds to form polysaccharides, existing in linear or branched structures [139]. Carbohydrates exist both inside and on the cell surface, creating the distinct glycocalyx layer. This layer plays a vital part in numerous biological recognition processes by engaging with carbohydrate-binding proteins [140]. Polysaccharides' primary adsorption mechanism is hydrogen bonding. For example, starch's ability to bond with iron oxide results from hydrogen bonding, which ultimately leads to the creation of chemical complexes [141,142]. The glycocalyx forms a dominant feature of freely moving cells in blood circulation and also of the composite cells located as lining in the case of blood vessels,

Advances in Healthcare and Nanoparticle Toxicology
Materials Research Foundations 171 (2024) 154-190

Materials Research Forum LLC
https://doi.org/10.21741/9781644903339-6

alimentary canal, urinary bladder, ureter, urethra, etc. The oligosaccharides that are present in glycocalyx act as cell markers and facilitate cell identification [143]. The interaction of NPs with glycocalyx is explained in the following subsection.

8.1 Interaction of nanoparticles with glycocalyx

The glycocalyx, a meshwork of glycoproteins and polysaccharides, forms a protective coat and regulates the interaction and translocation of materials that come in contact with endothelial cells. NPs or metabolites can only be absorbed through endothelial cells in locations where the erosion of the glycocalyx has occurred. Polystyrene nanospheres (50 nm) enter the human endothelial cells during cell culture experiments [144,145]. PEG amine or galactose-coated gold NPs translocate via endothelial glycocalyx in renal and brain cells. It's important to note that how coated gold NPs are absorbed into various tissue can be affected by the presence of glycocalyx [146]. The role of glycocalyx in the translocation of nanomaterials through the plasma membrane is a fascinating area of study. The interaction between glycosylated nanomaterials and the cellular glycocalyx, particularly with lipid anchors, involves the extracellular leaflet of the plasma membrane [147]. It's important to consider the different factors that impact the movement of nanomaterials across the plasma membrane, including the glycocalyx's chemical composition, thickness, ability to renew, density, and surface charge [148]. Recent research shows that glycopeptide dendrimers can prevent bacteria from sticking to the glycocalyx [149]. Some of the carbohydrates' interaction with NPs and their applications are discussed in the following sections.

8.2 Carbohydrate-modified gold nanoparticles

The use of glyconanoparticles (Carbohydrate-modified NPs) as a platform for studying carbohydrate-protein and carbohydrate-carbohydrate interactions is highly intriguing. Remarkably, most glyconanoparticles feature an inorganic core of nanoscale dimensions, along with a flexible organic layer, linker, or polymer that's either covalently or noncovalently conjugated with carbohydrates. Different materials like noble metal NPs, semiconductor QDs, or magnetic nanomaterial may be used for the core of the glyconanoparticles [150]. Gold NP-based colorimetric assays are highly effective in detecting a wide range of glycan-binding proteins, such as toxins and lectins [151]. Gold and silver NPs have been utilized in combination with mass spectroscopic techniques to enhance the sensitivity of biomolecule detection and characterization. Glycosyl transferase activity on oligo (ethylene glycol)-coated AuNPs has been monitored using mass spectrometry, as well as the direct analysis and epitope mapping of glyconanoparticle-bound proteins [152,153]. Penades and colleagues have successfully developed magnetic glyconanoparticles consisting of a 4nm magnetic core and a 1nm gold shell functionalized with carbohydrates. These NPs have been designed to locate endogenous neural precursor cells, which migrate toward a brain damage site. Monoclonal antibody Nilo2, which recognizes cell surface antigens on neuroblasts, has been coupled with magnetic glyconanoparticles (mGNPs). The Nilo2-mGNP complexes have enabled the in vivo identification of endogenous neural precursors at their niche and their migration to a lesion site, an induced brain tumor [154]. Water-soluble gold NPs coated with mannose-ferrocene conjugates have been successfully used for the development of electrochemical sensing, such as mannose-binding lectin concanavalin A. Gold NPs covered by a polymer with Tn-antigen glycans can potentially serve as synthetic cancer vaccines aiding in the generation of antibodies selective for aberrant mucin glycans [155]. Glyconanoparticles show great potential for delivering anticancer drugs to specific sites. Gold NPs with lactose and β-

Advances in Healthcare and Nanoparticle Toxicology
Materials Research Foundations 171 (2024) 154-190

Materials Research Forum LLC
https://doi.org/10.21741/9781644903339-6

cyclodextrin linkers interact with peanut agglutinin and human galectin-3 (Gal-3). Gal-3 is overexpressed in many tumors, making glyconanoparticles a promising targeted drug delivery system [156]. Gold glyconanoparticles equipped with a trivalent, α-2,6-thio-linked sialic acid–galactose derivative can detect influenza virus in just 30 minutes. Recent research showed that these NPs act as a plasmonic sensor and can distinguish between human and avian influenza [157].

8.3 Glycans conjugate with carbon nanostructures

The unique properties of CNTs and graphene combined with glycans make them competent materials for biosensors. By incorporating glycoconjugates into CNTs / graphene, these sensors can accurately detect medically significant targets through optical and electrochemical means. As a result, there are now a multitude of biosensor applications utilizing CNTs and graphene. These materials can be modified covalently or noncovalently. For biosensor applications, noncovalent methods are preferred to avoid disrupting the structure and losing electronic properties, achieved through the use of pyrene/porphyrin conjugates or glycopolymers/lipids composites [158–161]. CNTs are commonly used in Field-effect transistor (FET) devices for efficient and label-free biosensors. The development of functional sensors that can selectively interact with bacterial cells is greatly aided by glycans [162]. Chen et al. led a study in which an SWNT network was functionalized with specific glycoconjugates through conjugation with porphyrins. This resulted in the creation of specific sensors that target bacterial detection. These sensors respond to bacterial lectins that interact with the specific sugar [163]. Reuel et al. have successfully embedded SWNTs in a chitosan matrix, creating a unique biosensor array. These SWNT bundles act as individual sensors, utilizing local fluorescence to detect glycan-lectin-binding and monitor IgG affinity distributions. Their innovative approach has also proven effective in detecting hypermannosylation [164,165]. CNTs and graphene have shown strong interaction with enzymes and are effective in mediating electron transfer. This has led to their recent exploration for use in biofuel cells [166]. One of the favorable techniques is to incorporate GO into a silica sol-gel to form a fuel cell that is porous and free of membranes [167]. In a more recent development, Prasad et al. produced an enzymatic biofuel cell by utilizing laccase and glucose oxidase on a hybrid graphene-CNT scaffold [168].

The use of graphene-based nanomaterials in biomedical applications is truly remarkable and ranges from cell imaging, photothermal and photodynamic therapies and drug delivery. Study has shown that amylose can wrap around carbon nanotubes in a self-assembly process, which could potentially be used for biosensing in vivo [169,170]. Similarly, DNA helical oligomers have also been observed to wrap around SWNTs, providing water solubility. By conjugating with PEG, the near-infrared fluorescence properties of the nanotubes were stabilized for the detection of nitric oxide with a high degree of sensitivity [171]. Hexakis-adducts of C60 showing 12-6 mannoside residues have been accounted for as nanomolar inhibitors of viral infection intervened by DC-SIGN [172]. Additionally, chitosan -SWNTs conjugates provide porous, biodegradable, and biocompatible chitosan [173]. Association with hydroxyapatite allowed for the development of a composite matrix useful for cell proliferation.

8.4 Chitosan-based nanoparticle platform

Polysaccharides, when used for the construction of NP platforms, provide the advantage of biodegradability and create customized derivatives for the desired application. Some natural polysaccharides used to make NP platforms for diagnostics and therapy are chitosan, arabinogalactan, pullulan, and dextran [174]. Chitosan-based NPs selectively adsorb to the negatively charged tumor cell surface because of the amino groups present. This property is exploited to target drugs to the liver, spleen, lung, and colon [175]. PEGylation of chitosan improves the enhanced permeability and retention effect of NPs [176]. Loading of the therapeutic drug into the NPs is achieved either by direct conjugation to the chitosan molecules or by encapsulation. Chitosan-based NPs nanocarriers are utilized for the delivery of DNA molecules because positively charged chitosan promotes interaction with negatively charged NAs. These nanocarriers further benefit from the mucoadhesive properties and the permeation-enhancing capability of chitosan [177,178]. The abundance of amine groups in chitosan-based NP platforms also helps in lysosomal evasion. The efficiency of chitosan-NP conjugates is enhanced by certain modifications, such as the conjugation of amine-rich polymers and cell-penetrating peptides [179,180]. Chitosan-based NP platforms find their application in photodynamic therapy, too, as polysaccharides provide a biocompatible, solubilizing platform for the hydrophobic photosensitizers and efficiently target photosensitizers to tumor tissue [181,182].

Conclusion

Nanotechnology has opened new avenues in medicine. Researchers have intensively studied the association of NPs with biological systems. To decipher how NPs behave in biological systems, it is necessary to address the physicochemical properties of NPs, including their size, shape, coating, morphology, surface charge, hydrophobicity, chemical composition, and structure. The nature of biomolecules further defines the modes of NPs interaction. Studies determining the biocompatibility and cytotoxicity of NPs would help further the advancement of novel therapeutics. Nanomaterials have a superior advantage due to efficient penetration, retention, and specific targeting of tumor cells. Not surprisingly, NPs are utilized in immunotherapies and cancer treatments. NPs' relationship with biomolecules is also exploited for diagnostic and imaging purposes. A better understanding of nanomaterial-biomolecules' interaction and their behaviour in biological systems could harness their potential in nanomedicine.

Abbreviations

IgG	Immunoglobulin G
PC	Protein corona
NPs	Nanoparticles
IgM	Immunoglobulin M
pI	Isoelectric points
SWCNT	Single-walled carbon nanotubes
RES	Reticuloendothelial system
BSA	Bovine serum albumin
PEG	Poly (ethylene glycol)
Tf	Transferrin
SiNPs	Silica nanoparticles
EGFP	Green fluorescent protein

EGF1	Epidermal growth factor
GRP	Gastrin releasing peptide
Apo	Apolipoproteins
BBB	Blood-Brain barrier
CTAB	Cetyltrimethylammonium bromide
SPION	Superparamagnetic iron oxide nanoparticles
CNTs	Carbon nanotubes
MWCNTs	Multiwalled carbon nanotubes
GO	Graphene oxide
ssDNA	Single-stranded DNA
Tm	Melting time
dsDNA	Double-Stranded DNA
rGO	Reduced graphene oxide
CSP	Chemosensory proteins
PAMAM	Poly-amid-amine
PEI	Polyethylenimine
PAA	Polyallylamine
siRNA	Small interfering RNA
NAs	Nucleic acids
PEG	Polyethylene glycol
NGR	Asparagine-glycine-arginine
DOX	Doxorubicin
UCNPs	Upconversion Nanoparticles
NIR	Near-infrared radiation
GNPs	Graphene nanoplatelets
QDs	Quantum dots
MRI	Magnetic resonance imaging
LFDs	Lateral flow devices
MSNs	Mesoporous silica-based nanoparticles
DOTAP	1,2 Dioleoyl 3 Trimethylammoniopropane
IONPs	Iron oxide nanoparticles
ΔGdef	The energy cost for bilayer deformation
$\Delta\Omega$	Interparticle interaction
DMPC	1,2-dimyristoyl-sn-glycero-3-phosphocholine
RF	Radiofrequency
DPPC	1,2-dipalmitoyl-sn-glycero-3-phosphocholine
SECs	Surfactant-encapsulated clusters

POM	Polyoxometalate
DLPC	Dilauroyl phosphatidylcholine
DOPC	1,2-Dioleoyl-sn-glycero-3-phosphocholine
mGNPs	Magnetic glyconanoparticles
Gal-3	Human galectin-3
pDNA	Plasmid DNA
FET	Field-effect transistor

References

[1] R. Saini, S. Saini, R.S. Sugandha, Pharmacogenetics: The future medicine, J. Adv. Pharm. Technol. Res. 1 (2010) 423–424. https://doi.org/10.4103/0110-5558.76443.

[2] A.W. Hübler, O. Osuagwu, Digital quantum batteries: Energy and information storage in nanovacuum tube arrays, Complexity. 15 (2010) 48–55. https://doi.org/10.1002/cplx.20306.

[3] A. Verma, F. Stellacci, Effect of surface properties on nanoparticle-cell interactions, Small. 6 (2010) 12–21. https://doi.org/10.1002/smll.200901158.

[4] A.P. Ingle, N. Duran, M. Rai, Bioactivity, mechanism of action, and cytotoxicity of copper-based nanoparticles: A review, Appl. Microbiol. Biotechnol. 98 (2014) 1001–1009. https://doi.org/10.1007/s00253-013-5422-8.

[5] A. Albanese, P.S. Tang, W.C.W. Chan, The effect of nanoparticle size, shape, and surface chemistry on biological systems, Annu. Rev. Biomed. Eng. 14 (2012) 1–16. https://doi.org/10.1146/annurev-bioeng-071811-150124.

[6] A. Rahi, N. Sattarahmady, H. Heli, Toxicity of nanomaterials-physicochemical effects, Austin J Nanomedicine Nanotechnol. 2 (2014).

[7] I. Khan, K. Saeed, I. Khan, Nanoparticles: Properties, applications and toxicities, Arab. J. Chem. 12 (2019) 908–931. https://doi.org/10.1016/j.arabjc.2017.05.011.

[8] H. Heinz, C. Pramanik, O. Heinz, Y. Ding, R.K. Mishra, D. Marchon, R.J. Flatt, I. Estrela-Lopis, J. Llop, S. Moya, R.F. Ziolo, Nanoparticle decoration with surfactants: Molecular interactions, assembly, and applications, Surf. Sci. Rep. 72 (2017) 1–58. https://doi.org/10.1016/j.surfrep.2017.02.001.

[9] Y.K. Lahir, Impacts of Metal and Metal Oxide Nanoparticles on Reproductive Tissues and Spermatogenesis in Mammals., J. Exp. Zool. India. 21 (2018) 593–608.

[10]   G. Karp, J. Iwasa, W. Marshall, Karp's Cell and Molecular Biology, John Wiley \& Sons, 2020.

[11]   F. Chellat, Y. Merhi, A. Moreau, L. Yahia, Therapeutic potential of nanoparticulate systems for macrophage targeting, Biomaterials. 26 (2005) 7260–7275. https://doi.org/10.1016/j.biomaterials.2005.05.044.

[12]   M. Lundqvist, I. Sethson, B.H. Jonsson, Protein adsorption onto silica nanoparticles: Conformational changes depend on the particles' curvature and the protein stability, Langmuir. 20 (2004) 10639–10647. https://doi.org/10.1021/la0484725.

[13]   I. Lynch, K.A.A. Dawson, Protein-nanoparticle interactions, Nano Today. 3 (2008) 40–47. https://doi.org/10.1016/S1748-0132(08)70014-8.

[14]   T. Cedervall, I. Lynch, S. Lindman, T. Berggård, E. Thulin, H. Nilsson, K.A. Dawson, S. Linse, Understanding the nanoparticle-protein corona using methods to quntify exchange rates and affinities of proteins for nanoparticles, Proc. Natl. Acad. Sci. U. S. A. 104 (2007) 2050–2055. https://doi.org/10.1073/pnas.0608582104.

[15]   M. Lundqvist, J. Stigler, G. Elia, I. Lynch, T. Cedervall, K.A. Dawson, Nanoparticle size and surface properties determine the protein corona with possible implications for biological impacts, Proc. Natl. Acad. Sci. U. S. A. 105 (2008) 14265–14270. https://doi.org/10.1073/pnas.0805135105.

[16]   P.P. Karmali, D. Simberg, Interactions of nanoparticles with plasma proteins: Implication on clearance and toxicity of drug delivery systems, Expert Opin. Drug Deliv. 8 (2011) 343–357. https://doi.org/10.1517/17425247.2011.554818.

[17]   R. García-álvarez, M. Vallet-Regí, Hard and soft protein corona of nanomaterials: Analysis and relevance, Nanomaterials. 11 (2021) 888. https://doi.org/10.3390/nano11040888.

[18]   L. Vroman, A.L. Adams, G.C. Fischer, P.C. Munoz, Interaction of high molecular weight kininogen, factor XII, and fibrinogen in plasma at interfaces, Blood. 55 (1980) 156–159. https://doi.org/10.1182/blood.v55.1.156.bloodjournal551156.

[19]   M. Lundqvist, J. Stigler, T. Cedervall, T. Berggård, M.B. Flanagan, I. Lynch, G. Elia, K. Dawson, The evolution of the protein corona around nanoparticles: A test study, ACS Nano. 5 (2011) 7503–7509. https://doi.org/10.1021/nn202458g.

[20]   T.M. Göppert, R.H. Müller, Polysorbate-stabilized solid lipid nanoparticles as colloidal carriers for intravenous targeting of drugs to the brain: Comparison of plasma protein adsorption patterns, J. Drug Target. 13 (2005) 179–187. https://doi.org/10.1080/10611860500071292.

[21]   P. Aggarwal, J.B. Hall, C.B. McLeland, M.A. Dobrovolskaia, S.E. McNeil, Nanoparticle interaction with plasma proteins as it relates to particle biodistribution, biocompatibility and therapeutic efficacy, Adv. Drug Deliv. Rev. 61 (2009) 428–437. https://doi.org/10.1016/j.addr.2009.03.009.

[22]   V. Hirsch, C. Kinnear, M. Moniatte, B. Rothen-Rutishauser, M.J.D. Clift, A. Fink, Surface charge of polymer coated SPIONs influences the serum protein adsorption, colloidal stability and subsequent cell interaction in vitro, Nanoscale. 5 (2013) 3723–3732. https://doi.org/10.1039/c2nr33134a.

[23]   M.P. Monopoli, D. Walczyk, A. Campbell, G. Elia, I. Lynch, F. Baldelli Bombelli, K.A. Dawson, Physical-Chemical aspects of protein corona: Relevance to in vitro and in vivo biological impacts of nanoparticles, J. Am. Chem. Soc. 133 (2011) 2525–2534. https://doi.org/10.1021/ja107583h.

[24]   S. Tenzer, D. Docter, S. Rosfa, A. Wlodarski, J. Kuharev, A. Rekik, S.K. Knauer, C. Bantz, T. Nawroth, C. Bier, J. Sirirattanapan, W. Mann, L. Treuel, R. Zellner, M. Maskos, H. Schild, R.H. Stauber, Nanoparticle size is a critical physicochemical determinant of the

human blood plasma corona: A comprehensive quantitative proteomic analysis, ACS Nano. 5 (2011) 7155–7167. https://doi.org/10.1021/nn201950e.

[25]   M.A. Dobrovolskaia, A.K. Patri, J. Zheng, J.D. Clogston, N. Ayub, P. Aggarwal, B.W. Neun, J.B. Hall, S.E. McNeil, Interaction of colloidal gold nanoparticles with human blood: effects on particle size and analysis of plasma protein binding profiles, Nanomedicine Nanotechnology, Biol. Med. 5 (2009) 106–117. https://doi.org/10.1016/j.nano.2008.08.001.

[26]   V.H. Nguyen, B.J. Lee, Protein corona: A new approach for nanomedicine design, Int. J. Nanomedicine. 12 (2017) 3137–3151. https://doi.org/10.2147/IJN.S129300.

[27]   A. Gessner, A. Lieske, B.R. Paulke, R.H. Müller, Influence of surface charge density on protein adsorption on polymeric nanoparticles: Analysis by two-dimensional electrophoresis, Eur. J. Pharm. Biopharm. 54 (2002) 165–170. https://doi.org/10.1016/S0939-6411(02)00081-4.

[28]   S. Muzammil, J. Neves Cruz, R. Mumtaz, I. Rasul, S. Hayat, M.A. Khan, A.M. Khan, M.U. Ijaz, R.R. Lima, M. Zubair, Effects of Drying Temperature and Solvents on In Vitro Diabetic Wound Healing Potential of Moringa oleifera Leaf Extracts, Molecules. 28 (2023) 710. https://doi.org/10.3390/molecules28020710.

[29]   S. Lindman, I. Lynch, E. Thulin, H. Nilsson, K.A. Dawson, S. Linse, Systematic investigation of the thermodynamics of HSA adsorption to N-iso-propylacrylamide/N-tert-butylacrylamide copolymer nanoparticles. Effects of particle size and hydrophobicity, Nano Lett. 7 (2007) 914–920. https://doi.org/10.1021/nl062743+.

[30]   S.M. Moghimi, H.M. Patel, Tissue specific opsonins for phagocytic cells and their different affinity for cholesterol-rich liposomes, FEBS Lett. 233 (1988) 143–147. https://doi.org/10.1016/0014-5793(88)81372-3.

[31]   Z.J. Deng, G. Mortimer, T. Schiller, A. Musumeci, D. Martin, R.F. Minchin, Differential plasma protein binding to metal oxide nanoparticles, Nanotechnology. 20 (2009) 455101. https://doi.org/10.1088/0957-4484/20/45/455101.

[32]   M. Mahmoudi, S.E. Lohse, C.J. Murphy, A. Fathizadeh, A. Montazeri, K.S. Suslick, Variation of protein corona composition of gold nanoparticles following plasmonic heating, Nano Lett. 14 (2014) 6–12. https://doi.org/10.1021/nl403419e.

[33]   J. O'Brien, K.J. Shea, Tuning the Protein Corona of Hydrogel Nanoparticles: The Synthesis of Abiotic Protein and Peptide Affinity Reagents, Acc. Chem. Res. 49 (2016) 1200–1210. https://doi.org/10.1021/acs.accounts.6b00125.

[34]   M. Kosmulski, pH-dependent surface charging and points of zero charge. IV. Update and new approach, J. Colloid Interface Sci. 337 (2009) 439–448. https://doi.org/10.1016/j.jcis.2009.04.072.

[35]   S. Khan, A. Mukherjee, N. Chandrasekaran, Silver nanoparticles tolerant bacteria from sewage environment, J. Environ. Sci. 23 (2011) 346–352. https://doi.org/10.1016/S1001-0742(10)60412-3.

[36]   S.M. Moghimi, I.S. Muir, L. Illum, S.S. Davis, V. Kolb-Bachofen, Coating particles with a block co-polymer (poloxamine-908) suppresses opsonization but permits the activity of

dysopsonins in the serum, BBA - Mol. Cell Res. 1179 (1993) 157–165. https://doi.org/10.1016/0167-4889(93)90137-E.

[37]   A.B. Engin, M. Neagu, K. Golokhvast, A. Tsatsakis, Nanoparticles and endothelium: An update on the toxicological interactions, Farmacia. 63 (2015) 792–804.

[38]   C.C. Fleischer, C.K. Payne, Nanoparticle-cell interactions: Molecular structure of the protein corona and cellular outcomes, Acc. Chem. Res. 47 (2014) 2651–2659. https://doi.org/10.1021/ar500190q.

[39]   M.M. Yallapu, N. Chauhan, S.F. Othman, V. Khalilzad-Sharghi, M.C. Ebeling, S. Khan, M. Jaggi, S.C. Chauhan, Implications of protein corona on physico-chemical and biological properties of magnetic nanoparticles, Biomaterials. 46 (2015) 1–12. https://doi.org/10.1016/j.biomaterials.2014.12.045.

[40]   G. Maiorano, S. Sabella, B. Sorce, V. Brunetti, M.A. Malvindi, R. Cingolani, P.P. Pompa, Effects of cell culture media on the dynamic formation of protein-nanoparticle complexes and influence on the cellular response, ACS Nano. 4 (2010) 7481–7491. https://doi.org/10.1021/nn101557e.

[41]   R.B.M. de Almeida, D.B. Barbosa, M.R. do Bomfim, J.A.O. Amparo, B.S. Andrade, S.L. Costa, J.M. Campos, J.N. Cruz, C.B.R. Santos, F.H.A. Leite, M.B. Botura, Identification of a Novel Dual Inhibitor of Acetylcholinesterase and Butyrylcholinesterase: In Vitro and In Silico Studies, Pharmaceuticals. 16 (2023) 95. https://doi.org/10.3390/ph16010095.

[42]   I.Y. Podolski, Z.A. Podlubnaya, E.A. Kosenko, E.A. Mugantseva, E.G. Makarova, L.G. Marsagishvili, M.D. Shpagina, Y.G. Kaminsky, G. V. Andrievsky, V.K. Klochkov, Effects of hydrated forms of C 60 fullerene on amyloid β-peptide fibrillization in vitro andperformance of the cognitive task, J. Nanosci. Nanotechnol. 7 (2007) 1479–1485. https://doi.org/10.1166/jnn.2007.330.

[43]   S. Laurent, M.R. Ejtehadi, M. Rezaei, P.G. Kehoe, M. Mahmoudi, Interdisciplinary challenges and promising theranostic effects of nanoscience in Alzheimer's disease, RSC Adv. 2 (2012) 5008–5033. https://doi.org/10.1039/c2ra01374f.

[44]   S. Mirsadeghi, R. Dinarvand, M.H. Ghahremani, M.R. Hormozi-Nezhad, Z. Mahmoudi, M.J. Hajipour, F. Atyabi, M. Ghavami, M. Mahmoudi, Protein corona composition of gold nanoparticles/nanorods affects amyloid beta fibrillation process, Nanoscale. 7 (2015) 5004–5013. https://doi.org/10.1039/c4nr06009a.

[45]   E. Mahon, A. Salvati, F. Baldelli Bombelli, I. Lynch, K.A. Dawson, Designing the nanoparticle-biomolecule interface for "targeting and therapeutic delivery," J. Control. Release. 161 (2012) 164–174. https://doi.org/10.1016/j.jconrel.2012.04.009.

[46]   M. Mahmoudi, I. Lynch, M.R. Ejtehadi, M.P. Monopoli, F.B. Bombelli, S. Laurent, Protein- nanoparticle interactions: opportunities and challenges, Chem. Rev. 111 (2011) 5610–5637.

[47]   S. Zanganeh, R. Spitler, M. Erfanzadeh, A.M. Alkilany, M. Mahmoudi, Protein corona: opportunities and challenges, Int. J. Biochem. \& Cell Biol. 75 (2016) 143–147.

[48]   A. Salvati, A.S. Pitek, M.P. Monopoli, K. Prapainop, F.B. Bombelli, D.R. Hristov, P.M. Kelly, C. Åberg, E. Mahon, K.A. Dawson, Transferrin-functionalized nanoparticles lose their

targeting capabilities when a biomolecule corona adsorbs on the surface, Nat. Nanotechnol. 8 (2013) 137–143. https://doi.org/10.1038/nnano.2012.237.

[49]   V. Mirshafiee, M. Mahmoudi, K. Lou, J. Cheng, M.L. Kraft, Protein corona significantly reduces active targeting yield, Chem. Commun. 49 (2013) 2557–2559. https://doi.org/10.1039/c3cc37307j.

[50]   K.P. García, K. Zarschler, L. Barbaro, J.A. Barreto, W. O'Malley, L. Spiccia, H. Stephan, B. Graham, Zwitterionic-coated "stealth" nanoparticles for biomedical applications: Recent advances in countering biomolecular corona formation and uptake by the mononuclear phagocyte system, Small. 10 (2014) 2516–2529. https://doi.org/10.1002/smll.201303540.

[51]   G. Caracciolo, D. Pozzi, A.L. Capriotti, C. Cavaliere, F. Cardarelli, A. Bifone, G. Bardi, F. Salomone, A. Laganà, Cancer cell targeting of lipid gene vectors by protein corona, in: Tech. Proc. 2012 NSTI Nanotechnol. Conf. Expo, NSTI-Nanotech 2012, 2012: pp. 354–357.

[52]   M.A. Foote, Using nanotechnology to improve the characteristics of antineoplastic drugs: Improved characteristics of nab-paclitaxel compared with solvent-based paclitaxel, Biotechnol. Annu. Rev. 13 (2007) 345–357. https://doi.org/10.1016/S1387-2656(07)13012-X.

[53]   A. Cifuentes-Rius, H. De Puig, J.C.Y. Kah, S. Borros, K. Hamad-Schifferli, Optimizing the properties of the protein corona surrounding nanoparticles for tuning payload release, ACS Nano. 7 (2013) 10066–10074. https://doi.org/10.1021/nn404166q.

[54]   S. Behzadi, V. Serpooshan, R. Sakhtianchi, B. Müller, K. Landfester, D. Crespy, M. Mahmoudi, Protein corona change the drug release profile of nanocarriers: The "overlooked" factor at the nanobio interface, Colloids Surfaces B Biointerfaces. 123 (2014) 143–149. https://doi.org/10.1016/j.colsurfb.2014.09.009.

[55]   S.A. Alex, N. Chandrasekaran, A. Mukherjee, Impact of gold nanorod functionalization on biocorona formation and their biological implication, J. Mol. Liq. 248 (2017) 703–712. https://doi.org/10.1016/j.molliq.2017.10.119.

[56]   W. Hu, C. Peng, M. Lv, X. Li, Y. Zhang, N. Chen, C. Fan, Q. Huang, Protein corona-mediated mitigation of cytotoxicity of graphene oxide, ACS Nano. 5 (2011) 3693–3700. https://doi.org/10.1021/nn200021j.

[57]   F.E. Guaouguaou, N.E. Es-Safi, Cotula cinerea as a source of natural products with potential biological activities, in: J.N. Cruz (Ed.), Drug Discov. Des. Using Nat. Prod., Springer Nature Switzerland, Cham, 2023: pp. 465–500. https://doi.org/10.1007/978-3-031-35205-8_17.

[58]   A. Brown, T. Brown, Curtailing their negativity, Nat. Chem. 11 (2019) 501–503. https://doi.org/10.1038/s41557-019-0274-1.

[59]   C. Cha, S.R. Shin, N. Annabi, M.R. Dokmeci, A. Khademhosseini, Carbon-based nanomaterials: Multifunctional materials for biomedical engineering, ACS Nano. 7 (2013) 2891–2897. https://doi.org/10.1021/nn401196a.

[60]   M. Zheng, A. Jagota, E.D. Semke, B.A. Diner, R.S. McLean, S.R. Lustig, R.E. Richardson, N.G. Tassi, DNA-assisted dispersion and separation of carbon nanotubes, Nat. Mater. 2 (2003) 338–342. https://doi.org/10.1038/nmat877.

[61]   L. Feng, S. Zhang, Z. Liu, Graphene based gene transfection, Nanoscale. 3 (2011) 1252–1257. https://doi.org/10.1039/c0nr00680g.

[62]   G. V. Theodosopoulos, P. Bilalis, G. Sakellariou, Polymer Functionalized Graphene Oxide: A Versatile Nanoplatform for Drug/Gene Delivery, Curr. Org. Chem. 19 (2015) 1828–1837. https://doi.org/10.2174/1385272819666150526005714.

[63]   M.E. Hughes, E. Brandin, J.A. Golovchenko, Optical absorption of DNA-carbon nanotube structures, Nano Lett. 7 (2007) 1191–1194. https://doi.org/10.1021/nl062906u.

[64]   S. Alidori, K. Asqiriba, P. Londero, M. Bergkvist, M. Leona, D.A. Scheinberg, M.R. McDevitt, Deploying RNA and DNA with functionalized carbon nanotubes, J. Phys. Chem. C. 117 (2013) 5982–5992. https://doi.org/10.1021/jp312416d.

[65]   X. Li, Y. Peng, X. Qu, Carbon nanotubes selective destabilization of duplex and triplex DNA and inducing B-A transition in solution, Nucleic Acids Res. 34 (2006) 3670–3676. https://doi.org/10.1093/nar/gkl513.

[66]   W. Sun, J. Zhao, Z. Du, Density-functional-theory-based study of interaction of DNA/RNA nucleobases with hydroxyl- and carboxyl-functionalized armchair (6,6)CNT, Comput. Theor. Chem. 1102 (2017) 60–68. https://doi.org/10.1016/j.comptc.2017.01.001.

[67]   B. Nandy, M. Santosh, P.K. Maiti, Interaction of nucleic acids with carbon nanotubes and dendrimers, J. Biosci. 37 (2012) 457–474. https://doi.org/10.1007/s12038-012-9220-8.

[68]   Y. Wu, J.A. Phillips, H. Liu, R. Yang, W. Tan, Carbon nanotubes protect DNA strands during cellular delivery, ACS Nano. 2 (2008) 2023–2028. https://doi.org/10.1021/nn800325a.

[69]   H.H. Gürel, B. Salmankurt, Binding mechanisms of DNA/RNA nucleobases adsorbed on graphene under charging: First-principles van der Waals study, Mater. Res. Express. 4 (2017) 65401. https://doi.org/10.1088/2053-1591/aa6e67.

[70]   L. Tang, H. Chang, Y. Liu, J. Li, Duplex DNA/graphene oxide biointerface: From fundamental understanding to specific enzymatic effects, Adv. Funct. Mater. 22 (2012) 3083–3088. https://doi.org/10.1002/adfm.201102892.

[71]   S. He, B. Song, D. Li, C. Zhu, W. Qi, Y. Wen, L. Wang, S. Song, H. Fang, C. Fan, A craphene nanoprobe for rapid, sensitive, and multicolor fluorescent DNA analysis, Adv. Funct. Mater. 20 (2010) 453–459. https://doi.org/10.1002/adfm.200901639.

[72]   M. Liu, H. Zhao, S. Chen, H. Yu, X. Quan, Capture of double-stranded DNA in stacked-graphene: Giving new insight into the graphene/DNA interaction, Chem. Commun. 48 (2012) 564–566. https://doi.org/10.1039/c1cc16429e.

[73]   B. Zheng, C. Wang, C. Wu, X. Zhou, M. Lin, X. Wu, X. Xin, X. Chen, L. Xu, H. Liu, J. Zheng, J. Zhang, S. Guo, Nuclease activity and cytotoxicity enhancement of the DNA intercalators via graphene oxide, J. Phys. Chem. C. 116 (2012) 15839–15846. https://doi.org/10.1021/jp3050324.

[74]   A.M. Giuliodori, A. Brandi, S. Kotla, F. Perrozzi, R. Gunnella, L. Ottaviano, R. Spurio, A. Fabbretti, Development of a graphene oxide-based assay for the sequence-specific detection of double-stranded DNA molecules, PLoS One. 12 (2017) e0183952. https://doi.org/10.1371/journal.pone.0183952.

Materials Research Forum LLC
https://doi.org/10.21741/9781644903339-6

[75]   X. Zhao, A. Striolo, P.T. Cummings, C60 binds to and deforms nucleotides, Biophys. J. 89 (2005) 3856–3862. https://doi.org/10.1529/biophysj.105.064410.

[76]   S.K. Vittala, S.K. Saraswathi, J. Joseph, Self-Assembled Functional Fullerenes and DNA Hybrid Nanomaterials for Various Applications, Templated DNA Nanotechnol. (2019) 271–300. https://doi.org/10.1201/9780429428661-9.

[77]   F.S. Alves, J.N. Cruz, I.N. de Farias Ramos, D.L. do Nascimento Brandão, R.N. Queiroz, G.V.G.V. da Silva, G.V.G.V. da Silva, M.F. Dolabela, M.L. da Costa, A.S. Khayat, J. de Arimatéia Rodrigues do Rego, D. do Socorro Barros Brasil, Evaluation of Antimicrobial Activity and Cytotoxicity Effects of Extracts of Piper nigrum L. and Piperine, Separations. 10 (2023) 21. https://doi.org/10.3390/separations10010021.

[78]   T.H. Wang, Discerning single molecule interactions of DNA and quantum dots, Biotechnol. J. 8 (2013) 15–16. https://doi.org/10.1002/biot.201200309.

[79]   I.L. Medintz, H.T. Uyeda, E.R. Goldman, H. Mattoussi, Quantum dot bioconjugates for imaging, labelling and sensing, Nat. Mater. 4 (2005) 435–446. https://doi.org/10.1038/nmat1390.

[80]   Y. Zhang, T.H. Wang, Quantum dot enabled molecular sensing and diagnostics, Theranostics. 2 (2012) 631–654. https://doi.org/10.7150/thno.4308.

[81]   K. Li, W. Zhang, Y. Chen, Quantum dot binding to DNA: Single-molecule imaging with atomic force microscopy, Biotechnol. J. 8 (2013) 110–116. https://doi.org/10.1002/biot.201200155.

[82]   W. Chen, N.J. Turro, D.A. Tomalia, Using ethidium bromide to probe the interactions between DNA and dendrimers, Langmuir. 16 (2000) 15–19. https://doi.org/10.1021/la981429v.

[83]   H.H. Wong, N.R. Lemoine, Y. Wang, Oncolytic viruses for cancer therapy: Overcoming the obstacles, Viruses. 2 (2010) 78–106. https://doi.org/10.3390/v2010078.

[84]   C.E. Thomas, A. Ehrhardt, M.A. Kay, Progress and problems with the use of viral vectors for gene therapy, Nat. Rev. Genet. 4 (2003) 346–358. https://doi.org/10.1038/nrg1066.

[85]   D. Pantarotto, R. Singh, D. McCarthy, M. Erhardt, J.P. Briand, M. Prato, K. Kostarelos, A. Bianco, Functionalized carbon nanotubes for plasmid DNA gene delivery, Angew. Chemie - Int. Ed. 43 (2004) 5242–5246. https://doi.org/10.1002/anie.200460437.

[86]   L. Gao, L. Nie, T. Wang, Y. Qin, Z. Guo, D. Yang, X. Yan, Carbon nanotube delivery of the GFP gene into mammalian cells, ChemBioChem. 7 (2006) 239–242. https://doi.org/10.1002/cbic.200500227.

[87]   Y. Liu, D.C. Wu, W. De Zhang, X. Jiang, C. Bin He, T.S. Chung, S.H. Goh, K.W. Leong, Polyethylenimine-grafted multiwalled carbon nanotubes for secure noncovalent immobilization and efficient delivery of DNA, Angew. Chemie - Int. Ed. 44 (2005) 4782–4785. https://doi.org/10.1002/anie.200500042.

[88]   A. Nunes, N. Amsharov, C. Guo, J. Van Den Bossche, P. Santhosh, T.K. Karachalios, S.F. Nitodas, M. Burghard, K. Kostarelos, K.T. Al-Jamal, Hybrid polymer-grafted

multiwalled carbon nanotubes for in vitro gene delivery, Small. 6 (2010) 2281–2291. https://doi.org/10.1002/smll.201000864.

[89] M.P. Xiong, M. Laird Forrest, G. Ton, A. Zhao, N.M. Davies, G.S. Kwon, Poly(aspartate-g-PEI800), a polyethylenimine analogue of low toxicity and high transfection efficiency for gene delivery, Biomaterials. 28 (2007) 4889–4900. https://doi.org/10.1016/j.biomaterials.2007.07.043.

[90] K.T. Al-Jamal, F.M. Toma, A. Yilmazer, H. Ali-Boucetta, A. Nunes, M.A. Herrero, B. Tian, A. Eddaoui, W. Al-Jamal, A. Bianco, M. Prato, K. Kostarelos, Enhanced cellular internalization and gene silencing with a series of cationic dendron-multiwalled carbon nanotube:siRNA complexes, FASEB J. 24 (2010) 4354–4365. https://doi.org/10.1096/fj.09-141036.

[91] M.A. Herrero, F.M. Toma, K.T. Al-Jamal, K. Kostarelos, A. Bianco, T. Da Ros, F. Bano, L. Casalis, G. Scoles, M. Prato, Synthesis and characterization of a carbon nanotube-dendron series for efficient siRNA delivery, J. Am. Chem. Soc. 131 (2009) 9843–9848. https://doi.org/10.1021/ja903316z.

[92] N.W.S. Kam, Z. Liu, H. Dai, Functionalization of carbon nanotubes via cleavable disulfide bonds for efficient intracellular delivery of siRNA and potent gene silencing, J. Am. Chem. Soc. 127 (2005) 12492–12493. https://doi.org/10.1021/ja053962k.

[93] L. Wang, J. Shi, H. Zhang, H. Li, Y. Gao, Z. Wang, H. Wang, L. Li, C. Zhang, C. Chen, Z. Zhang, Y. Zhang, Synergistic anticancer effect of RNAi and photothermal therapy mediated by functionalized single-walled carbon nanotubes, Biomaterials. 34 (2013) 262–274. https://doi.org/10.1016/j.biomaterials.2012.09.037.

[94] R. Imani, W. Shao, S. Taherkhani, S.H. Emami, S. Prakash, S. Faghihi, Dual-functionalized graphene oxide for enhanced siRNA delivery to breast cancer cells, Colloids Surfaces B Biointerfaces. 147 (2016) 315–325. https://doi.org/10.1016/j.colsurfb.2016.08.015.

[95] N. Dinauer, S. Balthasar, C. Weber, J. Kreuter, K. Langer, H. Von Briesen, Selective targeting of antibody-conjugated nanoparticles to leukemic cells and primary T-lymphocytes, Biomaterials. 26 (2005) 5898–5906. https://doi.org/10.1016/j.biomaterials.2005.02.038.

[96] Y. Guo, H. Xu, Y. Li, F. Wu, Y. Li, Y. Bao, X. Yan, Z. Huang, P. Xu, Hyaluronic acid and Arg-Gly-Asp peptide modified Graphene oxide with dual receptor-targeting function for cancer therapy, J. Biomater. Appl. 32 (2017) 54–65. https://doi.org/10.1177/0885328217712110.

[97] L. Zhang, Z. Lu, Q. Zhao, J. Huang, H. Shen, Z. Zhang, Enhanced chemotherapy efficacy by sequential delivery of siRNA and anticancer drugs using PEI-grafted graphene oxide, Small. 7 (2011) 460–464. https://doi.org/10.1002/smll.201001522.

[98] L. Le Li, P. Wu, K. Hwang, Y. Lu, An exceptionally simple strategy for DNA-functionalized Up-conversion nanoparticles as biocompatible agents for nanoassembly, DNA delivery, and imaging, J. Am. Chem. Soc. 135 (2013) 2411–2414. https://doi.org/10.1021/ja310432u.

[99]   Z. Wang, J. Zhang, J.M. Ekman, P.J.A. Kenis, Y. Lu, DNA-mediated control of metal nanoparticle shape: One-pot synthesis and cellular uptake of highly stable and functional gold nanoflowers, Nano Lett. 10 (2010) 1886–1891. https://doi.org/10.1021/nl100675p.

[100]   A.L.C. de Souza, A. do Rego Pires, C.A.F. Moraes, C.H.C. de Matos, K.I.P. dos Santos, R.C. e Silva, S.P.C. Acuña, S. dos Santos Araújo, Chromatographic methods for separation and identification of bioactive compounds, in: J.N. Cruz (Ed.), Drug Discov. Des. Using Nat. Prod., Springer Nature Switzerland, Cham, 2023: pp. 153–176. https://doi.org/10.1007/978-3-031-35205-8_6.

[101]   A. Banerjee, T. Pons, N. Lequeux, B. Dubertret, Quantum dots--DNA bioconjugates: synthesis to applications, Interface Focus. 6 (2016) 20160064.

[102]   H.L. Jung, Z. Wang, J. Liu, Y. Lu, Highly sensitive and selective colorimetric sensors for uranyl (UO 22+): Development and comparison of labeled and label-free DNAzyme-gold nanoparticle systems, J. Am. Chem. Soc. 130 (2008) 14217–14226. https://doi.org/10.1021/ja803607z.

[103]   M.V. Yigit, D. Mazumdar, H.K. Kim, J.H. Lee, B. Odintsov, Y. Lu, Smart "turn-on" magnetic resonance contrast agents based on aptamer-functionalized superparamagnetic iron oxide nanoparticles, ChemBioChem. 8 (2007) 1675–1678. https://doi.org/10.1002/cbic.200700323.

[104]   J. Liu, D. Mazumdar, Y. Lu, A simple and sensitive "dipstick" test in serum based on lateral flow separation of aptamer-linked nanostructures, Angew. Chemie - Int. Ed. 45 (2006) 7955–7959. https://doi.org/10.1002/anie.200603106.

[105]   M. Rossetti, S. Ranallo, A. Idili, G. Palleschi, A. Porchetta, F. Ricci, Allosteric DNA nanoswitches for controlled release of a molecular cargo triggered by biological inputs, Chem. Sci. 8 (2017) 914–920. https://doi.org/10.1039/c6sc03404g.

[106]   L. Le Li, M. Xie, J. Wang, X. Li, C. Wang, Q. Yuan, D.W. Pang, Y. Lu, W. Tan, A vitamin-responsive mesoporous nanocarrier with DNA aptamer-mediated cell targeting, Chem. Commun. 49 (2013) 5823–5825. https://doi.org/10.1039/c3cc41072b.

[107]   J. Willem de Vries, S. Schnichels, J. Hurst, L. Strudel, A. Gruszka, M. Kwak, K.U. Bartz-Schmidt, M.S. Spitzer, A. Herrmann, DNA nanoparticles for ophthalmic drug delivery, Biomaterials. 157 (2018) 98–106. https://doi.org/10.1016/j.biomaterials.2017.11.046.

[108]   L. Abarca-Cabrera, P. Fraga-García, S. Berensmeier, Bio-nano interactions: binding proteins, polysaccharides, lipids and nucleic acids onto magnetic nanoparticles, Biomater. Res. 25 (2021) 1–18. https://doi.org/10.1186/s40824-021-00212-y.

[109]   I. Budin, N.K. Devaraj, Membrane assembly driven by a biomimetic coupling reaction, J. Am. Chem. Soc. 134 (2012) 751–753. https://doi.org/10.1021/ja2076873.

[110]   P. Bohley, Molecular Cell Biology, Macmillan, 1987. https://doi.org/10.1016/0307-4412(87)90114-2.

[111]   C. Contini, M. Schneemilch, S. Gaisford, N. Quirke, Nanoparticle--membrane interactions, J. Exp. Nanosci. 13 (2018) 62–81.

[112] C. Auría-Soro, T. Nesma, P. Juanes-Velasco, A. Landeira-Viñuela, H. Fidalgo-Gomez, V. Acebes-Fernandez, R. Gongora, M.J.A. Parra, R. Manzano-Roman, M. Fuentes, Interactions of nanoparticles and biosystems: Microenvironment of nanoparticles and biomolecules in nanomedicine, Nanomaterials. 9 (2019) 1365. https://doi.org/10.3390/nano9101365.

[113] Y.H. Lahir, P. Avti, Nanomaterials and Their Interactive Behavior with Biomolecules, Cells and Tissues, Bentham Science Publishers, 2020. https://doi.org/10.2174/97898114617811200101.

[114] M.F.H. Sarfraz, M. Zubair, B. Aslam, A. Ashraf, M.H. Siddique, S. Hayat, J.N. Cruz, S. Muzammil, M. Khurshid, M.F.H. Sarfraz, A. Hashem, T.M. Dawoud, G.D. Avila-Quezada, E.F. Abd_Allah, Comparative analysis of phyto-fabricated chitosan, copper oxide, and chitosan-based CuO nanoparticles: antibacterial potential against Acinetobacter baumannii isolates and anticancer activity against HepG2 cell lines, Front. Microbiol. 14 (2023) 1188743. https://doi.org/10.3389/fmicb.2023.1188743.

[115] M.R. Rasch, E. Rossinyol, J.L. Hueso, B.W. Goodfellow, J. Arbiol, B.A. Korgel, Hydrophobic gold nanoparticle self-assembly with phosphatidylcholine lipid: Membrane-loaded and janus vesicles, Nano Lett. 10 (2010) 3733–3739. https://doi.org/10.1021/nl102387n.

[116] G. Von White, Y. Chen, J. Roder-Hanna, G.D. Bothun, C.L. Kitchens, Structural and thermal analysis of lipid vesicles encapsulating hydrophobic gold nanoparticles, ACS Nano. 6 (2012) 4678–4685. https://doi.org/10.1021/nn2042016.

[117] S. Tatur, M. MacCarini, R. Barker, A. Nelson, G. Fragneto, Effect of functionalized gold nanoparticles on floating lipid bilayers, Langmuir. 29 (2013) 6606–6614. https://doi.org/10.1021/la401074y.

[118] J. Park, W. Lu, Interaction of nanoparticles with lipid layers, Phys. Rev. E - Stat. Nonlinear, Soft Matter Phys. 80 (2009) 941–944. https://doi.org/10.1103/PhysRevE.80.021607.

[119] G.D. Bothun, Y. Chen, A. Bose, Controlled release from membrane-decorated magnetoliposomes via electromagnetic heating, 20th Annu. Meet. North Am. Membr. Soc. 11th Int. Conf. Inorg. Membr. 2010, NAMS/ICIM 2010. 4 (2010) 167–168.

[120] B. Collin, E. Oostveen, O. V. Tsyusko, J.M. Unrine, Influence of natural organic matter and surface charge on the toxicity and bioaccumulation of functionalized ceria nanoparticles in Caenorhabditis elegans, Environ. Sci. Technol. 48 (2014) 1280–1289. https://doi.org/10.1021/es404503c.

[121] H. Li, H. Sun, W. Qi, M. Xu, L. Wu, Onionlike hybrid assemblies based on surfactant-encapsulated polyoxometalates, Angew. Chemie - Int. Ed. 46 (2007) 1300–1303. https://doi.org/10.1002/anie.200603934.

[122] M. Assis, M.O. Gonçalves, C.C. de Foggi, M. Burck, S. dos Passos Ramos, L.O. Libero, A.R.C. Braga, E. Longo, C.P. de Sousa, Applications of (nano)encapsulated natural products by physical and chemical methods, in: J.N. Cruz (Ed.), Drug Discov. Des. Using Nat. Prod., Springer Nature Switzerland, Cham, 2023: pp. 323–374. https://doi.org/10.1007/978-3-031-35205-8_11.

[123] H. Nabika, Y. Inomata, E. Itoh, K. Unoura, Activity of Keggin and Dawson polyoxometalates toward model cell membrane, RSC Adv. 3 (2013) 21271–21274. https://doi.org/10.1039/c3ra41522h.

[124] B. Jing, M. Hutin, E. Connor, L. Cronin, Y. Zhu, Polyoxometalate macroion induced phase and morphology instability of lipid membrane, Chem. Sci. 4 (2013) 3818–3826. https://doi.org/10.1039/c3sc51404h.

[125] B. Wang, L. Zhang, C.B. Sung, S. Granick, Nanoparticle-induced surface reconstruction of phospholipid membranes, Proc. Natl. Acad. Sci. U. S. A. 105 (2008) 18171–18175. https://doi.org/10.1073/pnas.0807296105.

[126] S. Hong, P.R. Leroueil, E.K. Janus, J.L. Peters, M.M. Kober, M.T. Islam, B.G. Orr, J.R. Baker, M.M. Banaszak Holl, Interaction of polycationic polymers with supported lipid bilayers and cells: Nanoscale hole formation and enhanced membrane permeability, Bioconjug. Chem. 17 (2006) 728–734. https://doi.org/10.1021/bc060077y.

[127] P.R. Leroueil, S.A. Berry, K. Duthie, G. Han, V.M. Rotello, D.Q. McNerny, J.R. Baker, B.G. Orr, M.M.B. Holl, Wide varieties of cationic nanoparticles induce defects in supported lipid bilayers, Nano Lett. 8 (2008) 420–424. https://doi.org/10.1021/nl0722929.

[128] J. Lin, H. Zhang, Z. Chen, Y. Zheng, Penetration of lipid membranes by gold nanoparticles: Insights into cellular uptake, cytotoxicity, and their relationship, ACS Nano. 4 (2010) 5421–5429. https://doi.org/10.1021/nn1010792.

[129] R.R. Arvizo, O.R. Miranda, M.A. Thompson, C.M. Pabelick, R. Bhattacharya, J. David Robertson, V.M. Rotello, Y.S. Prakash, P. Mukherjee, Effect of nanoparticle surface charge at the plasma membrane and beyond, Nano Lett. 10 (2010) 2543–2548. https://doi.org/10.1021/nl101140t.

[130] V. V. Ginzburg, S. Balijepalli, Modeling the thermodynamics of the interaction of nanoparticles with cell membranes, Nano Lett. 7 (2007) 3716–3722. https://doi.org/10.1021/nl072053l.

[131] G. Gopalakrishnan, C. Danelon, P. Izewska, M. Prummer, P.Y. Bolinger, I. Geissbühler, D. Demurtas, J. Dubochet, H. Vogel, Multifunctional lipid/quantum dot hybrid nanocontainers for controlled targeting of live cells, Angew. Chemie - Int. Ed. 45 (2006) 5478–5483. https://doi.org/10.1002/anie.200600545.

[132] P.J. Sintic, E. Wenbo, Z. Ou, J. Shao, J.A. McDonald, Z.L. Cai, K.M. Kadish, M.J. Crossley, J.R. Reimers, Control of the site and potential of reduction and oxidation processes in π-expanded quinoxalinoporphyrins (Physical Chemistry Chemical Physics (2008) 10, (268-280) DOI: 10.1039/b711320j), Phys. Chem. Chem. Phys. 10 (2008) 7328. https://doi.org/10.1039/b820726g.

[133] G.D. Bothun, Hydrophobic silver nanoparticles trapped in lipid bilayers: Size distribution, bilayer phase behavior, and optical properties, J. Nanobiotechnology. 6 (2008) 1–10. https://doi.org/10.1186/1477-3155-6-13.

[134] M. Breidenich, R.R. Netz, R. Lipowsky, The influence of non-anchored polymers on the curvature of vesicles, Mol. Phys. 103 (2005) 3169–3183. https://doi.org/10.1080/00268970500270484.

[135] M.R.R. De Planque, S. Aghdaei, T. Roose, H. Morgan, Electrophysiological characterization of membrane disruption by nanoparticles, ACS Nano. 5 (2011) 3599–3606. https://doi.org/10.1021/nn103320j.

[136] P.R. Leroueil, S. Hong, A. Mecke, J.R. Baker, B.G. Orr, M.M.B. Holl, Nanoparticle interaction with biological membranes: Does nanotechnology present a janus face?, Acc. Chem. Res. 40 (2007) 335–342. https://doi.org/10.1021/ar600012y.

[137] J. Chen, J.A. Hessler, K. Putchakayala, B.K. Panama, D.P. Khan, S. Hong, D.G. Mullen, S.C. DiMaggio, A. Som, G.N. Tew, A.N. Lopatin, J.R. Baker, M.M.B. Holl, B.G. Orr, Cationic nanoparticles induce nanoscale disruption in living cell plasma membranes, J. Phys. Chem. B. 113 (2009) 11179–11185. https://doi.org/10.1021/jp9033936.

[138] V. A, C. RD, E. JD, S. P, H. GW, A. M, D. AG, K. T, P. NH, P. JH, S. RL, S. PH, Essentials of Glycobiology [Internet], Cold Spring Harb. (2015) 823. https://pubmed.ncbi.nlm.nih.gov/27010055/.

[139] F. Assa, H. Jafarizadeh-Malmiri, H. Ajamein, N. Anarjan, H. Vaghari, Z. Sayyar, A. Berenjian, A biotechnological perspective on the application of iron oxide nanoparticles, Nano Res. 9 (2016) 2203–2225. https://doi.org/10.1007/s12274-016-1131-9.

[140] L. Skálová, Becker, WM, Kleinsmith, LJ, Hardin, J.: The World of the Cell., (2003).

[141] C.H. Veloso, L.O. Filippov, I. V. Filippova, S. Ouvrard, A.C. Araujo, Adsorption of polymers onto iron oxides: Equilibrium isotherms, J. Mater. Res. Technol. 9 (2020) 779–788. https://doi.org/10.1016/j.jmrt.2019.11.018.

[142] M. Zarei, J. Aalaie, Profiling of nanoparticle–protein interactions by electrophoresis techniques, Anal. Bioanal. Chem. 411 (2019) 79–96. https://doi.org/10.1007/s00216-018-1401-3.

[143] P.J. Russell, A Molecular Approach, 2, Cell. 2nd Ed. Sunderland, MA Sinauer Assoc. (2010) 1–5.

[144] B. Uhl, S. Hirn, R. Immler, K. Mildner, L. Möckl, M. Sperandio, C. Bräuchle, C.A. Reichel, D. Zeuschner, F. Krombach, The Endothelial Glycocalyx Controls Interactions of Quantum Dots with the Endothelium and Their Translocation across the Blood-Tissue Border, ACS Nano. 11 (2017) 1498–1508. https://doi.org/10.1021/acsnano.6b06812.

[145] L. Möckl, S. Hirn, A.A. Torrano, B. Uhl, C. Bräuchle, F. Krombach, The glycocalyx regulates the uptake of nanoparticles by human endothelial cells in vitro, Nanomedicine. 12 (2017) 207–217. https://doi.org/10.2217/nnm-2016-0332.

[146] R. Gromnicova, M. Kaya, I.A. Romero, P. Williams, S. Satchell, B. Sharrack, D. Male, Transport of gold nanoparticles by vascular endothelium from different human tissues, PLoS One. 11 (2016) e0161610. https://doi.org/10.1371/journal.pone.0161610.

[147] M.L. Huang, K. Godula, Nanoscale materials for probing the biological functions of the glycocalyx, Glycobiology. 26 (2016) 797–803. https://doi.org/10.1093/glycob/cww022.

[148] H. Bouwmeester, M. van der Zande, M.A. Jepson, Effects of food-borne nanomaterials on gastrointestinal tissues and microbiota, Wiley Interdiscip. Rev. Nanomedicine Nanobiotechnology. 10 (2018) e1481. https://doi.org/10.1002/wnan.1481.

[149] J. Šebestík, M. Reiniš, J. Ježek, Biomedical applications of peptide-, glyco- and glycopeptide dendrimers, and analogous dendrimeric structures, Springer Science \& Business Media, 2012. https://doi.org/10.1007/978-3-7091-1206-9.

[150] M. Marradi, F. Chiodo, I. García, S. Penadés, Glyconanoparticles as multifunctional and multimodal carbohydrate systems, Chem. Soc. Rev. 42 (2013) 4728–4745. https://doi.org/10.1039/c2cs35420a.

[151] A.J. Reynolds, A.H. Haines, D.A. Russell, Gold glyconanoparticles for mimics and measurement of metal Ion-mediated carbohydrate - carbohydrate interactions, Langmuir. 22 (2006) 1156–1163. https://doi.org/10.1021/la052261y.

[152] Y.J. Chen, S.H. Chen, Y.Y. Chien, Y.W. Chang, H.K. Liao, C.Y. Chang, M.D. Jan, K.T. Wang, C.C. Lin, Carbohydrate-encapsulated gold nanoparticles for rapid target-protein identification and binding-epitope mapping, ChemBioChem. 6 (2005) 1169–1173. https://doi.org/10.1002/cbic.200500023.

[153] N. Nagahori, S.I. Nishimura, Direct and efficient monitoring of glycosyltransferase reactions on gold colloidal nanoparticles by using mass spectrometry, Chem. - A Eur. J. 12 (2006) 6478–6485. https://doi.org/10.1002/chem.200501267.

[154] G. Elvira, I. García, M. Benito, J. Gallo, M. Desco, S. Penadés, J.A. Garcia-Sanz, A. Silva, Live Imaging of Mouse Endogenous Neural Progenitors Migrating in Response to an Induced Tumor, PLoS One. 7 (2012). https://doi.org/10.1371/journal.pone.0044466.

[155] A.L. Parry, N.A. Clemson, J. Ellis, S.S.R. Bernhard, B.G. Davis, N.R. Cameron, "Multicopy multivalent" glycopolymer-stabilized gold nanoparticles as potential synthetic cancer vaccines, J. Am. Chem. Soc. 135 (2013) 9362–9365. https://doi.org/10.1021/ja4046857.

[156] A. Aykaç, M.C. Martos-Maldonado, J.M. Casas-Solvas, I. Quesada-Soriano, F. García-Maroto, L. García-Fuentes, A. Vargas-Berenguel, B-Cyclodextrin-Bearing Gold Glyconanoparticles for the Development of Site Specific Drug Delivery Systems, Langmuir. 30 (2014) 234–242. https://doi.org/10.1021/la403454p.

[157] M.J. Marín, A. Rashid, M. Rejzek, S.A. Fairhurst, S.A. Wharton, S.R. Martin, J.W. McCauley, T. Wileman, R.A. Field, D.A. Russell, Glyconanoparticles for the plasmonic detection and discrimination between human and avian influenza virus, Org. Biomol. Chem. 11 (2013) 7101–7107. https://doi.org/10.1039/c3ob41703d.

[158] S. Liu, X. Guo, NPG Asia Mater. 4, e23 (2012), (2012).

[159] S. Kruss, A.J. Hilmer, J. Zhang, N.F. Reuel, B. Mu, M.S. Strano, Carbon nanotubes as optical biomedical sensors, Adv. Drug Deliv. Rev. 65 (2013) 1933–1950. https://doi.org/10.1016/j.addr.2013.07.015.

[160] D.R. Kauffman, A. Star, Electronically monitoring biological interactions with carbon nanotube field-effect transistors, Chem. Soc. Rev. 37 (2008) 1197–1206. https://doi.org/10.1039/b709567h.

[161] Y. Chen, A. Star, S. Vidal, Sweet carbon nanostructures: Carbohydrate conjugates with carbon nanotubes and graphene and their applications, Chem. Soc. Rev. 42 (2013) 4532–4542. https://doi.org/10.1039/c2cs35396b.

Materials Research Forum LLC
https://doi.org/10.21741/9781644903339-6

[162] A.M. Münzer, Z.P. Michael, A. Star, Carbon nanotubes for the label-free detection of biomarkers, ACS Nano. 7 (2013) 7448–7453. https://doi.org/10.1021/nn404544e.

[163] H. Vedala, Y. Chen, S. Cecioni, A. Imberty, S. Vidal, A. Star, Nanoelectronic detection of lectin-carbohydrate interactions using carbon nanotubes, Nano Lett. 11 (2011) 170–175. https://doi.org/10.1021/nl103286k.

[164] N.F. Reuel, J.H. Ahn, J.H. Kim, J. Zhang, A.A. Boghossian, L.K. Mahal, M.S. Strano, Transduction of glycan-lectin binding using near-infrared fluorescent single-walled carbon nanotubes for glycan profiling, J. Am. Chem. Soc. 133 (2011) 17923–17933. https://doi.org/10.1021/ja2074938.

[165] N.F. Reuel, B. Grassbaugh, S. Kruss, J.Z. Mundy, C. Opel, A.O. Ogunniyi, K. Egodage, R. Wahl, B. Helk, J. Zhang, Z.I. Kalcioglu, K. Tvrdy, D.O. Bellisario, B. Mu, S.S. Blake, K.J. Van Vliet, J.C. Love, K.D. Wittrup, M.S. Strano, Emergent properties of nanosensor arrays: Applications for monitoring IgG affinity distributions, weakly affined hypermannosylation, and colony selection for biomanufacturing, ACS Nano. 7 (2013) 7472–7482. https://doi.org/10.1021/nn403215e.

[166] M.T. Meredith, S.D. Minteer, Biofuel cells: Enhanced enzymatic bioelectrocatalysis, Annu. Rev. Anal. Chem. 5 (2012) 157–179. https://doi.org/10.1146/annurev-anchem-062011-143049.

[167] C. Liu, S. Alwarappan, Z. Chen, X. Kong, C.Z. Li, Membraneless enzymatic biofuel cells based on graphene nanosheets, Biosens. Bioelectron. 25 (2010) 1829–1833. https://doi.org/10.1016/j.bios.2009.12.012.

[168] K.P. Prasad, Y. Chen, P. Chen, Three-dimensional graphene-carbon nanotube hybrid for high-performance enzymatic biofuel cells, ACS Appl. Mater. Interfaces. 6 (2014) 3387–3393. https://doi.org/10.1021/am405432b.

[169] H. Hong, Y. Zhang, J.W. Engle, T.R. Nayak, C.P. Theuer, R.J. Nickles, T.E. Barnhart, W. Cai, In vivo targeting and positron emission tomography imaging of tumor vasculature with 66Ga-labeled nano-graphene, Biomaterials. 33 (2012) 4147–4156. https://doi.org/10.1016/j.biomaterials.2012.02.031.

[170] A. Star, D.W. Steuerman, J.R. Heath, J.F. Stoddart, Starched carbon nanotubes, Angew. Chemie - Int. Ed. 41 (2002) 2508–2512. https://doi.org/10.1002/1521-3773(20020715)41:14<2508::AID-ANIE2508>3.0.CO;2-A.

[171] N.M. Iverson, P.W. Barone, M. Shandell, L.J. Trudel, S. Sen, F. Sen, V. Ivanov, E. Atolia, E. Farias, T.P. McNicholas, N. Reuel, N.M.A. Parry, G.N. Wogan, M.S. Strano, In vivo biosensing via tissue-localizable near-infrared-fluorescent single-walled carbon nanotubes, Nat. Nanotechnol. 8 (2013) 873–880. https://doi.org/10.1038/nnano.2013.222.

[172] J. Luczkowiak, A. Muñoz, M. Sánchez-Navarro, R. Ribeiro-Viana, A. Ginieis, B.M. Illescas, N. Martín, R. Delgado, J. Rojo, Glycofullerenes inhibit viral infection, Biomacromolecules. 14 (2013) 431–437. https://doi.org/10.1021/bm3016658.

[173] J. Venkatesan, Z.J. Qian, B. Ryu, N. Ashok Kumar, S.K. Kim, Preparation and characterization of carbon nanotube-grafted-chitosan - Natural hydroxyapatite composite for

bone tissue engineering, Carbohydr. Polym. 83 (2011) 569–577.
https://doi.org/10.1016/j.carbpol.2010.08.019.

[174] S. Mizrahy, D. Peer, Polysaccharides as building blocks for nanotherapeutics, Chem. Soc. Rev. 41 (2012) 2623–2640. https://doi.org/10.1039/c1cs15239d.

[175] J.H. Park, G. Saravanakumar, K. Kim, I.C. Kwon, Targeted delivery of low molecular drugs using chitosan and its derivatives, Adv. Drug Deliv. Rev. 62 (2010) 28–41. https://doi.org/10.1016/j.addr.2009.10.003.

[176] F.Q. Hu, P. Meng, Y.Q. Dai, Y.Z. Du, J. You, X.H. Wei, H. Yuan, PEGylated chitosan-based polymer micelle as an intracellular delivery carrier for anti-tumor targeting therapy, Eur. J. Pharm. Biopharm. 70 (2008) 749–757. https://doi.org/10.1016/j.ejpb.2008.06.015.

[177] S. Mao, W. Sun, T. Kissel, Chitosan-based formulations for delivery of DNA and siRNA, Adv. Drug Deliv. Rev. 62 (2010) 12–27. https://doi.org/10.1016/j.addr.2009.08.004.

[178] X. Liu, K.A. Howard, M. Dong, M. Andersen, U.L. Rahbek, M.G. Johnsen, O.C. Hansen, F. Besenbacher, J. Kjems, The influence of polymeric properties on chitosan/siRNA nanoparticle formulation and gene silencing, Biomaterials. 28 (2007) 1280–1288. https://doi.org/10.1016/j.biomaterials.2006.11.004.

[179] W.F. Lai, M.C.M. Lin, Nucleic acid delivery with chitosan and its derivatives, J. Control. Release. 134 (2009) 158–168. https://doi.org/10.1016/j.jconrel.2008.11.021.

[180] J. Nguyen, F.C. Szoka, Nucleic acid delivery: The missing pieces of the puzzle?, Acc. Chem. Res. 45 (2012) 1153–1162. https://doi.org/10.1021/ar3000162.

[181] S. Trivedi, A. Paunikar, N. Raut, V. Belgamwar, Photodynamic therapy for cancer treatment, Photophysics Nanophysics Ther. 3 (2022) 89–114. https://doi.org/10.1016/B978-0-323-89839-3.00010-5.

[182] V. Simon, C. Devaux, A. Darmon, T. Donnet, E. Thiénot, M. Germain, J. Honnorat, A. Duval, A. Pottier, E. Borghi, L. Levy, J. Marill, Pp IX silica nanoparticles demonstrate differential interactions with in vitro tumor cell lines and in vivo mouse models of human cancers, Photochem. Photobiol. 86 (2010) 213–222. https://doi.org/10.1111/j.1751-1097.2009.00620.x.

Advances in Healthcare and Nanoparticle Toxicology
Materials Research Foundations 171 (2024) 191-216

Materials Research Forum LLC
https://doi.org/10.21741/9781644903339-7

Chapter 7

Interactions of Nanoparticles with Lipid and Cell Membranes

Rohit Awale[1], Nilesh Kulkarni[1], Saurabh Khadse[1*]

R.C. Patel Institute of Pharmaceutical Education & Research, Karvand Naka Shirpur, Dist. Dhule (MS) 425405 India

* khadse_s@rediffmail.com

Abstract

The interactions between nanoparticles and lipid or cell membranes are of paramount importance in the realms of nanomedicine and nanotoxicology. Nanoparticles, with their unique physicochemical properties, exhibit dynamic interactions upon contact with biological membranes. These interactions include adsorption, penetration, and potential disruption of lipid bilayers. Such interactions play a crucial role in drug delivery systems, enabling precise targeting and controlled release of therapeutic agents. However, these interactions also raise concerns about nanoparticle toxicity due to potential membrane damage. A deeper understanding of these intricate processes is essential for harnessing the benefits of nanoparticles while ensuring their safety in biomedical applications.

Keywords

Nanoparticles, Cell Membranes, Nanoparticle Interactions, Lipid Bilayers, Cellular Uptake, Permeability, Protein Interactions, Membrane Structure

Contents

1. Introduction

Nanotechnology has heralded a groundbreaking era in science, opening new avenues for targeted drug delivery, advanced diagnostics, and innovative therapeutic strategies [1]. At the forefront of these advancements lie nanoparticles, a minuscule entity with exceptional properties that have captured the imagination of researchers and medical professionals alike. Among the paramount considerations in harnessing their potential is understanding how nanoparticles interact with lipid and cell membranes, the gatekeepers governing cellular entry and response [2].

The lipid bilayer, comprising the cell membrane, forms an essential boundary between the intracellular milieu and the extracellular environment. Its selective permeability is vital for maintaining cellular homeostasis, regulating signal transduction, and safeguarding the integrity of the cell. Cellular uptake of nanoparticles refers to the process by which these nanoparticles are internalized by cells. Nanoparticles are very small particles, typically ranging from 1 to 100 nanometres in size. When nanoparticles interact with these delicate membranes, they can trigger a myriad of responses, shaping cellular uptake, bio-distribution, and therapeutic efficacy.

This chapter embarks on a comprehensive exploration of the interactions that nanoparticles perform with lipid and cell membranes. We will delve into the various factors that influence their interactions, such as nanoparticle size, surface charge, shape, and functionalization [3]. Additionally, we will discuss the role of membrane composition, fluidity, curvature and impact of biological environment in governing the behavior of nanoparticles [4, 5]. The engineering of nanomedicines has substantial difficulties due to the possibility that a nanoparticle's physicochemical characteristics may alter considerably after being exposed to biological environment [6]. By amalgamating knowledge from multiple disciplines, we aim to construct a holistic understanding of the diverse scenarios in which nanoparticles engage with cell membranes. This chapter will also shed light on the potential therapeutic applications of nanoparticle-membrane interactions. Harnessing these interactions can enable targeted drug delivery systems, enhance cellular uptake, and pave the way for innovative therapies in various diseases, including cancer, neurodegenerative disorders, and infectious diseases.

In this multidisciplinary field, researchers seek to unravel the underlying mechanisms that govern the interactions of nanoparticles with lipid and cell membranes. A thorough comprehension of these processes promises to revolutionize drug delivery systems, optimize nanomedicine strategies, and minimize potential toxicities. Nanoparticles must get through the cell plasma membrane in order to enter cells. Understanding the underlying mechanisms of nanoparticles is crucial because they govern their function, intracellular destiny, and biological response [7, 8, 9].

2. Cellular uptake of nanoparticles:

Numerous diverse pathways exist for nanoparticles to traverse a cell's plasma membrane when they are exposed to cells, whether *in-vivo* or *in-vitro*. These pathways can be broadly classified into two main categories: (i) uptake pathways based on endocytosis, and (ii) direct entry of nanoparticles into the cells [6].

(i) Endocytosis based uptake pathways:

This is one of the primary mechanisms for cellular uptake of nanoparticles. Endocytosis is a process in which cells engulf extracellular material by forming vesicles derived from the cell membrane. There are several endocytosis-based pathways that nanoparticles can exploit to travel

Advances in Healthcare and Nanoparticle Toxicology   Materials Research Forum LLC
Materials Research Foundations 171 (2024) 191-216   https://doi.org/10.21741/9781644903339-7

into cells. The specific pathway taken by nanoparticles largely depends on their physicochemical properties, surface modifications, and the type of cells they interact with. Some of the key endocytosis-based pathways for nanoparticle cellular uptake are as follows [10, 11].

These pathways can be categorized into five distinct classes based on their underlying mechanisms [10]:

- Clathrin Mediated Endocytosis
- Caveolin Mediated Endocytosis
- Clathrin and Caveolin Independent Endocytosis
- Phagocytosis
- Macropinocytosis

Numerous factors, such as nanoparticle dimensions, morphology, surface electrical charge, surface modifications, and the specific cell type, can impact the efficiency and route of cellular internalization. Researchers often modify these properties to enhance cellular uptake for specific applications, such as targeted drug delivery to specific tissues or organs.

Understanding the mechanisms of cellular uptake is crucial for designing effective nanoparticle-based therapies and minimizing potential adverse effects on cells and tissues. However, it's worth noting that nanoparticle cellular uptake is a complex and active area of research, and scientists continue to explore new approaches to improve the efficiency and specificity of nanoparticle uptake for various biomedical applications.

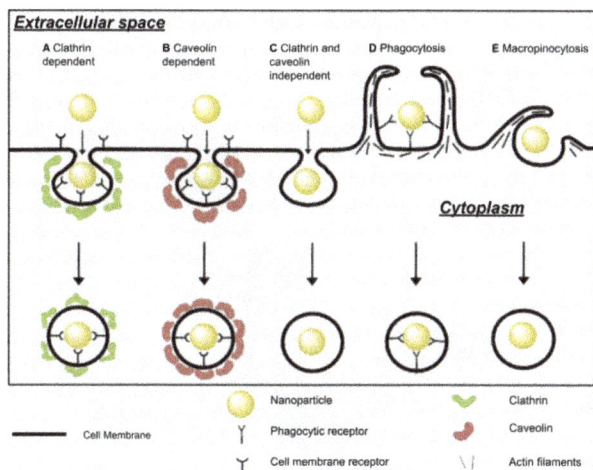

Figure 1.: Endocytosis based uptake pathways like a. Clathrin mediated endocytosis, b. Caveolin mediated endocytosis, c. Clathrin and caveolin independent endocytosis, d. Phagocytosis, e. Macropinocytosis

Advances in Healthcare and Nanoparticle Toxicology   Materials Research Forum LLC
Materials Research Foundations 171 (2024) 191-216   https://doi.org/10.21741/9781644903339-7

Clathrin Mediated Endocytosis (CME):

Clathrin mediated endocytosis (CME) is a highly regulated and well-studied process used by cells to internalize extracellular molecules, including nutrients, signalling molecules, and various types of nanoparticles. It plays a vital role in maintaining cell homeostasis, regulating cell surface receptor levels, and transporting materials from the cell surface to the intracellular compartments. Clathrin mediated endocytosis entails the creation of vesicles coated with clathrin molecules, which detach from the cell's outer membrane and carry their contents into the interior of the cell.

Elaborative Clathrin mediated endocytosis process is as follows:

The process of Clathrin mediated endocytosis begins with the binding of ligands (e.g., nanoparticles or other extracellular molecules) to specific cell surface receptors. These receptors recognize and interact with the ligands, leading to the activation of the endocytic process. Once the cell surface receptors are activated, a cluster of proteins called adaptors gathers at the inner surface of the plasma membrane. These adaptors connect to the cytoplasmic tails of the activated receptors and recruit clathrin, a protein that plays a key role in the formation of vesicles. Clathrin molecules assemble into a lattice-like structure, forming a "clathrin coat" on the inner side of the plasma membrane. As clathrin molecules continue to accumulate, they deform the plasma membrane, causing it to invaginate inward. This inward curvature results in the formation of a clathrin-coated pit on the cell surface, with the ligand-receptor complexes sequestered within it. The clathrin-coated pit deepens and eventually pinches off from the plasma membrane to form a clathrin-coated vesicle. With the aid of conformational changes from the GTPase enzyme dynamin, vesicles are squeezed off the membrane [12]. This vesicle contains the bound ligand-receptor complexes, effectively internalizing them within the cell. These vesicles scission off the membrane and normally go to endosomes with the aid of intracellular actin filaments [13, 14]. Once the Clathrin coated vesicle has formed, it must shed its clathrin coat to fuse with other intracellular compartments. This process, is called uncoating, involves the removal of clathrin from the vesicle by a variety of accessory proteins. After uncoating, the vesicle becomes uncoated and ready to fuse with its target compartment. The uncoated vesicle then undergoes fusion with early endosomes, which are a type of intracellular compartment. This fusion allows the cargo, including nanoparticles, to be released into the early endosome. Within the endosomal network, the cargo can be sorted into different intracellular compartments. Depending on the fate of the cargo, it may be directed for recycling back to the cell surface, targeted for degradation in lysosomes, or transported to specific intracellular organelles.

Overall, clathrin mediated endocytosis is a highly regulated and dynamic process essential for cellular uptake. Nanoparticles with diameters between 100 and 500 nm are entrapped in intracellular vesicles through the clathrin mediated endocytosis mechanism [15]. Its ability to specifically transport ligand-bound receptors and other cargo from the cell surface to intracellular compartments makes it an attractive pathway for targeted drug delivery and other biomedical applications. Researchers continue to investigate and manipulate this process to design efficient and specific nanoparticle-based therapies for various diseases.

Caveolae Mediated Endocytosis:

It is a specialized form of endocytosis used by cells to internalize extracellular molecules. This process involves the formation of small invaginations in the plasma membrane called caveolae, which play a role in transporting specific cargo from the cell surface to the cell interior. Caveolae are flask-shaped membrane structures with diameters of 50-100 nm that are stabilized by a caveolin protein-based coat [16]. Caveolae are rich in cholesterol and glycosphingolipids, and they have unique properties that distinguish them from other endocytic pathways. Caveolin based vesicles frequently go to the Golgi apparatus and the endoplasmic reticulum within the cell [17].

Detailed explanation of caveolae-mediated endocytosis is as follows:

The process of caveolae-mediated endocytosis is initiated when specific ligands (e.g., nanoparticles or other molecules) bind to their corresponding cell surface receptors associated with caveolae. This binding triggers the activation of signalling cascades that lead to the recruitment and clustering of the receptor-ligand complexes in the caveolar membrane. As the receptor-ligand complexes cluster in the caveolar membrane, the membrane begins to invaginate, resulting in the formation of a caveolar vesicle. This vesicle, known as a caveosome, contains the ligand-receptor complexes, effectively internalizing them within the cell. The caveosome is then internalized into the cell's interior through the scission of the caveolar neck from the plasma membrane. Once inside the cell, caveosomes can undergo maturation and fuse with other endocytic compartments, such as early endosomes and recycling endosomes. Within the endosomal network, the cargo in the caveosome can be sorted into different intracellular compartments. This sorting process dictates the fate of the cargo, which may include recycling back to the cell surface, delivery to specific intracellular organelles, or targeting for degradation in lysosomes.

Caveolae mediated endocytosis is considered to be an efficient and highly regulated endocytic pathway. It is particularly relevant for certain cell types, such as endothelial cells and adipocytes, which are enriched in caveolae and caveolins. This pathway is also associated with lipid raft-mediated signalling and plays a role in cellular processes like transcytosis (transporting substances across the endothelial barrier) and regulation of membrane tension.

Clathrin and Caveolin Independent Endocytosis:

Clathrin and Caveolin Independent Endocytosis are also known as Clathrin-independent/GPI-anchored protein-enriched endocytic compartment (CLIC/GEEC). It is a less characterized and relatively recently discovered endocytic pathway that operates independently of the Clathrin coated vesicles. It involves the internalization of nanoparticles and certain proteins, into specialized vesicles i.e., lipid rafts that are rich in glycosylphosphatidylinositol (GPI)-anchored proteins [18]. Furthermore, specific ligands, like cholera toxin B and SIV40, attach to lipid-enriched areas of the cell's plasma membrane, initiating endocytosis through lipid raft-mediated processes [19]. The pathway is known by different names, and CLIC and GEEC are sometimes used interchangeably to refer to this endocytic route.

Virus-like particles and various nanoparticle varieties can permeate the cellular plasma membrane and gain access to the cell interior without depending on the clathrin and caveolin-mediated routes (refer to Figure 1). An alternative proposed pathway for this clathrin and caveolin-independent cellular entry involves lipid rafts specialized domains within the plasma membrane abundant in cholesterol and sphingolipids, which initiate endocytic processes when appropriately stimulated [18]. Lipid raft driven endocytosis is a common mechanism observed in immunological contexts, where lymphocytes uptake and process interleukins [16]. Furthermore, particular ligands like cholera toxin B and SIV40 adhere to lipid-enriched regions on the cell's plasma membrane that engage in lipid raft-driven endocytosis [19]. Recent research has indicated the involvement of a lipid raft-based endocytosis route in the internalization of nanoparticles that have been engineered

Advances in Healthcare and Nanoparticle Toxicology   Materials Research Forum LLC
Materials Research Foundations 171 (2024) 191-216   https://doi.org/10.21741/9781644903339-7

with specific cell-penetrating peptides (CPPs) and nucleic acids. [20, 21]. The proposed consolidation of lipid raft, actin cytoskeleton, and cholera toxin subunit B (CTB)-mediated endocytosis could be succinctly referred to as the actin cytoskeleton and cholera toxin subunit B (CTB) pathways [22].

Detailed explanation of the Clathrin-independent/GPI-anchored protein-enriched endocytic compartment is as follows:

Glycosylphosphatidylinositol (GPI)-anchored proteins are a specific class of cell surface proteins that are attached to the outer leaflet of the plasma membrane through a GPI anchor. These proteins have a distinct lipid composition that makes them functionally different from other cell surface proteins. The CLIC/GEEC pathway is initiated when specific ligands or nanoparticles bind to their corresponding receptors on the cell surface. The binding of these nanoparticles triggers the formation of specialized vesicles that are enriched in GPI-anchored proteins. Following the receptor-ligand interaction, the specialized vesicles, called CLIC/GEEC vesicles, are internalized from the plasma membrane. These vesicles are distinct from the well-characterized clathrin-coated vesicles and do not involve clathrin in their formation. Once internalized, the CLIC/GEEC vesicles move within the cell and may fuse with other intracellular compartments, such as early endosomes or recycling endosomes. Within these compartments, the cargo can be sorted based on specific signals, leading to its subsequent intracellular trafficking.

The Clathrin-independent/GPI-anchored protein-enriched endocytic compartment is an area of active research, and its full molecular mechanism and physiological roles are still being elucidated. It is worth noting that the term "CLIC/GEEC" encompasses different pathways and mechanisms that may share common features but can also differ in some aspects, making it a complex and dynamic endocytic route.

Macropinocytosis:

It is a type of endocytosis that allows cells to engulf and internalize large volumes of extracellular fluid, along with any solutes present in the fluid, including nanoparticles through actin-stabilized plasma membrane extensions [23]. The term "macropinocytosis" is derived from "macropinosome," which refers to the large vesicles formed during this process. Macropinocytosis is distinct from other endocytic pathways, such as clathrin-mediated endocytosis and caveolae-mediated endocytosis, and it serves as a non-selective mechanism for the bulk uptake of extracellular material. Macropinosomes are leaking intracellular vesicles that might allow nanoparticles to escape before being degraded by lysosomes [24, 25].
Detailed explanation of the macropinocytosis process is as follows:

The initiation of macropinocytosis involves the activation of signaling pathways that lead to actin polymerization at the cell surface. This actin polymerization causes the plasma membrane to undergo extensive ruffling, resulting in the formation of large, transient membrane protrusions called macropinocytic cups [16, 26]. As the macropinocytic cups extend, they engulf a considerable volume of extracellular fluid and solutes, including nanoparticles present in the surrounding environment. Eventually, the edges of the cup come together, sealing off the fluid-filled vesicle, which is now referred to as a macropinosome. The macropinosomes then move into the cell's interior, where they undergo a maturation process. This involves changes in the macropinosome's membrane composition and acquisition of specific markers. The maturation

Advances in Healthcare and Nanoparticle Toxicology   Materials Research Forum LLC
Materials Research Foundations 171 (2024) 191-216   https://doi.org/10.21741/9781644903339-7

process can vary between different cell types and can lead to either degradation or recycling of the internalized nanoparticles.

Macropinocytosis is particularly important for certain cell types, such as immune cells like macrophages and dendritic cells. These cells utilize macropinocytosis to capture antigens from their surroundings, a process essential for immune surveillance and antigen presentation [27].

Macropinocytosis is considered to be a non-selective process, and it occurs continuously in many cell types, particularly in cells that have a high capacity for nutrient uptake or those actively involved in immune functions. It provides a means for cells to sample the extracellular environment and take up a wide range of extracellular materials, including nutrients, growth factors, pathogens, and nanoparticles.

Researchers are exploring the potential of macropinocytosis for nanoparticle-based drug delivery, as the pathway's non-selective nature could enable the uptake of nanoparticles without the need for specific targeting ligands [28, 29].

Phagocytosis:

It is a specialized form of endocytosis that involves the engulfment and internalization of large particles by certain specialized cells called phagocytes. The term "phagocytosis" is derived from the Greek words "phago," meaning "to eat," and "kytos," meaning "cell." Phagocytes play a crucial role in the immune system, where they engulf and destroy foreign invaders, such as bacteria, viruses, and cellular debris [30]. Immune cells, such as macrophages, dendritic cells, neutrophils, and B lymphocytes, engage in phagocytosis. Physical binding to phagocyte cell surface receptors often initiates nanoparticle phagocytosis. Fc receptors, mannose receptors, scavenger receptors, and complement receptors are examples of cell surface receptors [31, 32]. Due to the remarkable efficiency of phagocytosis in clearing opsonized nanoparticles, it poses a substantial hurdle in the development of successful nanomedicines. When nanoparticles are introduced into the bloodstream through intravenous administration, they usually undergo swift opsonization upon encountering blood [33, 34].

These opsonized nanoparticles are subsequently captured by macrophages and other phagocytic cells within the mononuclear phagocyte system (MPS) [6, 35]. As much as 99% of a nanoparticle bolus dose introduced systemically can be captured by the MPS [6, 33]. Additionally, tissue-resident macrophages, like those found in tumors, have demonstrated a greater propensity to uptake nanoparticles designed to target cancer cells compared to the malignant cells themselves [36]. To mitigate the sequestration of nanoparticles by the MPS, surface modifications for nanoparticles have been devised to reduce their opsonization [37].

Detailed explanation of the phagocytosis process is as follows:

Phagocytosis is initiated when phagocytes, such as macrophages and neutrophils, recognize and bind to particles that are targeted for removal. These particles can be pathogens, dead cells, cellular debris, or other foreign materials. Once bound to the particle, the phagocyte extends its plasma membrane around the particle, forming a phagocytic cup. This cup progressively engulfs the particle, enclosing it within a membrane-bound vesicle called a phagosome. As the phagocytic cup completely surrounds the particle, it closes off, and the phagosome is internalized within the phagocyte. The phagosome now contains the ingested particle but remains separate from the rest of the cell's interior. The phagosome then undergoes a maturation process during which it fuses with lysosomes, forming a phagolysosome. Lysosomes contain powerful enzymes that can degrade the contents of the phagosome, effectively destroying the engulfed particle. Once the particle is broken down within the phagolysosome, the resulting waste products are either recycled by the cell or expelled from the phagocyte through exocytosis.

(ii) Direct cytoplasmic delivery of nanoparticles

Usually, direct penetration of nanoparticles into the cytoplasm is not observed when they enter the cell through endocytosis. Nevertheless, alternative pathways for nanoparticle delivery, as illustrated in Figure 2, can facilitate this direct access to the cytoplasm. Through biochemical or physical mechanisms, nanoparticles can traverse the cell's plasma membrane to enter the cytoplasm directly. Once nanoparticles are dispersed within the cytoplasm, they can interact with subcellular organelles and intracellular structures, enabling them to induce specific biological responses and serve medical purposes effectively.

Direct cytoplasmic delivery of nanoparticles refers to the process of delivering nanoparticles directly into the cytoplasm of target cells. This delivery approach is particularly useful for therapeutic or research applications where the nanoparticles need to interact with specific cellular organelles or molecules within the cell's cytoplasm. Direct cytoplasmic delivery is often desired when the target site of action is within the cell's interior, as opposed to targeting molecules on the cell surface.

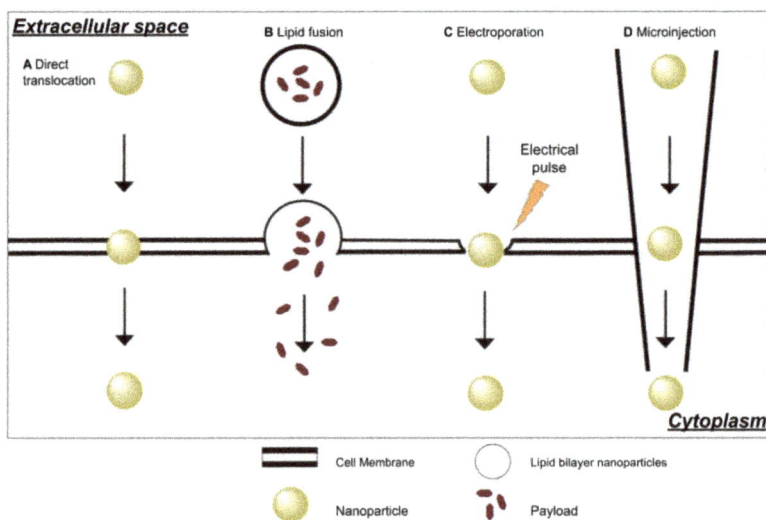

Fig. 2: Direct cytoplasmic delivery of nanoparticles like a) Cytoplasmic entry by direct translocation b) Cytoplasmic entry by lipid fusion, c) Electroporation, d) Microinjection

Advances in Healthcare and Nanoparticle Toxicology   Materials Research Forum LLC
Materials Research Foundations 171 (2024) 191-216   https://doi.org/10.21741/9781644903339-7

Several methods have been developed to achieve direct cytoplasmic delivery of nanoparticles. Here are some common techniques [6]:

- Cytoplasmic entry by direct translocation:
- Cytoplasmic entry by lipid fusion
- Electroporation
- Microinjection

Cytoplasmic entry by direct translocation:

Cytoplasmic entry by direct translocation of nanoparticles refers to a process where nanoparticles can cross the cell membrane without engaging in endocytosis or other traditional uptake pathways. Instead, these nanoparticles directly pass through the cell membrane, entering the cytoplasm and gaining access to the cell's interior. This mode of entry is particularly relevant for achieving efficient and rapid delivery of nanoparticles to the cytoplasm, bypassing potential entrapment or degradation in endosomes and lysosomes [38]. Computational models have been employed to simulate and clarify various aspects of nanoparticle diffusion through lipid bilayer membranes [39-42]. Yang et al. investigated through the utilization of computer simulations, he delves into the physical movement mechanisms of nanoparticles possessing diverse geometries, such as spheres, ellipsoids, rods, discs, and pushpin-like particles, as they traverse a lipid bilayer. Their findings underscore the pivotal roles played by both the particle's shape anisotropy and its initial orientation in governing the nature of the interaction between the particle and the lipid bilayer. The nanoparticle's capacity to breach the lipid bilayer is chiefly governed by two factors: the extent of contact between the particle and the lipid bilayer, and the curvature of the particle at the specific contact point. Interestingly, the particle's volume exerts an indirect influence on translocation, while particle rotation can introduce complexities into the penetration process [39]. These computational studies have yielded valuable insights, enabling nanomedicine researchers to translate theoretical information and modeling results into in vitro experimental observations. For instance, Hinde et. al. combined correlation microscopy techniques to demonstrate that polymeric nanoparticles, despite having the same surface chemistry, exhibited varying rates of movement across different cellular barriers, ultimately determining the location of drug release. They observed rods and worms, in contrast to micelles and vesicles, were capable of entering the nucleus through passive diffusion. They illustrate that achieving effective drug delivery through the primary cellular obstacle, the nuclear envelope, is crucial for enhancing the efficiency of doxorubicin and can be accomplished by utilizing nanoparticles with suitable shapes [43]. In a separate investigation conducted by Jewell et. al., it was observed that nanoparticles with a coating characterized by self-assembled domains arranged in a stripe-like pattern can transport various lengths and types of DNA cargo into cells without any detectable harm. Importantly, the manner in which these materials facilitate cellular entry depends on the specific organization of the outer layer composed of mercapto-1-undecanesulphonate ligands. Nanoparticles with uniform mercapto-1-undecanesulphonate ligand structures achieve DNA delivery through endocytic pathways, while nanoparticles with "striped" mercapto-1-undecanesulphonate 1-octanethiol coatings enable delivery through both endocytosis and direct penetration of the cell membrane [44].

Another notable strategy for achieving the direct translocation of nanoparticles through the cell's plasma membrane entails using cell-penetrating peptides (CPPs) as surface ligands on the nanoparticles. Cell-penetrating peptides (CPPs) are short peptides, typically consisting of fewer than 40 amino acids, that possess the capability to enter cells through diverse mechanisms, primarily involving endocytosis. Furthermore, these peptides can effectively assist in the intracellular transportation of bioactive cargos, which can be either covalently or noncovalently attached, such as low molecular weight drugs and nucleic acids, without causing any toxicity [45]. In Radstrom's study, compelling evidence is presented indicating that the primary mechanism by which CADY (a 20-residue peptide) associates non-covalently with siRNA (Small interfering RNA) to enter cells involves direct penetration through the cell membrane rather than utilizing the endosomal pathway. The research demonstrates that CADY:siRNA complexes do not coincide with most markers associated with endosomes and retain their full functionality even when subjected to inhibitors targeting the endosomal pathway. In contrast, the study reveals that CADY:siRNA complexes unmistakably trigger a temporary permeabilization of the cell membrane, which is swiftly restored due to the fluidity of the cell membrane. Consequently, the prevailing hypothesis suggests that the major route for CADY:siRNA complex cellular entry is via direct translocation through the cell membrane [46]. The relevance of these pathways appears to be regulated by factors such as (i) the type of nanoparticle to which cell-penetrating peptides are attached, and (ii) concentration levels of lipids and peptides in the cellular plasma membrane at a specific location [46-48].

Cytoplasmic entry by lipid fusion

Cytoplasmic entry by lipid fusion is a delivery strategy that involves the direct fusion of lipid-based nanoparticles or liposomes with membrane, which allows their cargo to be delivered into the cytoplasm. This mechanism of delivery takes advantage of the natural fusion properties of certain lipids and the cell membrane to achieve efficient and direct intracellular delivery of therapeutic agents or nanoparticles.

Lipid fusion refers to the mechanism through which specific lipid bilayer-coated structures merge with the plasma membrane of a cell. Detailed process of cytoplasmic entry by lipid fusion typically involves liposomes a lipid-based nanoparticles composed of one or more lipid bilayers enclosing aqueous core engineered to carry various payloads, including drugs, genes, or imaging agents, within their aqueous compartments. These lipid-based nanoparticles or liposomes are brought into contact with the cell membrane of the target cells. The lipids on the nanoparticle surface interact with the lipids in the cell membrane, creating a favorable environment for fusion. Under specific conditions, like presence of fusogenic peptides, changes in temperature, pH, the lipids on the nanoparticle surface and the cell membrane undergo structural rearrangements, leading to the fusion of the two lipid bilayers. This process allows the nanoparticle or liposome to merge with the cell membrane. Upon fusion with the cell membrane, the contents of the nanoparticle or liposome, including the therapeutic cargo, are directly released into the cytoplasm of the cell. This bypasses the need for endocytosis, allowing for rapid and efficient delivery of the cargo to its intracellular target.

In a study by Kube et al., Fusogenic liposomes, a molecular transport system, were employed for the delivery of proteins. When liposomes containing water-soluble proteins came into contact with mammalian cells, the liposomal membrane seamlessly merged with the cell's plasma membrane, ensuring the unharmed transport of liposomal contents into the cell's cytoplasm, free from any degradation. To investigate the critical factors involved in this proteofection process, various aspects were examined, including the formation of complexes between fusogenic liposomes and

Advances in Healthcare and Nanoparticle Toxicology   Materials Research Forum LLC
Materials Research Foundations 171 (2024) 191-216   https://doi.org/10.21741/9781644903339-7

the target proteins, as well as the Protein size (2.3 kDa (fluorescently tagged LifeAct) over 27 kDa (EGFP) to 240 kDa (R-phycoerythrin) and zeta potential of the resulting fusogenic proteoliposomes. The delivery of proteins into the cells was assessed using fluorescence microscopy and flow cytometry. Notably, proteins such as R-phycoerythrin, EGFP, Dendra2 as well as NTF2-AlexaFluor488 and LifeAct-FITC peptides, were effectively introduced into mammalian cells with high levels of efficiency [49]. In a study by He et al. engineered lipid-based liquid crystalline nanoparticles, referred to as "nano-Transformers," which can undergo structural changes in the acidic intracellular conditions and efficiently deliver siRNA in cancer therapy. These nano-Transformers exhibit favourable siRNA loading efficiency while maintaining low levels of cytotoxicity. These nano-Transformers triggers fusion with endosomal membranes, resulting in the simple separation of siRNA from the nanocarriers and its immediate release into the cytoplasm. Their research demonstrates that transfection with nano-Transformers loaded with cyclin-dependent kinase 1 (CDK1)-siRNA achieves a remarkable upto 95% reduction in the corresponding mRNA levels in vitro and significantly hinders tumor growth in vivo without eliciting any immunogenic response [50].

Kim and colleagues observed a reduction in the levels of a proinflammatory marker (IRF5) in macrophages. This reduction facilitated the clearance of Staphylococcus aureus pneumonia through phagocytosis and improved the survival of mice that were infected. Recent research investigating the interaction between lipid layers has shown that nanoparticles with gold core with amphiphilic organic shell shows a behaviour which is size dependent when it comes to lipid fusion at the lipid-lipid interface [51].

Electroporation:

It is a technique used to introduce molecules, such as nucleic acids, proteins, or nanoparticles, into cells by applying brief electric pulses to the cells. These electric pulses create temporary pores in the cell membrane, allowing the molecules to pass through and enter the cell's cytoplasm. The creation of membrane pores using electroporation can be precisely regulated by adjusting the parameters of the electrical pulse, such as pulse duration and voltage. This precise control ensures that the newly formed pores do not adversely affect the viability of the cell [52]. Electroporation is a widely used method in molecular biology, genetics, and biotechnology for the delivery of foreign materials into cells.

The process of electroporation typically involves harvesting cells to be electroporated growth medium or buffer. Cells are then suspended in an isotonic buffer or solution to ensure their stability during the electroporation process. Molecules to be delivered, such as DNA, RNA, proteins, or nanoparticles, are mixed with the cell suspension. The molecules may carry genetic information or other cargoes intended for specific cellular functions or research purposes. Cell and molecule mixture is then placed in a specialized electroporation cuvette or chamber. When an electric field is applied to the cells, the membrane temporarily becomes permeable due to the formation of transient pores, allowing the molecules to pass through. The size and number of pores depend on the strength and duration of the electric pulse. Molecules that were mixed with the cells are now able to enter the cell's cytoplasm through the temporary pores in the membrane. After the electric pulse is removed, the cell membrane re-seals, closing the temporary pores. The cells can then be processed further for the expression or action of the delivered molecules.

Kim and associates created Novel T1 magnetic resonance imaging (MRI) contrast agents by encapsulating hollow manganese oxide nanoparticles within mesoporous silica shells. In this innovative approach, adipose-derived mesenchymal stem cells (MSCs) were effectively labelled using electroporation. This labelling method resulted in significantly shorter T1 relaxation times of water protons in tissue, compared to the conventional approach of direct incubation without electroporation [53]. Hobo and associated created nanoparticles based on lipids which successfully transferred siRNA via electroporation using an electrical pulse of around 200 volts. This resulted in the silencing of co-inhibitory molecules PD-L1 and PD-L2 expression on dendritic cells derived from human monocytes.

Electroporation offers several advantages has a) high efficacy ensuring that a significant proportion of cells take up the desired molecules, b) non-viral unlike viral delivery methods, electroporation does not require the use of viruses, reducing the risk of viral-induced cytotoxicity or unwanted immune responses. c) Versatile: Electroporation can be used with a wide range of cell types, including plant cells, mammalian cells, bacteria and yeast. d) Non-Integrated Delivery: In the case of DNA delivery, electroporation often results in non-integrated plasmid DNA, allowing transient expression of genes without altering the host cell's genome.

Electroporation is commonly employed in gene editing techniques, such as CRISPR/Cas9, gene expression studies, cell-based assays, and transfection experiments. Researchers continuously explore and optimize electroporation parameters to achieve improved delivery efficiency and minimize cell damage while maximizing the successful introduction of molecules into target cells.

Microinjection:

It is a direct cytoplasmic delivery technique that involves the physical injection of nanoparticles or other molecules directly into the cytoplasm of individual cells using a fine microneedle [54]. This method allows for precise and controlled delivery of cargo to specific cells, making it a valuable tool in research and various biomedical applications.
Process of microinjection as a direct cytoplasmic delivery method:

The nanoparticles or cargo to be delivered are prepared and loaded into a solution suitable for microinjection. This may involve functionalizing the nanoparticles with targeting ligands or surface modifications to improve their intracellular delivery. Microinjection requires specialized equipment, including a micromanipulator to control the movement of the microneedle, a microscope for visualization, and a microinjector to generate the required pressure for injection. Cells that will receive the nanoparticle delivery are selected and placed in a culture dish or on a petri dish with suitable media to keep them viable during the injection process. The microneedle, which is typically a thin glass or quartz pipette pulled to form a fine tip, is filled with the nanoparticle solution or cargo to be injected. The microneedle is carefully inserted into the selected cell's cytoplasm while being visually guided using a microscope. The micromanipulator allows precise control over the needle's position and movement. Once the microneedle is inside the cell's cytoplasm, a small volume of the nanoparticle solution is gently injected. The nanoparticles are released within the cytoplasm, enabling them to engage with the cellular constituents. After the injection is complete, the microneedle is carefully removed from the cell, and the injected cell is allowed to recover and continue its normal biological processes.

Candeloro and colleagues conducted a study to examine how Ag and Fe_3O_4 nanoparticles impact HeLa cells where they introduced these nanoparticles into the cells through microinjection and then used Raman spectroscopy to analyze the cells shortly after incubation. They found distinct behaviors in the cells exposed to nanoparticles compared to the control cells. This divergence suggests that the cells may be responding to the nanoparticles through mechanisms associated with the onset of oxidative stress [55]. Further investigations using microinjection have revealed that the delivery of genes mediated by nanoparticles can be obstructed when lysosomal sequestration is initiated as a result of autophagy processes [56].

Microinjection is a highly precise technique that enables researchers to deliver nanoparticles directly into specific cells of interest, making it particularly valuable for experiments requiring intracellular delivery of cargo with minimal disturbance to the cell. It has found diverse applications in various fields, encompassing introduction of genetic material (e.g., DNA, RNA, proteins) into cells, delivering nanoparticles for imaging or drug delivery, and studying cellular responses to specific molecules or stimuli.

However, microinjection is a labor-intensive process and may not be suitable for large-scale delivery or high-throughput experiments. Additionally, the injection process can cause some cellular damage, and not all cell types are amenable to microinjection.

As technology advances, other methods such as electroporation, lipofection, and nanoparticle surface modifications continue to be developed to improve the efficiency and scalability of direct cytoplasmic delivery of nanoparticles. These alternative techniques offer different advantages and challenges, allowing researchers to choose the most appropriate method for their specific research needs.

3.   Mediating nanoparticle uptake through material design

Apart from methods by which the nanoparticles can be internalized by cells i.e. what cell does to the nanoparticle, we can also modify the nanoparticles in order to assist the internalization process. This involves tailoring the characteristics of nanoparticles to enhance their cellular internalization and uptake by target cells. By carefully engineering the properties of nanoparticles, researchers can improve their interactions with cell membranes and endocytic pathways, thereby increasing their uptake efficiency and specificity. Material design considerations can influence various aspects of nanoparticle uptake, including cellular binding, endocytosis, and intracellular trafficking. Some key strategies for mediating nanoparticle uptake through material design include:

3.1   Surface Chemistry

The surface chemistry of nanoparticles plays a crucial role in determining their engagement with the cellular membrane. Through the modification of the nanoparticle's surface with precise ligands like small compounds, peptides or antibodies, it is possible to design nanoparticles capable of attaching selectively to cell surface receptors, promoting receptor-mediated endocytosis, this is called as "active targeting" [57-60]. The justification for employing this surface modification approach lies in the potential of targeting ligands to amplify a nanoparticle's engagement with cells. This engagement can trigger subsequent cellular signalling pathways that result in a desired biological outcome, such as cell apoptosis, or it can bolster the uptake of nanoparticles by cells, facilitating the transportation of therapeutic and diagnostic payloads inside the cell. To develop nanoparticles for active targeting, multiple design aspects need careful consideration and optimization to attain efficient targeting. These factors encompass aspects such as the length of the target ligand, the density of target ligands, hydrophobic characteristics, and avidity.

Nanoparticles lacking distinct surface-targeting ligands are termed "passive targeting"

nanoparticles. Passive targeting signifies that the interactions between nanoparticles and cells are not selective, meaning these interactions can promote nanoparticle uptake in both healthy and diseased cells. Many passive targeting nanotherapeutics have been approved by regulatory authorities.

Both passive and active targeting nanoparticles face a challenge in that their intentionally designed surface properties can undergo alterations when exposed to a biological environment. For example, nanoparticles decorated with targeting ligands may undergo alterations in their targeting effectiveness when introduced into biological fluids such as the bloodstream. One primary factor contributing to this phenomenon is the development of a protein corona on the nanoparticle's surface due to the adsorption of proteins from the serum. This formation of a protein corona transforms the nanoparticles' initially engineered synthetic characteristics into a biological identity, frequently leading to notable effects on the interactions between nanoparticles and cells [61, 62].

Various factors can influence the creation and characteristics of nanoparticle protein coronae. These factors include the temperature during incubation, the choice of protein or serum sources, whether the plasma or serum is derived from humans or animals, and localized temperature fluctuations in the case of plasmonic nanoparticles. These variations can potentially lead to the development of distinct and customized protein coronae around the nanoparticles. When exposed to human plasma, nanoparticles come into contact with substantial quantities of proteins, resulting in an enlargement of their size and potentially enhancing their subsequent uptake by immune cells. Furthermore, targeting ligands may become obscured within the protein corona, possibly leading to a decrease or even the complete loss of their precise targeting functionality. To address this problem Tonigold and colleagues demonstrated their ability to employ a pre-adsorption technique for binding targeting antibodies onto the nanocarrier's surface. The antibodies pre-adsorbed in this manner retain their functionality and are not entirely displaced or concealed by the biomolecular corona. In contrast, antibodies that are directly coupled to the polystyrene nanoparticles are more susceptible to being influenced or masked by this protective shielding [63].

3.2 Size and Shape

The size and shape of nanoparticles influence their cellular uptake. Small nanoparticles can more easily diffuse through the extracellular matrix and cell membrane, while larger nanoparticles may be taken up through phagocytosis by certain cell types, such as macrophages. Additionally, elongated or rod-shaped nanoparticles have shown enhanced cellular uptake compared to spherical nanoparticles due to their improved contact area with the cell membrane. A study by Agarwala et. al. demonstrated that when considering shape independently of volume, charge, and material composition, under typical in vitro conditions, mammalian epithelial and immune cells exhibit a preference for the internalization of hydrophilic nanoparticles with a disc-like shape, negative charge, and high aspect ratios, as opposed to nanorods and nanodiscs with lower aspect ratios. In this study polyethylene glycol diacrylate based discoidal [220-nm diameter (d)×100-nm height (h), 325-nmd×100-nm h, and 80-nmd×70-nmh] and cuboidal (rod-shaped) nanoparticles (100×100×400 nm and 100×100×800 nm) were prepared having zeta potential of about−57 mV. Uptake studies were performed on HeLa cells, and particle internalization was qualitatively assessed using confocal microscopy. Nanodiscs of intermediate aspect ratios, however, are preferred by endothelial cells. Strangely, larger hydrogel nanodiscs and nanorods are internalised more effectively than their smaller counterparts, in contrast to nanospheres. The rate, effectiveness, and processes of uptake are all cell type- and shape-dependent. Both epithelial and endothelial cells can internalise nanoparticles through the process of micropinocytosis, but only epithelial cells can do so through the caveolae-mediated pathway. As opposed to epithelial cells, human umbilical vein endothelial cells adopt clathrin-mediated uptake for all shapes and have a substantially better uptake efficiency. [64]. In order to determine how the size and amount of folate decoration on polymeric nanoparticles affected the cellular absorption kinetics and choice of endocytic routes in retinal pigment epithelium (RPE) cells, Wang and colleagues conducted a study. They found that nanoparticles were ingested by ARPE-19 cells, a human RPE cell line, through receptor-mediated endocytosis, and that the rate of absorption increased as the particle size reduced from 250 nm to 50 nm. Additionally, it was found that clathrin- and caveolae-mediated endocytosis pathways were used to internalise both 50 nm and 120 nm folate-decorated nanoparticles. However, the caveolae-mediated pathway was the only route by which the 250 nm folate-decorated nanoparticles were internally absorbed [65]. In a separate investigation, Chang and colleagues explored the impact of size variations in self-assembled block copolymer spherical micelles and vesicles on their uptake by human colon carcinoma cells. Their findings indicated that smaller micelles were rapidly internalized, whereas larger micelles were taken up at a slower pace. However, over the course of a few hours, all cells internalized aggregates to a comparable extent [66].

3.3   Surface Charge

The charge on nanoparticles can have a significant impact on lipid and cell membranes due to electrostatic interactions and other physicochemical effects. An further crucial factor to take into account is surface charge. Nanoparticles with positive charge might interact more favourably cell membranes which are negatively charged, boosting cellular absorption. These interactions can influence the stability, permeability, and behaviour of membranes. Scientists can manipulate synthetic nanoparticles to have positive, negative, or no surface charge and the charge can be measured by zeta potential exhibited by the colloidal species.

The nanoparticle zeta potential changes with the environmental conditions. In one of the study, Walkey et al. showed effect on the gold collide before and after incubation in pooled human plasma [67]. In another study Dobrovolskaia, et. al. observed upon incubating 30-nm and 50-nm colloidal gold in human plasma, an approximately twofold increase in the hydrodynamic size of the nanoparticles was noted however, the presence of adsorbed plasma proteins resulted in a reduction in the overall particle charge [68]. Tenzer et.al. discussed about, rapid formation of coronas which are highly complex protein that attach to the nanoparticle's surface to create a layer in physiological environment. Tenzer et.al. employed label-free snapshot proteomics to generate quantitative, time-dependent profiles of human plasma coronas that developed on silica and polystyrene nanoparticles with differing sizes and surface modifications. He also found coronas contained nearly 300 different proteins. He also discussed the effect of corona formation on early exposure-related factors, including hemolysis, platelet activation, nanoparticle internalization, and endothelial cell mortality [69].

Conclusion

Nanoparticles can be synthesized using a variety of inorganic and organic materials, each possessing distinctive physical, chemical, and biological attributes, which render them valuable for medical applications. Nanoparticles interact with many tissues and cells once they are introduced into the body. Although achieving precise delivery of nanoparticles to specific tissues and cells presents formidable challenges, field of nanomedicine holds the potential to revolutionize diagnostic and therapeutic approaches.

Nevertheless, the advancement of nanomedicine faces hurdles stemming from biological and physical obstacles. Study of interactions between nanoparticles and lipid and cell membranes has revealed a complex interplay of physical, chemical, and biological processes. Nanoparticles can adsorb onto lipid bilayers, leading to changes in membrane properties such as fluidity and permeability. To address these challenges and facilitate the clinical adoption of nanomedicines, there is a pressing need for more comprehensive and quantitative investigations that delve into the intricate mechanisms of interactions between nanoparticles and biological systems. Such studies are vital for developing innovative solutions that surmount these barriers.

Extensive preclinical and clinical research within scientific literature details the broad array of medical applications associated with both organic and inorganic nanoparticles. These applications encompass vaccines, antimicrobial agents, medical imaging, diagnostic tools, and encompass diverse strategies for addressing cancer and other chronic disease [70-98]. In an effort to overcome intrinsic biological obstacles, recent research has presented the idea of responsive nanoparticles, which can be modified for specific goals in cancer care. [99-106]. Furthermore, ongoing research has explored the realm of nanoparticle-based immunoengineering and immunotherapy strategies.

In summary, the study of nanoparticle interactions with lipid and cell membranes is a multidisciplinary field that bridges materials science, biology, and medicine.

Future Directions

Interactions of nanoparticles with lipid and cell membranes is a dynamic and crucial field of research with a promising future. As we look ahead, several exciting directions and potential developments can be envisioned. Researchers will continue to refine and develop cutting-edge techniques for characterizing the interactions between nanoparticles and lipid/cell membranes to provide a deeper understanding of these complex interactions at the nanoscale. Researchers will focus on designing nanoparticles with specific properties optimized for interactions with lipid and cell membranes by tuning surface chemistry, size, shape, and charge to enhance their targeting and therapeutic potential and reducing side effects. Researchers may explore innovative therapies that leverage the unique properties of nanoparticles, such as photothermal therapy, gene delivery, and immunotherapy. These approaches hold promise for addressing some of the most challenging medical conditions. Collaboration between researchers from diverse fields, including nanotechnology, biology, chemistry, and medicine, will continue to drive progress in understanding and utilizing full potential of nanoparticle. Interdisciplinary teams will be essential for tackling complex challenges.

In summary, prospective avenues for researching interactions involving nanoparticles with lipid and cell membranes holds great promise for addressing critical health, environmental, and technological challenges. This interdisciplinary field will continue to evolve, leading to innovative solutions and transformative advancements in various domains.

References

[1]B.Y.S. Kim, J.T. Rutka, W.C.W. Chan, Nanomedicine, N. Engl. J. Med. 363 (2010) 2434–2443. https://doi.org/10.1056/nejmra0912273.

[2]C. von Roemeling, W. Jiang, C.K. Chan, I.L. Weissman, B.Y.S. Kim, Breaking Down the Barriers to Precision Cancer Nanomedicine, Trends Biotechnol. 35 (2017) 159–171. https://doi.org/10.1016/j.tibtech.2016.07.006.

[3]P.C. Ke, S. Lin, W.J. Parak, T.P. Davis, F. Caruso, A Decade of the Protein Corona, ACS Nano. 11 (2017) 11773–11776. https://doi.org/10.1021/acsnano.7b08008.

[4]J. Hühn, C. Carrillo-Carrion, M.G. Soliman, C. Pfeiffer, D. Valdeperez, A. Masood, I. Chakraborty, L. Zhu, M. Gallego, Z. Yue, M. Carril, N. Feliu, A. Escudero, A.M. Alkilany, B. Pelaz, P. Del Pino, W.J. Parak, Erratum: Selected Standard Protocols for the Synthesis, Phase Transfer, and Characterization of Inorganic Colloidal Nanoparticles (Chem. Mater. (2017) 29:1 (399−461) DOI: 10.1021/acs.chemmater.6b04738), Chem. Mater. 33 (2021) 4830. https://doi.org/10.1021/acs.chemmater.1c01764.

[5]M. Mahmoudi, N. Bertrand, H. Zope, O.C. Farokhzad, Emerging understanding of the protein corona at the nano-bio interfaces, Nano Today. 11 (2016) 817–832. https://doi.org/10.1016/j.nantod.2016.10.005.

[6]N.D. Donahue, H. Acar, S. Wilhelm, Concepts of nanoparticle cellular uptake, intracellular trafficking, and kinetics in nanomedicine, Adv. Drug Deliv. Rev. 143 (2019) 68–96. https://doi.org/10.1016/j.addr.2019.04.008.

[7]A. Panariti, G. Miserocchi, I. Rivolta, The effect of nanoparticle uptake on cellular behavior: Disrupting or enabling functions?, Nanotechnol. Sci. Appl. 5 (2012) 87–100. https://doi.org/10.2147/NSA.S25515.

[8]M.J.D. Clift, C. Brandenberger, B. Rothen-Rutishauser, D.M. Brown, V. Stone, The uptake and intracellular fate of a series of different surface coated quantum dots in vitro, Toxicology. 286 (2011) 58–68. https://doi.org/10.1016/j.tox.2011.05.006.

[9]J. Chen, Z. Yu, H. Chen, J. Gao, W. Liang, Transfection efficiency and intracellular fate of polycation liposomes combined with protamine, Biomaterials. 32 (2011) 1412–1418. https://doi.org/10.1016/j.biomaterials.2010.09.074.

[10]   D.J. Irvine, M.C. Hanson, K. Rakhra, T. Tokatlian, Synthetic Nanoparticles for Vaccines and Immunotherapy, Chem. Rev. 115 (2015) 11109–11146. https://doi.org/10.1021/acs.chemrev.5b00109.

[11]   P.A. Gleeson, The role of endosomes in innate and adaptive immunity, Semin. Cell Dev. Biol. 31 (2014) 64–72. https://doi.org/10.1016/j.semcdb.2014.03.002.

[12]   J.P. Mattila, A. V. Shnyrova, A.C. Sundborger, E.R. Hortelano, M. Fuhrmans, S. Neumann, M. Müller, J.E. Hinshaw, S.L. Schmid, V.A. Frolov, A hemi-fission intermediate

Advances in Healthcare and Nanoparticle Toxicology　　　　　　　Materials Research Forum LLC
Materials Research Foundations 171 (2024) 191-216　　　　https://doi.org/10.21741/9781644903339-7

links two mechanistically distinct stages of membrane fission, Nature. 524 (2015) 109–113.
https://doi.org/10.1038/nature14509.

[13]　P. Decuzzi, M. Ferrari, The receptor-mediated endocytosis of nonspherical particles,
Biophys. J. 94 (2008) 3790–3797. https://doi.org/10.1529/biophysj.107.120238.

[14]　A.S. Robertson, E. Smythe, K.R. Ayscough, Functions of actin in endocytosis, Cell. Mol.
Life Sci. 66 (2009) 2049–2065. https://doi.org/10.1007/s00018-009-0001-y.

[15]　M. Kaksonen, A. Roux, Mechanisms of clathrin-mediated endocytosis, Nat. Rev. Mol.
Cell Biol. 19 (2018) 313–326. https://doi.org/10.1038/nrm.2017.132.

[16]　S.D. Conner, S.L. Schmid, Regulated portals of entry into the cell, Nature. 422 (2003)
37–44. https://doi.org/10.1038/nature01451.

[17]　B. Yameen, W. Il Choi, C. Vilos, A. Swami, J. Shi, O.C. Farokhzad, Insight into
nanoparticle cellular uptake and intracellular targeting, J. Control. Release. 190 (2014) 485–
499. https://doi.org/10.1016/j.jconrel.2014.06.038.

[18]　P. Lajoie, I.R. Nabi, Regulation of raft-dependent endocytosis, J. Cell. Mol. Med. 11
(2007) 644–653. https://doi.org/10.1111/j.1582-4934.2007.00083.x.

[19]　D.J.F. Chinnapen, H. Chinnapen, D. Saslowsky, W.I. Lencer, Rafting with cholera toxin:
Endocytosis and trafficking from plasma membrane to ER, FEMS Microbiol. Lett. 266 (2007)
129–137. https://doi.org/10.1111/j.1574-6968.2006.00545.x.

[20]　C. Foerg, U. Ziegler, J. Fernandez-Carneado, E. Giralt, R. Rennert, A.G. Beck-Sickinger,
H.P. Merkle, Decoding the entry of two novel cell-penetrating peptides in HeLa cells: Lipid
raft-mediated endocytosis and endosomal escape, Biochemistry. 44 (2005) 72–81.
https://doi.org/10.1021/bi048330+.

[21]　R.B.M. de Almeida, D.B. Barbosa, M.R. do Bomfim, J.A.O. Amparo, B.S. Andrade, S.L.
Costa, J.M. Campos, J.N. Cruz, C.B.R. Santos, F.H.A. Leite, M.B. Botura, Identification of a
Novel Dual Inhibitor of Acetylcholinesterase and Butyrylcholinesterase: In Vitro and In Silico
Studies, Pharmaceuticals. 16 (2023) 95. https://doi.org/10.3390/ph16010095.

[22]　Y. Jiang, R. Tang, B. Duncan, Z. Jiang, B. Yan, R. Mout, V.M. Rotello, Direct cytosolic
delivery of siRNA using nanoparticle-stabilized nanocapsules, Angew. Chemie - Int. Ed. 54
(2015) 506–510. https://doi.org/10.1002/anie.201409161.

[23]　M.C. Kerr, R.D. Teasdale, Defining macropinocytosis, Traffic. 10 (2009) 364–371.
https://doi.org/10.1111/j.1600-0854.2009.00878.x.

[24]　J.S. Wadia, R. V. Stan, S.F. Dowdy, Transducible TAT-HA fusogenic peptide enhances
escape of TAT-fusion proteins after lipid raft macropinocytosis, Nat. Med. 10 (2004) 310–
315. https://doi.org/10.1038/nm996.

[25]　K.T. Love, K.P. Mahon, C.G. Levins, K.A. Whitehead, W. Querbes, J.R. Dorkin, J. Qin,
W. Cantley, L.L. Qin, T. Racie, M. Frank-Kamenetsky, K.N. Yip, R. Alvarez, D.W.Y. Sah,
A. De Fougerolles, K. Fitzgerald, V. Koteliansky, A. Akinc, R. Langer, D.G. Anderson,
Lipid-like materials for low-dose, in vivo gene silencing, Proc. Natl. Acad. Sci. U. S. A. 107
(2010) 1864–1869. https://doi.org/10.1073/pnas.0910603106.

[26]   J. Mercer, A. Helenius, Virus entry by macropinocytosis, Nat. Cell Biol. 11 (2009) 510–520. https://doi.org/10.1038/ncb0509-510.

[27]   M. Diken, S. Kreiter, A. Selmi, C.M. Britten, C. Huber, Ö. Türeci, U. Sahin, Selective uptake of naked vaccine RNA by dendritic cells is driven by macropinocytosis and abrogated upon DC maturation, Gene Ther. 18 (2011) 702–708. https://doi.org/10.1038/gt.2011.17.

[28]   S. Hirosue, I.C. Kourtis, A.J. van der Vlies, J.A. Hubbell, M.A. Swartz, Antigen delivery to dendritic cells by poly(propylene sulfide) nanoparticles with disulfide conjugated peptides: Cross-presentation and T cell activation, Vaccine. 28 (2010) 7897–7906. https://doi.org/10.1016/j.vaccine.2010.09.077.

[29]   J. Cullis, D. Siolas, A. Avanzi, S. Barui, A. Maitra, D. Bar-Sagi, Macropinocytosis of Nab-paclitaxel drives macrophage activation in pancreatic cancer, Cancer Immunol. Res. 5 (2017) 182–190. https://doi.org/10.1158/2326-6066.CIR-16-0125.

[30]   A. Martínez-Riaño, E.R. Bovolenta, P. Mendoza, C.L. Oeste, M.J. Martín-Bermejo, P. Bovolenta, M. Turner, N. Martínez-Martín, B. Alarcón, Antigen phagocytosis by B cells is required for a potent humoral response, EMBO Rep. 19 (2018). https://doi.org/10.15252/embr.201846016.

[31]   F. Chen, G. Wang, J.I. Griffin, B. Brenneman, N.K. Banda, V.M. Holers, D.S. Backos, L. Wu, S.M. Moghimi, D. Simberg, Complement proteins bind to nanoparticle protein corona and undergo dynamic exchange in vivo, Nat. Nanotechnol. 12 (2017) 387–393. https://doi.org/10.1038/nnano.2016.269.

[32]   R. Tavano, L. Gabrielli, E. Lubian, C. Fedeli, S. Visentin, P. Polverino De Laureto, G. Arrigoni, A. Geffner-Smith, F. Chen, D. Simberg, G. Morgese, E.M. Benetti, L. Wu, S.M. Moghimi, F. Mancin, E. Papini, C1q-Mediated Complement Activation and C3 Opsonization Trigger Recognition of Stealth Poly(2-methyl-2-oxazoline)-Coated Silica Nanoparticles by Human Phagocytes, ACS Nano. 12 (2018) 5834–5847. https://doi.org/10.1021/acsnano.8b01806.

[33]   Y.N. Zhang, W. Poon, A.J. Tavares, I.D. McGilvray, W.C.W. Chan, Nanoparticle–liver interactions: Cellular uptake and hepatobiliary elimination, J. Control. Release. 240 (2016) 332–348. https://doi.org/10.1016/j.jconrel.2016.01.020.

[34]   J. Lazarovits, Y.Y. Chen, E.A. Sykes, W.C.W. Chan, Nanoparticle-blood interactions: The implications on solid tumour targeting, Chem. Commun. 51 (2015) 2756–2767. https://doi.org/10.1039/c4cc07644c.

[35]   K.M. Tsoi, S.A. Macparland, X.Z. Ma, V.N. Spetzler, J. Echeverri, B. Ouyang, S.M. Fadel, E.A. Sykes, N. Goldaracena, J.M. Kaths, J.B. Conneely, B.A. Alman, M. Selzner, M.A. Ostrowski, O.A. Adeyi, A. Zilman, I.D. McGilvray, W.C.W. Chan, Mechanism of hard-nanomaterial clearance by the liver, Nat. Mater. 15 (2016) 1212–1221. https://doi.org/10.1038/nmat4718.

[36]   Q. Dai, S. Wilhelm, D. Ding, A.M. Syed, S. Sindhwani, Y. Zhang, Y.Y. Chen, P. Macmillan, W.C.W. Chan, Quantifying the Ligand-Coated Nanoparticle Delivery to Cancer Cells in Solid Tumors, ACS Nano. 12 (2018) 8423–8435. https://doi.org/10.1021/acsnano.8b03900.

[37]   C.D. Walkey, J.B. Olsen, H. Guo, A. Emili, W.C.W. Chan, Nanoparticle size and surface chemistry determine serum protein adsorption and macrophage uptake, J. Am. Chem. Soc. 134 (2012) 2139–2147. https://doi.org/10.1021/ja2084338.

[38]   R.C. Van Lehn, P.U. Atukorale, R.P. Carney, Y.S. Yang, F. Stellacci, D.J. Irvine, A. Alexander-Katz, Effect of particle diameter and surface composition on the spontaneous fusion of monolayer-protected gold nanoparticles with lipid bilayers, Nano Lett. 13 (2013) 4060–4067. https://doi.org/10.1021/nl401365n.

[39]   K. Yang, Y.Q. Ma, Computer simulation of the translocation of nanoparticles with different shapes across a lipid bilayer, Nat. Nanotechnol. 5 (2010) 579–583. https://doi.org/10.1038/nnano.2010.141.

[40]   R.B.M. de Almeida, D.B. Barbosa, M.R. do Bomfim, J.A.O. Amparo, B.S. Andrade, S.L. Costa, J.M. Campos, J.N. Cruz, C.B.R. Santos, F.H.A. Leite, M.B. Botura, Identification of a Novel Dual Inhibitor of Acetylcholinesterase and Butyrylcholinesterase: In Vitro and In Silico Studies, Pharmaceuticals. 16 (2023) 95. https://doi.org/10.3390/ph16010095.

[41]   B. Song, H. Yuan, S. V. Pham, C.J. Jameson, S. Murad, Nanoparticle permeation induces water penetration, ion transport, and lipid flip-flop, Langmuir. 28 (2012) 16989–17000. https://doi.org/10.1021/la302879r.

[42]   S. Pogodin, M. Werner, J.U. Sommer, V.A. Baulin, Nanoparticle-induced permeability of lipid membranes, ACS Nano. 6 (2012) 10555–10561. https://doi.org/10.1021/nn3028858.

[43]   E. Hinde, K. Thammasiraphop, H.T.T. Duong, J. Yeow, B. Karagoz, C. Boyer, J.J. Gooding, K. Gaus, Pair correlation microscopy reveals the role of nanoparticle shape in intracellular transport and site of drug release, Nat. Nanotechnol. 12 (2017) 81–89. https://doi.org/10.1038/nnano.2016.160.

[44]   C.M. Jewell, J.M. Jung, P.U. Atukorale, R.P. Carney, F. Stellacci, D.J. Irvine, Oligonucleotide delivery by cell-penetrating "striped" nanoparticles, Angew. Chemie - Int. Ed. 50 (2011) 12312–12315. https://doi.org/10.1002/anie.201104514.

[45]   D.M. Copolovici, K. Langel, E. Eriste, Ü. Langel, Cell-penetrating peptides: Design, synthesis, and applications, ACS Nano. 8 (2014) 1972–1994. https://doi.org/10.1021/nn4057269.

[46]   A. Rydström, S. Deshayes, K. Konate, L. Crombez, K. Padari, H. Boukhaddaoui, G. Aldrian, M. Pooga, G. Divita, Direct translocation as major cellular uptake for CADY self-assembling peptide-based nanoparticles, PLoS One. 6 (2011) e25924. https://doi.org/10.1371/journal.pone.0025924.

[47]   F.S. Alves, J.N. Cruz, I.N. de Farias Ramos, D.L. do Nascimento Brandão, R.N. Queiroz, G.V. da Silva, G.V. da Silva, M.F. Dolabela, M.L. da Costa, A.S. Khayat, J. de Arimatéia Rodrigues do Rego, D. do Socorro Barros Brasil, Evaluation of Antimicrobial Activity and Cytotoxicity Effects of Extracts of Piper nigrum L. and Piperine, Separations. 10 (2023) 21. https://doi.org/10.3390/separations10010021.

[48]   W.B. Kauffman, T. Fuselier, J. He, W.C. Wimley, Mechanism Matters: A Taxonomy of Cell Penetrating Peptides, Trends Biochem. Sci. 40 (2015) 749–764. https://doi.org/10.1016/j.tibs.2015.10.004.

[49]   S. Kube, N. Hersch, E. Naumovska, T. Gensch, J. Hendriks, A. Franzen, L. Landvogt, J.P. Siebrasse, U. Kubitscheck, B. Hoffmann, R. Merkel, A. Csiszár, Fusogenic liposomes as nanocarriers for the delivery of intracellular proteins, Langmuir. 33 (2017) 1051–1059. https://doi.org/10.1021/acs.langmuir.6b04304.

[50]   S. He, W. Fan, N. Wu, J. Zhu, Y. Miao, X. Miao, F. Li, X. Zhang, Y. Gan, Lipid-Based Liquid Crystalline Nanoparticles Facilitate Cytosolic Delivery of siRNA via Structural Transformation, Nano Lett. 18 (2018) 2411–2419. https://doi.org/10.1021/acs.nanolett.7b05430.

[51]   B. Kim, H.B. Pang, J. Kang, J.H. Park, E. Ruoslahti, M.J. Sailor, Immunogene therapy with fusogenic nanoparticles modulates macrophage response to Staphylococcus aureus, Nat. Commun. 9 (2018). https://doi.org/10.1038/s41467-018-04390-7.

[52]   P.U. Atukorale, Z.P. Guven, A. Bekdemir, R.P. Carney, R.C. Van Lehn, D.S. Yun, P.H. Jacob Silva, D. Demurtas, Y.S. Yang, A. Alexander-Katz, F. Stellacci, D.J. Irvine, Structure-Property Relationships of Amphiphilic Nanoparticles That Penetrate or Fuse Lipid Membranes, Bioconjug. Chem. 29 (2018) 1131–1140. https://doi.org/10.1021/acs.bioconjchem.7b00777.

[53]   G. Saulis, R. Saule, Size of the pores created by an electric pulse: Microsecond vs millisecond pulses, Biochim. Biophys. Acta - Biomembr. 1818 (2012) 3032–3039. https://doi.org/10.1016/j.bbamem.2012.06.018.

[54]   L. Damalakiene, V. Karabanovas, S. Bagdonas, M. Valius, R. Rotomskis, Intracellular distribution of nontargeted quantum dots after natural uptake and microinjection, Int. J. Nanomedicine. 8 (2013) 555–568. https://doi.org/10.2147/IJN.S39658.

[55]   P. Candeloro, L. Tirinato, N. Malara, A. Fregola, E. Casals, V. Puntes, G. Perozziello, F. Gentile, M.L. Coluccio, G. Das, C. Liberale, F. De Angelis, E. Di Fabrizio, Nanoparticle microinjection and Raman spectroscopy as tools for nanotoxicology studies, Analyst. 136 (2011) 4402–4408. https://doi.org/10.1039/c1an15313g.

[56]   T.F. Martens, K. Remaut, J. Demeester, S.C. De Smedt, K. Braeckmans, Intracellular delivery of nanomaterials: How to catch endosomal escape in the act, Nano Today. 9 (2014) 344–364. https://doi.org/10.1016/j.nantod.2014.04.011.

[57]   K. Cho, X. Wang, S. Nie, Z. Chen, D.M. Shin, Therapeutic nanoparticles for drug delivery in cancer, Clin. Cancer Res. 14 (2008) 1310–1316. https://doi.org/10.1158/1078-0432.CCR-07-1441.

[58]   N. Bertrand, J. Wu, X. Xu, N. Kamaly, O.C. Farokhzad, Cancer nanotechnology: The impact of passive and active targeting in the era of modern cancer biology, Adv. Drug Deliv. Rev. 66 (2014) 2–25. https://doi.org/10.1016/j.addr.2013.11.009.

[59]   J.D. Byrne, T. Betancourt, L. Brannon-Peppas, Active targeting schemes for nanoparticle systems in cancer therapeutics, Adv. Drug Deliv. Rev. 60 (2008) 1615–1626. https://doi.org/10.1016/j.addr.2008.08.005.

[60]   K. Shahane, M. Kshirsagar, S. Tambe, D. Jain, S. Rout, M.K.M. Ferreira, S. Mali, P. Amin, P.P. Srivastav, J. Cruz, R.R. Lima, An Updated Review on the Multifaceted

Therapeutic Potential of Calendula officinalis L., Pharmaceuticals. 16 (2023) 611. https://doi.org/10.3390/ph16040611

[61]   G. Caracciolo, O.C. Farokhzad, M. Mahmoudi, Biological Identity of Nanoparticles In Vivo: Clinical Implications of the Protein Corona, Trends Biotechnol. 35 (2017) 257–264. https://doi.org/10.1016/j.tibtech.2016.08.011.

[62]   E. Polo, M. Collado, B. Pelaz, P. Del Pino, Advances toward More Efficient Targeted Delivery of Nanoparticles in Vivo: Understanding Interactions between Nanoparticles and Cells, ACS Nano. 11 (2017) 2397–2402. https://doi.org/10.1021/acsnano.7b01197.

[63]   M. Tonigold, J. Simon, D. Estupiñán, M. Kokkinopoulou, J. Reinholz, U. Kintzel, A. Kaltbeitzel, P. Renz, M.P. Domogalla, K. Steinbrink, I. Lieberwirth, D. Crespy, K. Landfester, V. Mailänder, Pre-adsorption of antibodies enables targeting of nanocarriers despite a biomolecular corona, Nat. Nanotechnol. 13 (2018) 862–869. https://doi.org/10.1038/s41565-018-0171-6.

[64]   R. Agarwal, V. Singh, P. Jurney, L. Shi, S. V. Sreenivasan, K. Roy, Mammalian cells preferentially internalize hydrogel nanodiscs over nanorods and use shape-specific uptake mechanisms, Proc. Natl. Acad. Sci. U. S. A. 110 (2013) 17247–17252. https://doi.org/10.1073/pnas.1305000110.

[65]   W.L.L. Suen, Y. Chau, Size-dependent internalisation of folate-decorated nanoparticles via the pathways of clathrin and caveolae-mediated endocytosis in ARPE-19 cells, J. Pharm. Pharmacol. 66 (2014) 564–573. https://doi.org/10.1111/jphp.12134.

[66]   T. Chang, M.S. Lord, B. Bergmann, A. MacMillan, M.H. Stenzel, Size effects of self-assembled block copolymer spherical micelles and vesicles on cellular uptake in human colon carcinoma cells, J. Mater. Chem. B. 2 (2014) 2883–2891. https://doi.org/10.1039/c3tb21751e.

[67]   C.D. Walkey, J.B. Olsen, F. Song, R. Liu, H. Guo, D.W.H. Olsen, Y. Cohen, A. Emili, W.C.W. Chan, Protein corona fingerprinting predicts the cellular interaction of gold and silver nanoparticles, ACS Nano. 8 (2014) 2439–2455. https://doi.org/10.1021/nn406018q.

[68]   M.A. Dobrovolskaia, A.K. Patri, J. Zheng, J.D. Clogston, N. Ayub, P. Aggarwal, B.W. Neun, J.B. Hall, S.E. McNeil, Interaction of colloidal gold nanoparticles with human blood: effects on particle size and analysis of plasma protein binding profiles, Nanomedicine Nanotechnology, Biol. Med. 5 (2009) 106–117. https://doi.org/10.1016/j.nano.2008.08.001.

[69]   S. Tenzer, D. Docter, J. Kuharev, A. Musyanovych, V. Fetz, R. Hecht, F. Schlenk, D. Fischer, K. Kiouptsi, C. Reinhardt, K. Landfester, H. Schild, M. Maskos, S.K. Knauer, R.H. Stauber, Rapid formation of plasma protein corona critically affects nanoparticle pathophysiology, Nat. Nanotechnol. 8 (2013) 772–781. https://doi.org/10.1038/nnano.2013.181.

[70]   T. Tokatlian, B.J. Read, C.A. Jones, D.W. Kulp, S. Menis, J.Y.H. Chang, J.M. Steichen, S. Kumari, J.D. Allen, E.L. Dane, A. Liguori, M. Sangesland, D. Lingwood, M. Crispin, W.R. Schief, D.J. Irvine, Innate immune recognition of glycans targets HIV nanoparticle immunogens to germinal centers, Science (80-.). 363 (2019) 649–654. https://doi.org/10.1126/science.aat9120.

[71]   K.D. Moynihan, R.L. Holden, N.K. Mehta, C. Wang, M.R. Karver, J. Dinter, S. Liang, W. Abraham, M.B. Melo, A.Q. Zhang, N. Li, S. Le Gall, B.L. Pentelute, D.J. Irvine, Enhancement of peptide vaccine immunogenicity by increasing lymphatic drainage and boosting serum stability, Cancer Immunol. Res. 6 (2018) 1025–1038. https://doi.org/10.1158/2326-6066.CIR-17-0607.

[72]   K. Niikura, T. Matsunaga, T. Suzuki, S. Kobayashi, H. Yamaguchi, Y. Orba, A. Kawaguchi, H. Hasegawa, K. Kajino, T. Ninomiya, K. Ijiro, H. Sawa, Gold nanoparticles as a vaccine platform: Influence of size and shape on immunological responses in vitro and in vivo, ACS Nano. 7 (2013) 3926–3938. https://doi.org/10.1021/nn3057005.

[73]   G. Barhate, M. Gautam, S. Gairola, S. Jadhav, V. Pokharkar, Quillaja saponaria extract as mucosal adjuvant with chitosan functionalized gold nanoparticles for mucosal vaccine delivery: Stability and immunoefficiency studies, Int. J. Pharm. 441 (2013) 636–642. https://doi.org/10.1016/j.ijpharm.2012.10.033.

[74]   G. Barhate, M. Gautam, S. Gairola, S. Jadhav, V. Pokharkar, Enhanced mucosal immune responses against tetanus toxoid using novel delivery system comprised of chitosan-functionalized gold nanoparticles and botanical adjuvant: Characterization, immunogenicity, and stability assessment, J. Pharm. Sci. 103 (2014) 3448–3456. https://doi.org/10.1002/jps.24161.

[75]   A.G. Torres, A.E. Gregory, C.L. Hatcher, H. Vinet-Oliphant, L.A. Morici, R.W. Titball, C.J. Roy, Protection of non-human primates against glanders with a gold nanoparticle glycoconjugate vaccine, Vaccine. 33 (2015) 686–692. https://doi.org/10.1016/j.vaccine.2014.11.057.

[76]   M.H. Sarfraz, M. Zubair, B. Aslam, A. Ashraf, M.H. Siddique, S. Hayat, J.N. Cruz, S. Muzammil, M. Khurshid, M.F. Sarfraz, A. Hashem, T.M. Dawoud, G.D. Avila-Quezada, E.F. Abd_Allah, Comparative analysis of phyto-fabricated chitosan, copper oxide, and chitosan-based CuO nanoparticles: antibacterial potential against Acinetobacter baumannii isolates and anticancer activity against HepG2 cell lines, Front. Microbiol. 14 (2023) 1188743. https://doi.org/10.3389/fmicb.2023.1188743.

[77]   D. Wu, W. Fan, A. Kishen, J.L. Gutmann, B. Fan, Evaluation of the antibacterial efficacy of silver nanoparticles against Enterococcus faecalis biofilm, J. Endod. 40 (2014) 285–290. https://doi.org/10.1016/j.joen.2013.08.022.

[78]   V. Dhand, L. Soumya, S. Bharadwaj, S. Chakra, D. Bhatt, B. Sreedhar, Green synthesis of silver nanoparticles using Coffea arabica seed extract and its antibacterial activity, Mater. Sci. Eng. C. 58 (2016) 36–43. https://doi.org/10.1016/j.msec.2015.08.018.

[79]   D. Dinesh, K. Murugan, P. Madhiyazhagan, C. Panneerselvam, P. Mahesh Kumar, M. Nicoletti, W. Jiang, G. Benelli, B. Chandramohan, U. Suresh, Mosquitocidal and antibacterial activity of green-synthesized silver nanoparticles from Aloe vera extracts: towards an effective tool against the malaria vector Anopheles stephensi?, Parasitol. Res. 114 (2015) 1519–1529. https://doi.org/10.1007/s00436-015-4336-z.

[80]   S. Agnihotri, S. Mukherji, S. Mukherji, Size-controlled silver nanoparticles synthesized over the range 5-100 nm using the same protocol and their antibacterial efficacy, RSC Adv. 4 (2014) 3974–3983. https://doi.org/10.1039/c3ra44507k.

[81] R.A. Ismail, G.M. Sulaiman, S.A. Abdulrahman, T.R. Marzoog, Antibacterial activity of magnetic iron oxide nanoparticles synthesized by laser ablation in liquid, Mater. Sci. Eng. C. 53 (2015) 286–297. https://doi.org/10.1016/j.msec.2015.04.047.

[82] O.T. Bruns, T.S. Bischof, D.K. Harris, D. Franke, Y. Shi, L. Riedemann, A. Bartelt, F.B. Jaworski, J.A. Carr, C.J. Rowlands, M.W.B. Wilson, O. Chen, H. Wei, G.W. Hwang, D.M. Montana, I. Coropceanu, O.B. Achorn, J. Kloepper, J. Heeren, P.T.C. So, D. Fukumura, K.F. Jensen, R.K. Jain, M.G. Bawendi, Next-generation in vivo optical imaging with short-wave infrared quantum dots, Nat. Biomed. Eng. 1 (2017). https://doi.org/10.1038/s41551-017-0056.

[83] X. Sun, X. Huang, J. Guo, W. Zhu, Y. Ding, G. Niu, A. Wang, D.O. Kiesewetter, Z.L. Wang, S. Sun, X. Chen, Self-illuminating 64Cu-Doped CdSe/ZnS nanocrystals for in vivo tumor imaging, J. Am. Chem. Soc. 136 (2014) 1706–1709. https://doi.org/10.1021/ja410438n.

[84] R.M. Clauson, M. Chen, L.M. Scheetz, B. Berg, B. Chertok, Size-Controlled Iron Oxide Nanoplatforms with Lipidoid-Stabilized Shells for Efficient Magnetic Resonance Imaging-Trackable Lymph Node Targeting and High-Capacity Biomolecule Display, ACS Appl. Mater. Interfaces. 10 (2018) 20281–20295. https://doi.org/10.1021/acsami.8b02830.

[85] S.W. Chou, Y.H. Shau, P.C. Wu, Y.S. Yang, D. Bin Shieh, C.C. Chen, In vitro and in vivo studies of fept nanoparticles for dual modal CT/MRI molecular imaging, J. Am. Chem. Soc. 132 (2010) 13270–13278. https://doi.org/10.1021/ja1035013.

[86] Y. Lu, Y.J. Xu, G.B. Zhang, D. Ling, M.Q. Wang, Y. Zhou, Y.D. Wu, T. Wu, M.J. Hackett, B.H. Kim, H. Chang, J. Kim, X.T. Hu, L. Dong, N. Lee, F. Li, J.C. He, L. Zhang, H.Q. Wen, B. Yang, S.H. Choi, T. Hyeon, D.H. Zou, Iron oxide nanoclusters for T 1 magnetic resonance imaging of non-human primates article, Nat. Biomed. Eng. 1 (2017) 637–643. https://doi.org/10.1038/s41551-017-0116-7.

[87] K.D. Wegner, Z. Jin, S. Lindén, T.L. Jennings, N. Hildebrandt, Quantum-dot-based förster resonance energy transfer immunoassay for sensitive clinical diagnostics of low-volume serum samples, ACS Nano. 7 (2013) 7411–7419. https://doi.org/10.1021/nn403253y.

[88] K.L. Viola, J. Sbarboro, R. Sureka, M. De, M.A. Bicca, J. Wang, S. Vasavada, S. Satpathy, S. Wu, H. Joshi, P.T. Velasco, K. Macrenaris, E.A. Waters, C. Lu, J. Phan, P. Lacor, P. Prasad, V.P. Dravid, W.L. Klein, Towards non-invasive diagnostic imaging of early-stage Alzheimer's disease, Nat. Nanotechnol. 10 (2015) 91–98. https://doi.org/10.1038/nnano.2014.254.

[89] J. Kim, M.J. Biondi, J.J. Feld, W.C.W. Chan, Clinical Validation of Quantum Dot Barcode Diagnostic Technology, ACS Nano. 10 (2016) 4742–4753. https://doi.org/10.1021/acsnano.6b01254.

[90] L. Fan, H. Qi, J. Teng, B. Su, H. Chen, C. Wang, Q. Xia, Identification of serum miRNAs by nano-quantum dots microarray as diagnostic biomarkers for early detection of non-small cell lung cancer, Tumor Biol. 37 (2016) 7777–7784. https://doi.org/10.1007/s13277-015-4608-3.

[91] S.K. Libutti, G.F. Paciotti, A.A. Byrnes, H.R. Alexander, W.E. Gannon, M. Walker, G.D. Seidel, N. Yuldasheva, L. Tamarkin, Phase I and pharmacokinetic studies of CYT-6091, a novel PEGylated colloidal gold-rhTNF nanomedicine, Clin. Cancer Res. 16 (2010) 6139–6149. https://doi.org/10.1158/1078-0432.CCR-10-0978.

Materials Research Forum LLC
https://doi.org/10.21741/9781644903339-7

[92]   K. Maier-Hauff, F. Ulrich, D. Nestler, H. Niehoff, P. Wust, B. Thiesen, H. Orawa, V. Budach, A. Jordan, Efficacy and safety of intratumoral thermotherapy using magnetic iron-oxide nanoparticles combined with external beam radiotherapy on patients with recurrent glioblastoma multiforme, J. Neurooncol. 103 (2011) 317–324. https://doi.org/10.1007/s11060-010-0389-0.

[93]   S. Zanganeh, G. Hutter, R. Spitler, O. Lenkov, M. Mahmoudi, A. Shaw, J.S. Pajarinen, H. Nejadnik, S. Goodman, M. Moseley, L.M. Coussens, H.E. Daldrup-Link, Iron oxide nanoparticles inhibit tumour growth by inducing pro-inflammatory macrophage polarization in tumour tissues, Nat. Nanotechnol. 11 (2016) 986–994. https://doi.org/10.1038/nnano.2016.168.

[94]   Y.S. Yang, R.P. Carney, F. Stellacci, D.J. Irvine, Enhancing radiotherapy by lipid nanocapsule-mediated delivery of amphiphilic gold nanoparticles to intracellular membranes, ACS Nano. 8 (2014) 8992–9002. https://doi.org/10.1021/nn502146r.

[95]   L.C. Cheng, J.H. Huang, H.M. Chen, T.C. Lai, K.Y. Yang, R.S. Liu, M. Hsiao, C.H. Chen, L.J. Her, D.P. Tsai, Seedless, silver-induced synthesis of star-shaped gold/silver bimetallic nanoparticles as high efficiency photothermal therapy reagent, J. Mater. Chem. 22 (2012) 2244–2253. https://doi.org/10.1039/c1jm13937a.

[96]   L.M. Kranz, M. Diken, H. Haas, S. Kreiter, C. Loquai, K.C. Reuter, M. Meng, D. Fritz, F. Vascotto, H. Hefesha, C. Grunwitz, M. Vormehr, Y. Hüsemann, A. Selmi, A.N. Kuhn, J. Buck, E. Derhovanessian, R. Rae, S. Attig, J. Diekmann, R.A. Jabulowsky, S. Heesch, J. Hassel, P. Langguth, S. Grabbe, C. Huber, Ö. Türeci, U. Sahin, Systemic RNA delivery to dendritic cells exploits antiviral defence for cancer immunotherapy, Nature. 534 (2016) 396–401. https://doi.org/10.1038/nature18300.

[97]   M.H. Schwenk, Ferumoxytol: A new intravenous iron preparation for the treatment of iron deficiency anemia in patients with chronic kidney disease, Pharmacotherapy. 30 (2010) 70–79. https://doi.org/10.1592/phco.30.1.70.

[98]   M. Lu, M.H. Cohen, D. Rieves, R. Pazdur, FDA report: Ferumoxytol for intravenous iron therapy in adult patients with chronic kidney disease, Am. J. Hematol. 85 (2010) 315–319. https://doi.org/10.1002/ajh.21656.

[99]   Y. Zhang, Y. Lu, F. Wang, S. An, Y. Zhang, T. Sun, J. Zhu, C. Jiang, ATP/pH Dual Responsive Nanoparticle with d-[des-Arg10]Kallidin Mediated Efficient In Vivo Targeting Drug Delivery, Small. 13 (2017). https://doi.org/10.1002/smll.201602494.

[100] H.S. El-Sawy, A.M. Al-Abd, T.A. Ahmed, K.M. El-Say, V.P. Torchilin, Stimuli-Responsive Nano-Architecture Drug-Delivery Systems to Solid Tumor Micromilieu: Past, Present, and Future Perspectives, ACS Nano. 12 (2018) 10636–10664. https://doi.org/10.1021/acsnano.8b06104.

[101] S. Kashyap, N. Singh, B. Surnar, M. Jayakannan, Enzyme and thermal dual responsive amphiphilic polymer core-shell nanoparticle for doxorubicin delivery to cancer cells, Biomacromolecules. 17 (2016) 384–398. https://doi.org/10.1021/acs.biomac.5b01545.

[102] M. Grzelczak, L.M. Liz-Marzán, R. Klajn, Stimuli-responsive self-assembly of nanoparticles, Chem. Soc. Rev. 48 (2019) 1342–1361. https://doi.org/10.1039/c8cs00787j.

[103] G. Zhu, F. Zhang, Q. Ni, G. Niu, X. Chen, Efficient Nanovaccine Delivery in Cancer Immunotherapy, ACS Nano. 11 (2017) 2387–2392. https://doi.org/10.1021/acsnano.7b00978.

[104] W. Jiang, C.A. Von Roemeling, Y. Chen, Y. Qie, X. Liu, J. Chen, B.Y.S. Kim, Designing nanomedicine for immuno-oncology, Nat. Biomed. Eng. 1 (2017). https://doi.org/10.1038/s41551-017-0029.

[105] R.S. Riley, C.H. June, R. Langer, M.J. Mitchell, Delivery technologies for cancer immunotherapy, Nat. Rev. Drug Discov. 18 (2019) 175–196. https://doi.org/10.1038/s41573-018-0006-z.

[106] R. Zhang, M.M. Billingsley, M.J. Mitchell, Biomaterials for vaccine-based cancer immunotherapy, J. Control. Release. 292 (2018) 256–276. https://doi.org/10.1016/j.jconrel.2018.10.008.

Advances in Healthcare and Nanoparticle Toxicology
Materials Research Foundations 171 (2024) 217-250

Materials Research Forum LLC
https://doi.org/10.21741/9781644903339-8

Chapter 8

Nanoparticle Interactions with Endothelial Cells

Jibanananda Mishra[1]*, Jiban Jyoti Panda[2]

[1]School of Biosciences, RIMT University, Mandi Gobindgarh, Punjab 147301, India.

[2]Chemical Biology Unit, Institute of Nano Science and Technology, Mohali, Punjab 140306, India

* mjiban@gmail.com

Abstract

Nanoparticles (NPs) have earned significant attention for their prospective applications in various disciplines, including medicine. Discerning the interactions between NPs and biological systems is critical for the development of safe and effective nanomedicines. The interactions between NPs and endothelial cells play a pivotal role in drug delivery, diagnostics, and therapeutic interventions. This article aims on the complex interaction between NPs and endothelial cells, which configure the inner lining of the blood vessels. This chapter also discusses various mechanisms of NP-endothelial cell interactions, their implications in vascular biology, and the potential applications of these interactions in nanomedicine.

Keywords

Nanoparticles, Endothelial Cells, Interactions, Uptake, Therapeutic Applications

Contents

1.   Introduction

Endothelial cells, constituting the single-layer lining in the blood vessels and capillaries, hold a pivotal role in upholding vascular homeostasis, managing immune responses, and regulating blood flow [1,2]. These intricate functions can be disrupted, endangering a range of cardiovascular conditions like atherosclerosis, thrombosis, and hypertension [3]. In recent years, the integration of nanotechnology and biomedical research has widened innovative avenues in diagnostics, imaging, and targeted therapy. In this context, comprehending the intricate interactions between nanoparticles (NPs) and endothelial cells has emerged as a critical endeavor. Not only does it offer insights into vascular biology's complex mechanisms, but it also shapes the development of novel therapeutic strategies. NPs can be strategically engineered to ferry therapeutic agents, posing an alluring opportunity for precision drug delivery to the endothelial cells [4]. Furthermore, investigating NP-endothelial cell interactions facilitates the design of strategies for targeted drug dispensation to specific vascular sites. This approach minimizes systemic side effects and augments treatment efficiency. For instance, functionalized NPs exhibit promise in delivering anti-inflammatory agents to inflamed endothelium in atherosclerosis, as demonstrated by various researchers [5]. Besides, the distinctive optical, magnetic, and radiological properties of NPs position them as excellent candidates for advanced diagnostic imaging. Discerning how these NPs are internalized by endothelial cells and assembled inside the vascular microenvironment guides the development of contrast agents for refined imaging modalities. The research conducted by various workers has highlighted the potential of quantum dot NPs for high-resolution endothelial cell imaging [6]-[7].

Moreover, NPs exert influence on diverse aspects of endothelial cell function, including nitric oxide (NO) production, maintenance of physiological barrier integrity, and modulation of inflammatory responses. This knowledge is critical for devising interventions that counter endothelial dysfunction, a hallmark of numerous cardiovascular conditions. Studies exemplified by Yu et al. (2021) [8], Orlando et al. (2016) [8], and Muller et al. (2017) [8] have delved into the role of gold NPs in mitigating oxidative stress and inflammation in endothelial cells. The growing usage of NPs in biomedical applications accentuates the importance of assessing potential toxicity. Scrutinizing NP-endothelial cell interactions deciphers the mechanisms that trigger cellular responses and cytotoxicity. A number of studies have highlighted the influence of NP size and surface chemistry in determining their impact on endothelial cells [9,10]. The designing of NPs to engage in specific interactions with endothelial cells paves the way for personalized medicine. Unveiling how individual disparities in endothelial biology influence NP uptake and responses empowers the development of patient-tailored therapeutic strategies. Several studies illustrate the potential of personalized NP-based approaches in cardiovascular disease treatment [11]-[12].

The thorough examination of NP interactions with endothelial cells not only boosts our understanding of vascular biology but also pioneers innovative applications in cardiovascular medicine. The capacity to engineer NPs for targeted drug delivery, diagnostics, and the modulation of endothelial function stands poised to revolutionize cardiovascular disease management. The growing body of research underscores the indisputable significance of unraveling the intricate dance between NPs and endothelial cells.

2. Endothelial cells and their role in vascular function

Endothelial cells establish a dynamic interface between circulating blood and surrounding tissues. Their significance extends beyond being mere physical barriers, as they perform a multifaceted job in the regulation of vascular function, maintenance of homeostasis, and influence over various physiological processes. Notably, endothelial cells exhibit remarkable heterogeneity in both morphology and function, adapting to the distinct demands of diverse vascular beds [13]. They create a continuous monolayer lining the luminal surface of blood vessels, characterized by distinct structural features like intercellular junctions, glycocalyx, and specialized membrane domains. Of significance are tight junctions and adherens junctions, crucial for maintaining vascular integrity and barrier function, thereby preventing excessive fluid and solute leakage into surrounding tissues [14,15].

Serving as gatekeepers, endothelial cells control the exchange of gases, nutrients, and immune cells among blood and tissues. The tight regulation of vascular permeability is crucial to maintaining tissue fluid homeostasis and averting edema. This modulation of permeability occurs through paracellular and transcellular pathways, a process guided by complex signaling networks involving VEGF (vascular endothelial growth factors), angiopoietins, and NO [16]. Dysregulation of vascular permeability is linked to conditions like acute lung injury and diabetic retinopathy. Besides, endothelial cells play crucial roles in regulating vascular tone and blood flow. Through the production of vasoactive molecules such as NO, endothelin-1 (ET-1), and prostacyclin, they influence smooth muscle contraction and relaxation. NO produced by endothelial nitric oxide synthase (eNOS), fosters vasodilation and counteracts platelet aggregation, contributing significantly to vascular health. It's important to note that the dysfunction of endothelium-dependent vasodilation is implicated in hypertension and atherosclerosis [17].

Furthermore, endothelial cells actively participate in immune responses by coordinating leukocyte adhesion, extravasation, and tissue infiltration. Upon exposure to inflammatory signals,

Advances in Healthcare and Nanoparticle Toxicology   Materials Research Forum LLC
Materials Research Foundations 171 (2024) 217-250   https://doi.org/10.21741/9781644903339-8

endothelial activation occurs, leading to the upregulation of adhesion molecules like selectins and integrins. This enables leukocyte adherence to the endothelial surface and their subsequent transmigration into the tissues, a process necessary for combating infections and repairing injuries. Chronic endothelial activation, however, contributes to the pathogenesis of conditions like atherosclerosis and other inflammatory diseases [18]. Moreover, the involvement of endothelial cells in angiogenesis, the process of making new blood vessels from preexisting ones, is pivotal. Upon stimulation by pro-angiogenic factors like VEGF, endothelial cells undergo proliferation, migration, and formation of tubes, which are essential for tissue repair, wound healing, and adaptation to physiological changes. Yet, it's important to acknowledge that uncontrolled angiogenesis is implicated in pathological states such as tumor growth and retinopathy [19]-[20].

3. Physicochemical properties of NPs and their influence on endothelial interactions

The intercommunication between NPs and endothelial cells are intricately influenced by the physicochemical properties of the NPs, which encompasses factors like size, shape, surface charge, surface chemistry, and hydrophobicity/hydrophilicity. This profound influence on cellular interactions underscores the significance of comprehending how these attributes collectively shape interactions with endothelial cells, ultimately guiding the architecture of NPs optimized for targeted drug delivery, imaging, and therapeutic interventions [21]-[22].

The size and shape of NPs exert considerable impact on cellular interactions. Smaller NPs capitalize on their elevated surface area-to-volume ratio, resulting in augmented cellular uptake. However, an excessive reduction in size may trigger indiscriminate uptake by endothelial cells, raising concerns about potential toxicity. The geometric shape of NPs emerges as another pivotal factor, influencing cellular internalization mechanisms and binding affinities. For example, rod-shaped NPs chart distinct cellular uptake pathways compared to their spherical counterparts [23]-[24].

An astute selection of size and shape not only augments cellular internalization efficiency but also curtails unintended consequences. The surface charge of NPs, an outcome of functional group presence, fosters electrostatic interactions with negatively charged endothelial cell surfaces. Positively charged NPs exhibit heightened affinity for cellular membranes, often translating to elevated uptake. Conversely, negatively charged NPs may elicit repulsive interactions, thus influencing their internalization dynamics [25]-[26]. This facet of surface charge not only shapes uptake efficiency but also nuances the intracellular trajectory of NPs within endothelial cells.

Surface chemistry, delineating the array of functional groups on the NP surface, emerges as a pivotal modulator of targeting specificity and cellular interactions. Strategic functionalization with ligands that recognize endothelial cell surface receptors facilitates targeted delivery. An illustrative instance is the conjugation of antibodies against specific endothelial markers, which enhances NP uptake by the desired cell type [27,28]. This targeted approach has illuminated avenues in studies concerning tumor angiogenesis and cardiovascular ailments.

Hydrophobicity or hydrophilicity augments the panorama of NP-cell interactions. Hydrophobic NPs might adhere non-specifically to cell membranes, potentially disrupting cellular responses or triggering cytotoxicity. In contrast, hydrophilic NPs display augmented colloidal stability and reduced aggregation in physiological milieus, thereby enhancing their affinity for endothelial cells [29]-[30]. In the realm of biological fluids, NPs swiftly adsorb proteins, culminating in the formation of a protein corona on their surfaces. The composition of this protein corona substantially reshapes NPs' interactions with endothelial cells, exerting implications for cellular uptake, intracellular trafficking, and immune recognition. A nuanced comprehension of protein corona dynamics and its cascading effects on endothelial interactions is indispensable for prognosticating NP behavior within intricate biological contexts [31].

Thus, the physicochemical properties of NPs play a pivotal role in determining their interactions with endothelial cells. Properly tailoring these properties enables the design of NPs with enhanced cellular uptake, targeted delivery, and controlled intracellular trafficking. The optimization of these properties holds great promise for advancing nanoparticle-based therapies and diagnostic tools in various vascular-related diseases.

4.   Endothelial cell receptor-mediated uptake of NPs

Endothelial cell receptor-mediated uptake of NPs involves specific molecular interactions between NPs and cell surface receptors, leading to internalization and subsequent cellular effects [32],[33],[34]. This process is essential for targeted drug delivery, imaging, and therapeutic interventions. Understanding the mechanisms underlying receptor-mediated uptake is crucial for designing NPs that efficiently navigate the complex cellular landscape and achieve desired outcomes.

Endothelial cells have been found to express a variety of surface receptors that can recognize specific ligands presented on NPs. These ligands can be antibodies, peptides, proteins, or other molecular moieties that bind to endothelial receptors with high affinity and specificity. Integrins, selectins, and receptors involved in endocytosis (e.g., scavenger receptors) are commonly targeted for receptor-mediated uptake [35,36]. Receptor-mediated endocytosis is a primary mechanism for NP internalization by endothelial cells, this process is important for the delivery of therapeutic antibodies and other proteins linked to NPs. Kulkarni et. al. (2020) [8] observed that the nanomechanical comportment of cancer and endothelial cells differed when treated with targeted gold-NPs, indicating receptor-dependent and receptor-independent endocytosis processes, respectively. Caveolae, which are small invaginations in the plasma membrane, perform a significant task in NP internalization by endothelial cells [37]. Furthermore, the role of clathrin- and caveolin-mediated endocytosis in nanoparticle uptake has been explored. Clathrin-mediated endocytosis, caveolae-mediated endocytosis, and macropinocytosis are well-studied pathways involved in receptor-mediated NP uptake. Clathrin-coated pits and caveolae are specialized invaginations in the plasma membrane that facilitate the internalization of ligand-bound NPs [38]. Muro et al., 2003 studied the internalization of anti-cell adhesion molecule (CAM) NPs by endothelial cells. They found that multivalent anti-CAM NPs in the size range of 100-300 nm are internalized through CAM-mediated endocytosis [8]. This CAM-mediated endocytosis is distinct from clathrin- and caveolin-mediated endocytosis, phagocytosis, and micropinocytosis.

The size of NPs can influence receptor-mediated uptake. Smaller NPs may access specific cellular compartments more efficiently, while larger NPs can interact with multiple receptors, potentially enhancing their uptake. Furthermore, ligand density on NPs also impacts their binding affinity to receptors and subsequent internalization. Therefore, optimization of these parameters can enhance nanoparticle targeting and uptake [39]-[40].

Wang et al. (2009) investigated the size and dynamics of caveolae, which are involved in NP uptake, in the endothelial cells. They observed that caveolae could internalize multiple smaller NPs, indicating that caveolae are acquiescent structures with the ability to increase their size [37]. This indicates that the size of NPs can affect their uptake through caveolae-mediated pathways. Santos et al. (2011) compared the NP uptake efficiency of various cell lines and found that the internalization of negatively charged polystyrene NPs had a maximum size limit of 200 nm in nonphagocytic B16 cells [41]. This suggests that the size of NPs can influence their internalization by specific cell lines. Napierska et al. (2012) investigated the size-dependent effect of amorphous silica NPs on monocyte adhesion to endothelial cells. They observed a significant upsurge in monocyte adhesion for 18 and 54 nm particles, indicating that the size of NPs can affect their interaction with endothelial cells [42].

Upon internalization, NPs enter the endocytic pathway and may undergo various intracellular trafficking events. They can be sorted to lysosomes for degradation or directed to specific subcellular compartments for functional effects. Some NPs can escape lysosomal degradation, leading to sustained effects within the cell. Thus, understanding these trafficking patterns is important for controlling the intracellular fate of NPs and maximizing their therapeutic potential [43]. Receptor-mediated uptake of NPs offers a targeted approach for drug delivery. By engineering NPs with ligands specific to disease-associated endothelial receptors, therapeutic agents can be selectively delivered to desired vascular sites. This approach has shown promise in various diseases, including cancer, atherosclerosis, and inflammatory disorders [44]-[45].

The receptor-mediated uptake of NPs by endothelial cells is a complex and highly regulated process with significant implications for targeted therapeutics and diagnostics. By harnessing the specific interactions between NPs and endothelial receptors, researchers can design innovative strategies to improve drug delivery efficiency, enhance imaging techniques, and advance our ability to address various vascular-related diseases.

5.   Transport and trafficking of NPs across endothelial barriers

The transport and trafficking of NPs across endothelial barriers involve a complex process influenced by various factors which dictate their distribution within tissues and their interactions with cells. Understanding these mechanisms is important for optimizing NP-based therapies, targeted drug delivery, and imaging applications. This comprehensive overview explores the intricate journey of NPs across endothelial barriers and their implications for biomedical applications.

5.1   Transcytosis

Transcytosis is a fundamental mechanism by which NPs traverse endothelial barriers. This process involves the sequential uptake of NPs at the luminal surface, their transport across the endothelial cell, and their subsequent release at the basolateral surface. Two main pathways, clathrin-mediated and caveolae-mediated transcytosis, have been identified to date involved in this process. Clathrin-mediated transcytosis involves the formation of clathrin-coated vesicles, while caveolae-mediated transcytosis utilizes specialized lipid rafts called caveolae [46,47].

5.2 Intracellular trafficking

Upon internalization, NPs navigate through complex intracellular trafficking routes. They can be trafficked to early endosomes, late endosomes, lysosomes, or recycling endosomes, depending on the NP properties and cellular context. The pH, enzymatic activity, and organelle interactions influence the fate of NPs within the cell. Therefore, controlling intracellular trafficking is crucial for achieving desired therapeutic outcomes while minimizing undesirable effects [48]-[49],[24].

5.3 Endosomal escape

Many NPs face a challenge post-internalization – avoiding degradation within the endo-lysosomal pathway. Efficient endosomal escape is necessary for NPs to exert their intended effects within the cell. Some NPs possess inherent escape mechanisms, while others may require additional strategies, such as pH-responsive coatings or membrane-disrupting peptides, to facilitate their release into the cytosol [50]-[51].

5.4 Extracellular vesicle-mediated transport

Extracellular vesicles (EVs), involving exosomes and microvesicles, have surfaced as vehicles for intercellular communication and potential carriers of NPs. Endothelial cells release EVs that can transport encapsulated NPs to neighboring cells, influencing vascular and tissue responses. EV-mediated transport offers a novel avenue for enhancing nanoparticle delivery and targeting [52]-[53].

5.5 Implications for therapeutic applications

The transport and trafficking of NPs across endothelial barriers have direct implications for therapeutic applications. NPs can be engineered to exploit these mechanisms for targeted drug delivery to specific cell types or tissues. The use of ligands that recognize endothelial receptors can enhance transcytosis efficiency, allowing for effective delivery of therapeutic cargo to desired sites [54],[49].

Overall, the transport and trafficking of NPs across endothelial barriers represent a sophisticated interplay of cellular processes that govern NP distribution and effects. Advances in understanding these processes have far-reaching implications for designing targeted drug delivery systems, optimizing therapeutic efficacy, and unlocking the full potential of nanoparticle-based interventions in various diseases.

6. Nanoparticle-induced endothelial cell signaling pathways

NP interactions with endothelial cells can trigger intricate signaling pathways that profoundly influence cellular responses, ranging from inflammation to oxidative stress. These pathways play an important role in determining the ultimate impact of NPs on vascular health and disease. This exploration provides insight into the various cell signaling pathways activated by NPs in endothelial cells, shedding light on their implications for therapeutic applications.

6.1 Inflammatory signaling pathways

NPs can instigate inflammatory responses in endothelial cells through pathways implicating nuclear factor-kappa B (NF-κB) and mitogen-activated protein kinases (MAPKs) like p38, ERK, and JNK. Activation of these pathways leads to upregulation of pro-inflammatory cytokines (e.g., TNF-α, IL-6) and adhesion molecules (e.g., ICAM-1, VCAM-1), fostering leukocyte adhesion and promoting inflammation [55]-[56].

6.2   Oxidative stress pathways

NP exposure can induce oxidative stress within endothelial cells, steering an imbalance between ROS generation and antioxidant defenses. ROS-mediated activation of redox-sensitive transcription factors, like nuclear factor erythroid 2-related factor 2 (Nrf2), can trigger the expression of cytoprotective genes. However, excessive ROS production can result in cellular damage and dysfunction [57].

6.3   Nitric oxide (NO) signaling

NPs can modulate endothelial NO synthase (eNOS) action and NO manufacture, which plays a significant role in modulating vascular tone and homeostasis. Depending on NP properties, they can either augment or suppress NO production, influencing vasodilation and contributing to endothelial dysfunction in certain cases [58,59].

6.4   Endoplasmic reticulum (ER) stress pathways

ER stress is triggered when unfolded or misfolded proteins get accumulated within the ER. NPs can disrupt ER homeostasis and induce ER stress, giving rise to activation of the unfolded protein response (UPR) pathway. Various reports suggest that prolonged ER stress can lead to endothelial dysfunction, apoptosis, and inflammation [55,60,61].

6.5   Autophagy pathways

Nanoparticle exposure can also modulate autophagy, a cellular process that upholds homeostasis by degrading impaired organelles and proteins. Autophagy can have both protective and detrimental effects on endothelial cells, depending on the context. Autophagy dysregulation can contribute to endothelial dysfunction and disease progression [62]-[63].

6.6   Implications for therapeutic applications

Understanding the signaling pathways triggered by NPs in endothelial cells has direct implications for therapeutic interventions. Targeting specific pathways can help mitigating unwanted side effects while harnessing beneficial responses. For example, NPs can be designed to activate protective pathways such as Nrf2-mediated antioxidant defenses to counteract NP-induced oxidative stress [64].

NP-induced signaling pathways in endothelial cells are pivotal determinants of vascular responses and can dictate the therapeutic outcomes of NP-based interventions. Therefore, a comprehensive understanding of these pathways is essential for designing NPs that harness beneficial signaling while minimizing adverse effects, paving the way for innovative strategies to combat vascular diseases.

Advances in Healthcare and Nanoparticle Toxicology   Materials Research Forum LLC
Materials Research Foundations 171 (2024) 217-250   https://doi.org/10.21741/9781644903339-8

7.   Effects of NPs on endothelial barrier integrity

The endothelial barrier, crafted by a monolayer of endothelial cells lining blood vessels, plays a vital role in maintaining vascular homeostasis by controlling the exchange of nutrients, solutes, and immune cells between the bloodstream and surrounding tissues. The integrity of this barrier is vital for preventing excessive fluid leakage and maintaining tissue fluid balance. NPs have been increasingly investigated for their potential impact on endothelial barrier function, with implications for various physiological and pathological processes. Several studies have explored how NPs influence endothelial barrier integrity, emphasizing the mechanisms involved and their clinical implications.

7.1   Disruption of tight junctions and adherens junctions

Tight junctions and adherens junctions are intercellular structures that ensure the physical integrity of endothelial cell layers. NPs, depending on their physicochemical properties, can compromise the assembly and stability of these junctions. This disruption leads to increased permeability, allowing the passage of larger molecules and cells between endothelial cells [65]. Mechanistically, NPs can affect the distribution and expression of junctional proteins like occludin, claudins, and VE-cadherin, thus impairing junctional stability [66].

7.2   Increased permeability and vascular leakage

The disruption of endothelial junctions by NPs results in increased permeability of the endothelial barrier. This heightened permeability can bring about vascular leakage, which contributes to tissue edema and inflammation. Studies have shown that exposure to certain NPs can increase albumin leakage across endothelial barriers, indicative of barrier dysfunction [67]-[68]. Increased permeability can also facilitate the infiltration of immune cells into tissues, exacerbating inflammatory responses.

7.3   Inflammatory responses and cytokine release

NPs can trigger inflammatory signaling pathways in endothelial cells, leading to the release of pro-inflammatory cytokines, chemokines, and adhesion molecules. These molecules contribute to endothelial activation and leukocyte adhesion, which can further disrupt barrier integrity. Inflammatory responses induced by NPs can be mediated by pathways involving NF-κB, MAPKs, and ROS production [69].

7.4   Endothelial dysfunction and oxidative stress

NPs can induce oxidative stress in endothelial cells, result in endothelial malfunction. Elevated ROS levels can impair nitric oxide (NO) production, a key regulator of vascular tone and barrier function. The disruption of NO-mediated signaling contributes to vascular dysfunction and increased permeability[70] .Oxidative stress can also cause the activation of inflammatory pathways, perpetuating a cycle of barrier disruption and inflammation [71].

7.5   Nanoparticle properties and barrier effects

The effects of NPs on endothelial barrier integrity are highly dependent on their physicochemical properties. Size, surface charge/chemistry, and concentration play crucial roles in determining whether NPs enhance or disrupt the barrier function. Smaller NPs with specific surface

Advances in Healthcare and Nanoparticle Toxicology   Materials Research Forum LLC
Materials Research Foundations 171 (2024) 217-250   https://doi.org/10.21741/9781644903339-8

characteristics may penetrate the endothelial barrier more readily, affecting barrier integrity differently compared to larger NPs [31],[72]-[73].

7.6   Clinical implications

The effects of NPs on endothelial barrier integrity have significant clinical implications. In cardiovascular diseases, barrier dysfunction can contribute to atherosclerosis, hypertension, and edema. In cancer, NP-induced barrier disruption can facilitate drug delivery to tumors but also promote metastasis by enhancing tumor cell transmigration. Designing NPs that can selectively modulate barrier function may hold promise for therapeutic interventions.

In summary, the effects of NPs on endothelial barrier integrity encompass complex interactions and signaling pathways that impact vascular health and disease. Understanding these effects at the molecular and cellular levels is essential for designing nanoparticle-based therapies that maximize therapeutic benefits while minimizing potential disruptions to barrier function.

8.   NP and endothelial inflammation

Endothelial inflammation is a critical process in the pathogenesis of various vascular ailments, including atherosclerosis, hypertension, and inflammatory disorders. NPs, owing to their unique physicochemical properties, have garnered significant attention for their potential to modulate endothelial inflammation. Various studies shed light on the complex mechanisms underlying interactions between NPs and endothelial cells that drive inflammation.

8.1   Inflammatory pathways activation

NPs can activate inflammatory signaling pathways in endothelial cells, eliciting the discharge of pro-inflammatory cytokines, chemokines, and adhesion molecules. The NF-κB pathway, a central regulator of inflammation, is frequently implicated in nanoparticle-induced endothelial inflammation. NPs activate NF-κB by inducing its nuclear translocation and subsequent gene expression, leading to a pro-inflammatory environment [74]-[75].

8.2   ROS Generation and oxidative stress

NPs can induce the generation of ROS in endothelial cells, resulting in oxidative stress. ROS production is a common consequence of NP interactions with cellular components, including membranes and organelles. Elevated ROS levels contribute to endothelial inflammation by activating redox-sensitive transcription factors and promoting inflammatory gene expression [76].

8.3   Endothelial activation and adhesion molecule upregulation

NPs can incite endothelial cell activation, characterized by augmented expression of adhesion molecules like intercellular adhesion molecule-1 (ICAM-1) and vascular cell adhesion molecule-1 (VCAM-1). These molecules facilitate leukocyte adhesion and extravasation, key events in the inflammatory response. NP-induced adhesion molecule upregulation contributes to the recruitment of immune cells to inflamed endothelium [77].

Advances in Healthcare and Nanoparticle Toxicology
Materials Research Foundations 171 (2024) 217-250

Materials Research Forum LLC
https://doi.org/10.21741/9781644903339-8

8.4 Modulation of inflammatory mediators

NPs can influence the secretion of various inflammatory mediators by endothelial cells. These include cytokines (e.g., TNF-α, IL-6), chemokines (e.g., MCP-1), and growth factors. The release of these mediators contributes to the establishment of an inflammatory microenvironment that supports immune cell recruitment and activation [78].

8.5 NP properties and inflammatory responses

The impact of NPs on endothelial inflammation is highly reliant on their physicochemical properties. Nanoparticle size, surface charge/chemistry can all influence their interactions with endothelial cells and subsequent inflammatory responses. Different types of NPs, such as metal-based, polymer-based, and lipid-based NPs, may elicit distinct inflammatory profiles [79],[22].

8.6 Implications for disease and therapy

Understanding the interactions between NPs and endothelial inflammation has implications for both disease development and therapeutic interventions. NPs can exacerbate inflammation in various vascular diseases, contributing to disease progression. Conversely, NPs engineered with anti-inflammatory properties can be utilized to mitigate endothelial inflammation and promote vascular health [80]-[81].

The interactions between NPs and endothelial cells are intricately linked to the regulation of inflammation. These interactions can cause the activation of inflammatory signaling pathways, ROS generation, and adhesion molecule upregulation, ultimately promoting endothelial inflammation. Understanding these complex mechanisms is pivotal for designing NP-based interventions that either suppress or harness these inflammatory responses for therapeutic benefit in vascular diseases and beyond.

9. NP and endothelial dysfunction

Endothelial dysfunction, a hallmark of numerous cardiovascular and vascular diseases, is described by impaired regulation of vascular tone, inflammation, oxidative stress, and barrier function. Emerging evidence suggests that NPs, due to their unique physicochemical properties, can play a significant role in triggering and exacerbating endothelial dysfunction. Recent research findings probe into the complex connections between NPs and endothelial dysfunction, enlightening the underlying mechanisms.

9.1 Impaired NO signaling

Endothelial dysfunction often involves a decline in NO bioavailability, leading to vasoconstriction, platelet aggregation, and inflammation. NPs can influence NO synthesis and bioavailability through multiple mechanisms. For instance, some NPs can interfere with eNOS activity or modulate NO scavenging by ROS, disrupting NO-mediated vasodilation and vascular homeostasis [82].

9.2 Oxidative stress and redox imbalance

NPs can perturb the balance between ROS production and antioxidant defenses, leading to oxidative stress. The interaction between NPs and cellular components can trigger ROS generation, causing oxidative damage to lipids, proteins, and DNA within endothelial cells. Oxidative stress amplifies endothelial dysfunction by promoting inflammation, cell apoptosis, and altered intracellular signaling pathways [83].

9.3 Inflammatory responses and immune activation

NPs can induce inflammatory responses in endothelial cells by activating signaling pathways such as NF-κB and MAPKs. These pathways trigger the release of pro-inflammatory cytokines, chemokines, and adhesion molecules, contributing to endothelial activation and immune cell recruitment. Persistent inflammation is a hallmark of endothelial dysfunction and is linked with vascular pathologies [84,85].

Duan et al. (2013) examined the toxic effects of silica NPs on endothelial cells and observed that the mechanisms involved DNA damage response (DDR) through the Chk1-dependent G2/M checkpoint signaling pathway. This study suggests that exposure to silica NPs could potentially give rise to the development of cardiovascular diseases[86]. Han et al. (2013) focused on titanium dioxide (TiO2) NPs and their effects on vascular endothelial cells. The study demonstrated that TiO2 NPs increased the augmentation of mRNA and protein levels of vascular cell adhesion molecule-1 (VCAM-1) and mRNA levels of monocyte chemoattractant protein-1 (MCP-1). The authors also identified several cellular signaling pathways involved in the inflammatory response, including NF-κB, oxidative stress, Akt, ERK, JNK, and p38. Inhibitors for these pathways attenuated the TiO2 NP-induced expression of MCP-1 and VCAM-1 genes, as well as the activation of NF-κB. These findings suggest that TiO2 NPs can stimulate endothelial inflammatory responses through redox-sensitive cellular signaling pathways [55]. Furthermore, Zhu et al., (2020) investigated the activation of inflammatory responses by cell-penetrating NPs and found that these NPs could activate the inflammasome, an important constituent of the innate immune system. This activation led to enhanced antibody production, suggesting that engineered NPs could be used to potentiate the innate immune response for immunotherapy purposes [87].

9.4 Disruption of barrier integrity

Endothelial barrier dysfunction is a central feature of endothelial dysfunction. NPs, owing to their interactions with endothelial cells, can compromise the integrity of tight junctions and adherens junctions, leading to increased vascular permeability and tissue edema. This barrier disruption contributes to the extravasation of immune cells and potentially exacerbates inflammatory responses[88]-[89],[65]. Liu et al. (2018) showed that NPs in the 20 nm size range can be internalized by endothelial cells through caveolae/raft-mediated endocytosis, leading to an upregulation of intracellular calcium levels. This increase in intracellular calcium triggers myosin light chain kinase, causing actomyosin contraction and subsequent endothelial barrier impairment [90].

9.5 Cellular apoptosis and senescence

NPs can induce cellular apoptosis and senescence in endothelial cells, further contributing to endothelial dysfunction. Apoptosis of endothelial cells impairs the integrity of the endothelial layer, promoting atherosclerosis and other vascular diseases. Senescent endothelial cells exhibit altered secretory profiles, contributing to inflammation and dysfunction of neighboring cells [91],[42].

In a study conducted by Duan et al., 2014, the relationship between NanoSiO2-induced autophagic activity and endothelial disruption was confirmed. The researchers conducted various assessments in primary human umbilical vein endothelial cells (HUVECs) to investigate the biological behavior and toxic effects of nanoparticles on the vasculature. They observed cellular uptake of NPs, ultrastructural changes, monodansylcadaverine (MDC) staining, microtubule-associated protein 1 light chain 3β (LC3) conversion, NO and NOS system, and proinflammatory cytokine expression. These findings suggest that NPs, such as NanoSiO2, can provoke autophagy and endothelial dysfunction through the PI3K/Akt/mTOR signaling pathway [91]. Cao et al., 2014 investigated the effects of nano-sized carbon black (CB) exposure on endothelial dysfunction. The researchers found that exposure to CB NPs caused oxidative stress, endothelial dysfunction, and lipid accumulation in HUVECs and macrophages. The cytotoxicity of CB NPs was particularly evident in monocytes and macrophages. These findings suggest that oxidative stress and inflammation may be central mechanisms in the development of endothelial dysfunction and foam cell formation induced by NPs [92]. Furthermore, statins have been shown to inhibit endothelial senescence and improve endothelial function. In a study by Ota et al., 2010, the induction of eNOS, SIRT1, and catalase by statins was found to inhibit endothelial senescence through the Akt pathway. The activation of the Akt pathway plays a crucial role in promoting endothelial cell survival and preventing senescence. These findings suggest that statins may have a protective effect on endothelial function and can potentially counteract the detrimental effects of NPs on endothelial cells[93].

9.6 NP properties and endothelial dysfunction

NPs have distinctive physicochemical properties that allow them to interact with the endothelial layer and modulate endothelial permeability, a phenomenon known as NP-modulated endothelial leakiness (NanoEL) [94]. The influence of NPs on endothelial dysfunction is intricately linked to their physicochemical properties. NP size, surface charge/chemistry, and composition can modulate their interactions with endothelial cells and subsequent effects on cellular function. Each type of NP may elicit distinct responses, depending on its unique properties [89],[95,96]. For example, silica NPs have been shown to induce oxidative stress and alter mitochondrial dynamics, while polymeric NPs have been developed for targeted siRNA delivery to endothelial cells [58]. Fucoidan NPs have demonstrated antioxidant activity and protective effects on endothelial cell dysfunction in diabetic rats [97–99].

9.7 Implications for therapy and disease

Understanding the connections between NPs and endothelial dysfunction has implications for both disease progression and therapeutic interventions. NPs with anti-inflammatory, antioxidant, or vasodilatory properties can be designed to counteract endothelial dysfunction. Conversely, NPs that exacerbate endothelial dysfunction should be carefully considered in therapeutic design to minimize unwanted effects [82].

The intricate interplay between NPs and endothelial dysfunction underscores the importance of understanding their multifaceted effects on vascular health. By elucidating the underlying mechanisms, researchers can design NPs that either mitigate or exploit these interactions for therapeutic purposes. The potential to modulate endothelial dysfunction through NPs holds promise for advancing strategies to combat cardiovascular and vascular diseases.

10. Therapeutic applications of NPs in targeting endothelial cells

Targeting endothelial cells using NPs holds significant potential for treating various vascular and cardiovascular diseases as these NPs can be precisely designed to deliver various therapeutic agents, modulate cellular signaling, and enhance imaging capabilities, offering innovative strategies for improving patient outcomes. Further, NPs can be functionalized with ligands that recognize endothelial cell surface receptors, allowing for selective targeting of these cells. Such targeted delivery strategies enable site-specific therapeutic interventions and minimize off-target effects. For instance, in anti-angiogenic cancer therapies, NPs conjugated with anti-VEGF antibodies can specifically inhibit angiogenesis by disrupting the signaling pathways crucial for endothelial cell growth and survival [100]-[101]. Besides, NPs can be utilized for controlled drug delivery, improving the efficacy and safety of treatments. These NPs can encapsulate therapeutic agents, protecting them from degradation and facilitating their targeted release. Lipid-based NPs, polymeric NPs, and lipid-polymer hybrid NPs are examples of platforms that have been investigated for drug delivery to endothelial cells. These platforms can be engineered to release drugs in response to specific triggers, such as changes in pH or enzymatic activity within the cellular environment [102].

In the context of regenerative medicine, NPs can enhance endothelialization and tissue repair. Engineered NPs can be integrated into scaffolds for tissue engineering, promoting the adhesion, proliferation, and migration of endothelial cells. NPs can also be loaded with growth factors or cytokines that facilitate angiogenesis and tissue regeneration, offering new avenues for treating ischemic diseases and promoting wound healing [103].

Beyond drug delivery, NPs have gained attention for their potential to modulate cellular signaling pathways. For instance, NPs carrying small interfering RNA (siRNA) can target endothelial cell-specific genes, enabling gene silencing and regulation of key signaling pathways involved in inflammation, angiogenesis, and vascular permeability [104]. This approach holds promise for developing treatments that can attenuate aberrant endothelial cell functions in vascular diseases.

In the field of imaging, NPs offer enhanced contrast and improved visualization of endothelial cells for diagnostic purposes. Quantum dots, gold NPs, and magnetic NPs have been explored as contrast agents for various imaging modalities, including fluorescence imaging, computed tomography (CT), and magnetic resonance imaging (MRI). These imaging agents can enable early detection of endothelial dysfunction, aiding in disease diagnosis and monitoring [105].

NPs offer versatile and targeted approaches for addressing endothelial cell-related disorders and diseases. The ability to precisely deliver therapeutic agents, modulate signaling pathways, and enhance imaging capabilities makes NPs a powerful tool for advancing precision medicine strategies in vascular and cardiovascular therapeutics. As research in this field continues to evolve, NPs hold promise for revolutionizing the way we diagnose, treat, and manage vascular diseases.

11. Safety considerations and biocompatibility of NPs for endothelial applications

The utilization of NPs for targeting endothelial cells in therapeutic applications offers immense potential, but ensuring their safety and biocompatibility is paramount. As NPs interact with biological systems, understanding their potential adverse effects and optimizing their biocompatibility becomes essential. This section discusses the crucial safety considerations and biocompatibility aspects associated with the use of NPs for endothelial cell applications, supported by relevant references.

11.1 NP composition and surface properties

The composition and surface properties of NPs significantly influence their biocompatibility. Careful selection of materials and surface coatings can mitigate adverse effects. Biocompatible coatings, such as polymers or biomolecules, can improve NP stability, reduce nonspecific interactions, and enhance their circulation time within the bloodstream[106]-[107].

11.2 In vitro and in vivo toxicity assessments

Systematic toxicity evaluations are vital to judge the safety of NPs for endothelial cell applications. in vitro studies involving endothelial cell cultures can provide insights into NP-cell interactions, cytotoxicity, and potential mechanisms of toxicity. Animal studies, particularly using relevant models, offer insights into NP biodistribution, clearance, and long-term effects [108].

11.3 Immunological responses

NPs can trigger immune responses that influence their biocompatibility. The immune system's recognition of NPs can give rise to complement activation, opsonization, and macrophage uptake. Modulating nanoparticle size, surface charge, and coating can influence immune responses and enhance their acceptance by the immune system [109,110].

11.4 Endothelial barrier integrity

Maintaining endothelial barrier integrity is crucial for vascular health. NPs should not disrupt tight junctions and adherens junctions, which control barrier function. Studies evaluating nanoparticle-induced changes in vascular permeability and endothelial monolayer integrity are essential to ensure safe endothelial targeting [111].

11.5 Hemocompatibility and blood compatibility

NPs interacting with the bloodstream must possess hemocompatibility to prevent adverse interactions with blood components, such as platelets and red blood cells. Hemocompatibility tests assess NP-induced hemolysis, platelet activation, and clotting events to evaluate their blood compatibility [112].

11.6 Chronic exposure and long-term effects

Nanoparticle safety assessments should consider potential long-term effects of chronic exposure. Chronic NP exposure may lead to accumulation, altered cellular responses, and potential chronic toxicity. Studying the effects of repeated exposure over extended periods is essential for assessing safety in long-term therapeutic applications[113].

11.7 Regulatory approval and clinical translation

Before clinical translation, NP-based therapies targeting endothelial cells must undergo rigorous preclinical evaluation to ensure their safety and efficacy. Regulatory agencies evaluate data on nanoparticle toxicity, biocompatibility, and long-term effects to assess the potential risks and benefits of clinical applications [114]].

Advances in Healthcare and Nanoparticle Toxicology   Materials Research Forum LLC
Materials Research Foundations 171 (2024) 217-250   https://doi.org/10.21741/9781644903339-8

Ensuring the safety and biocompatibility of NPs for endothelial cell applications is crucial for realizing their therapeutic potential. By considering nanoparticle composition, surface properties, immune responses, and long-term effects, researchers can design NPs that minimize adverse interactions and maximize therapeutic benefits, advancing safe and effective targeted therapies for vascular and cardiovascular diseases.

12. Emerging technologies for studying NP-endothelial cell interactions

The study of NP-endothelial cell interactions is a rapidly evolving field, driven by the need to understand the complex mechanisms underlying NP behavior at the cellular level. Emerging technologies and techniques provide insights into the dynamic interactions between NPs and endothelial cells, shedding light on cellular responses, uptake mechanisms, and intracellular fate. This section elaborates extensively on the innovative methods and approaches that are shaping our understanding of NP-endothelial cell interactions, supported by relevant references.

12.1 Advanced imaging techniques

Advanced imaging techniques have revolutionized modern medicine by facilitating the examination of physiological incidents at the molecular level. Techniques such as stimulated emission depletion (STED) microscopy and stochastic optical reconstruction microscopy (STORM) enable visualization of NP-cell interactions at resolutions beyond the diffraction limit, providing detailed insights into subcellular localization and dynamics [115]-135]. Beside, cryo-electron microscopy allows imaging of NP-endothelial cell interactions in their native state, providing high-resolution structural information on cellular uptake mechanisms, intracellular trafficking, and NP localization [116,117].

One area of focus in imaging research is the manufacturing of nanoparticulate molecular imaging agents, which have the capability to image the specificity noninvasively, PK (pharmacokinetic) profiles, biodistribution, and therapeutic efficacy of novel compounds. While there are various types of NPs used for biomedical purposes, lipid-based NP platforms, such as liposomal NPs, micelles, and microemulsions, have gained popularity due to their ease of preparation, functionalization, and ability to combine multiple amphiphilic moieties. These platforms allow for the inclusion of various imaging agents like fluorescent molecules, chelated metals, and nanocrystals [118]. Inorganic nanocrystals, including iron oxide, gold, and quantum dots, as well as natural NPs and hybrid nanostructures, have also been applied in biomedical imaging [118]. The multimodal features of various contrast agent platforms have proven to be useful for validation purposes and discerning particle-target interactions at different level. These platforms permit the combination of multiple imaging techniques, providing a comprehensive view of the target [118]. The functionalization and modulation of PK profile of the NPs have enabled the imaging of key processes in cancer and cardiovascular diseases. The multimodal characteristics of different contrast agent platforms have proven to be valuable for understanding NP-target interactions.

Furthermore, advanced imaging techniques and visualization of NP-endothelial cell interactions have become crucial. Nanostructured materials, with their unique properties such as superparamagnetism and fluorescence, have emerged as promising agents for bioimaging, diagnostics, and therapy [119]. These materials, due to their small size comparable to biomolecules, can be functionalized and conjugated with targeting moieties to achieve specific targeted imaging and therapeutic properties [119]. One example of a targeted NP system is the use

Materials Research Forum LLC
https://doi.org/10.21741/9781644903339-8

of ultrasmall silica NPs functionalized with anti-human epidermal growth factor receptor 2 (HER2) single-chain variable fragments. This system exhibits great tumor-targeting proficiency and efficient renal clearance, making it a flexible platform for imaging and drug delivery [119]. The ability to control the biodistribution of NPs is important for attaining target specificity. By designing NPs that conglomerate the specificity of antibodies with favorable particle biodistribution profiles, researchers have been able to overcome the challenge of limited accumulation in tumor tissues by passive targeting [119]. In the context of imaging techniques, nanomaterials have shown great potential for various imaging modalities. However, their application in X-ray computed tomography (CT) imaging has been relatively limited compared to other techniques such as magnetic resonance imaging (MRI) and fluorescence imaging. This is primarily due to the challenges in preparing nanoparticulate CT contrast agents. Research in this area has lagged behind, but there is growing interest in developing nanomaterials specifically for CT imaging, as they offer advantages such as high X-ray attenuation and tunable properties [119].

12.2   Single-cell analysis

Single-cell RNA sequencing (scRNA-seq) is an efficient technique that enables the profiling of transcriptomes in individual cells. This technology has revolutionized our understanding of cellular heterogeneity and has assisted the identification of distinct cell populations and subtypes by capturing the gene expression profiles of individual cells [142]. This technique provides insights into the cellular response to NP exposure, revealing gene expression changes and identifying specific pathways activated upon interaction [143-145]. In the context of interaction of NPs with endothelial cells, scRNA-seq can provide valuable insights into the transcriptional changes and cellular responses induced by NP exposure. By profiling the transcriptomes of individual endothelial cells prior and post-nanoparticle treatment, researchers can identify differentially expressed genes, signaling pathways, and cellular processes involved in the interaction. This information can contribute to a better understanding of the molecular mechanisms underlying NP-induced effects on endothelial cells. However, careful data analysis and integration with other cellular modalities are necessary to fully understand the complex biological processes at play [142, 146-147].

12.3   Microfluidics systems

Microfluidic platforms mimic the vascular microenvironment and enable real-time observation of NP interactions with endothelial cells under controlled flow conditions. These systems offer insights into transport, adhesion, and transmigration processes [148-149]. Microfluidic platforms have surfaced as promising tools for studying the interactions between NPs and endothelial cells in a controlled and physiologically relevant environment. These platforms mimic the vascular microenvironment and enable real-time observation of nanoparticle interactions with endothelial cells. One example of a microfluidic platform is the blood-brain-barrier (BBB) model developed by Wang et al., 2016 [150]. This model closely mimics the physiological barrier functions of the BBB and has been used for drug permeability screening. The microfluidic platform is designed based on the residence time of blood in the brain tissues of human, letting medium recirculation at physiologically relevant flow rates. This platform has been shown to provide in vivo like barrier properties and can be integrated with other organ modules to simulate multi-organ interactions on drug response [150]. Another example is the lung-on-a-chip model developed by Huh et al., 2010 [151]. This biomimetic microsystem reconstitutes the essential functional alveolar-capillary interface of the human lung. It has been used to study the responses of the lung to bacteria and inflammatory cytokines, as well as the toxic and inflammatory effects of nanoparticles. The lung-on-a-chip model has revealed that cyclic mechanical strain heightens the toxic and inflammatory responses of the lung to silica NPs [151].

Mechanical strain has been found to enhance the uptake and transport of NPs by epithelial and endothelial cells in microfluidic platforms. demonstrated that physiological breathing in a lung-on-a-chip model and in whole mouse lungs have similar effects on nanoparticle absorption. These results feature the importance of integrating mechanical strain in organ-on-a-chip microdevices to better mimic tissue-tissue interfaces critical to organ function. These microdevices have the ability to expand the proficiencies of cell culture models and offer affordable substitutes to animal and clinical studies for drug screening and toxicology applications [151].

Thus, microfluidic platforms provide a valuable tool for studying nanoparticle interactions with endothelial cells. The BBB-on-a-chip and lung-on-a-chip models are examples of microfluidic platforms that closely mimic the physiological functions of the BBB and the lung, respectively. These platforms enable real-time observation of nanoparticle interactions and can be used for drug permeability screening and nanotoxicology studies. Incorporating mechanical strain in these microdevices enhances the uptake and transport of nanoparticles, providing a more realistic representation of tissue-tissue interfaces critical to organ function [150-151].

12.4   Organ-on-a-chip (OCC) systems

On the other hand, OCC systems replicate the organ-level function and interactions, allowing researchers to study NP effects on endothelial cells within a physiologically relevant context. This approach provides insights into systemic effects and intercellular crosstalk [152-153].

OOC systems have emerged as a promising technology for studying organ-level function and interactions in a controlled and physiologically relevant manner. These systems leverage microfabrication and microfluidics approaches to generate cell culture microenvironments that imitate the tissue-tissue interfaces, chemical gradients, and mechanical microenvironments of living organs [151]. One application of OOC systems is the study of the effects of NPs on endothelial cells. Endothelial cells play a crucial role in the transport of NPs in the bloodstream and their interaction with the vascular system. The transport of NPs in vivo is influenced by the flow rate of the bloodstream, which can influence their overall biodistribution. in vitro assays often focus on cancer cells and overlook the interaction between NPs and endothelial cells. However, recent studies have shown that up to 97% of NPs found in tumors enter through trans-endothelial processes [154]. Therefore, it is important to characterize the NP-endothelial cell interaction to better understand their behavior in vivo [154]. The uptake of NPs by endothelial cells can be influenced by various factors, including flow shear and the surface properties of NPs [154]. Studies have shown that rising flow velocity decreases nanoparticle uptake by endothelial cells [154]. However, modifying NP surfaces with endothelial-cell-binding ligands can partially restore uptake, suggesting that functionalizing NPs can enhance their resistance to flow effects. This finding highlights the potential of OOC systems to investigate NP-endothelial cell interactions under flow conditions [154]. Furthermore, the cytotoxicity of NPs on endothelial cells has been investigated using OOC systems. It has been observed that NPs, including alloy NPs, can cause toxic effects on endothelial cells [155]. The concentration, size, and surface properties of NPs, as well as the characteristics of the exposed tissue, can influence their cytotoxicity (Hahn et al., 2012).

For example, Cobalt and Nickel NPs have been found to cause the highest cytotoxicity on endothelial cells [155]. The toxic potential of alloy NPs like those found in cardiovascular implants, needs to be determined to ensure their safety [155]. In conclusion, OOC systems provide a valuable platform for studying NP effects on endothelial cells. These systems allow researchers to mimic the organ-level function and interactions, enabling the investigation of nanoparticle-endothelial cell interactions under flow conditions. They also facilitate the assessment of NP cytotoxicity on endothelial cells, which is crucial for evaluating the safety of NPs used in various applications, including drug delivery and cardiovascular implants [155]. By providing a more physiologically relevant model, OOC systems contribute to the development of effective in vitro disease models and reduce the reliance on animal studies [155-156].

12.5   Label-free sensing techniques

Surface Plasmon Resonance (SPR)-based assays enable label-free real-time monitoring of NP interactions with endothelial cells, revealing binding kinetics, affinity, and cellular responses without the need for exogenous labels [157],[72]. Label-free real-time monitoring of NP interactions with endothelial cells is an important area of research in nanotechnology. Nanomaterials, which are engineered structures with at least one dimension of 100 nanometers or less, have unique properties that make them suitable for various applications, including drug delivery, imaging, and therapy. However, it is crucial to understand the potential toxic effects of these materials on biological systems and the environment [69].

One approach to studying NP interactions with endothelial cells is through the use of gold nanorods. Gold nanorods have been shown to absorb and scatter strongly in the near-infrared (NIR) region, making them suitable for molecular imaging as well as photothermal therapy. In a study by Huang et al. (2006), gold nanorods were synthesized and conjugated to anti-epidermal growth factor receptor (anti-EGFR) monoclonal antibodies. These antibody-conjugated nanorods were then incubated with both nonmalignant and malignant oral epithelial cell lines. The results of this study showed that the anti-EGFR antibody-conjugated nanorods specifically bound to the surface of the malignant cells with a higher affinity due to the overexpression of EGFR on the cytoplasmic membrane of these cells. This specific binding allowed for the visualization and diagnosis of malignant cells from nonmalignant cells using dark field microscopy. Furthermore, when exposed to continuous red laser at 800 nm, the malignant cells required less laser energy to be photothermally destroyed compared to the nonmalignant cells [158]. This study demonstrates the potential of gold nanorods as contrast agents for both molecular imaging and photothermal cancer therapy. The ability to monitor NP interactions with endothelial cells in real-time and label-free provides valuable insights into the behavior and potential toxicity of nanomaterials. By understanding these interactions, researchers can develop safer and more effective nanomaterials for various applications.

12.6   Multi-omics approaches

Integrating proteomic and metabolomic analyses allows the comprehensive profiling of protein and metabolic changes in endothelial cells upon NP exposure. This approach reveals altered pathways and cellular responses at a systems level [159-161].

Comprehensive profiling of protein and metabolic changes in endothelial cells upon NP exposure approach provides valuable perceptions into the mechanisms underlying the harmful effects of

Advances in Healthcare and Nanoparticle Toxicology   Materials Research Forum LLC
Materials Research Foundations 171 (2024) 217-250   https://doi.org/10.21741/9781644903339-8

NPs on endothelial cells and their potential implications for cardiovascular diseases. One study by Mao et al., 2019 aimed to profile metabolic changes relevant to heterogeneous vascular endothelial functions in individuals at extreme cardiovascular risk [162]. This study utilized an untargeted metabolomics approach based on gas chromatography time-of-flight/mass spectrometry (GC-TOF/MS) to analyze the serum of asymptomatic individuals. The researchers sought to investigate potential mechanisms engaged in heterogeneous atherogenesis from metabolic clues. Another study by applied metabolic systems analysis to investigate endothelial dysfunction induced by lipopolysaccharides (LPS) in sepsis endothelial cell culture model in vitro [163]. The researchers used metabolomics to define the metabotypes coupled with augmented endothelial permeability and glycocalyx loss post inflammatory stimuli. They also utilized transcriptomic data to parametrize a metabolic model and analyze plasma metabolomics data from sepsis patients, providing insights into endothelial metabolism in these patients. Furthermore, a study investigated the toxic effects of silica NPs on endothelial cells [101]. The researchers observed dose- and time-dependent cytotoxicity in HUVECs exposed to silica NPs. They found that silica NPs induced ROS generation, oxidative damage, and apoptosis in HUVECs. The study also revealed that silica NPs were internalized into endothelial cells and caused DNA damage response via the Chk1-dependent G2/M checkpoint signaling pathway. Overall, these studies highlight the importance of integrating proteomic and metabolomic analyses to gain a comprehensive understanding of the protein and metabolic changes in endothelial cells upon nanoparticle exposure. This approach provides valuable insights into the mechanisms underlying endothelial dysfunction and the potential implications for cardiovascular diseases. Further research in this field is needed to elucidate the specific pathways and molecular mechanisms involved in nanoparticle-induced endothelial cell toxicity [86].

12.7 Computational modeling and simulations

Computational simulations offer insights into the atomic-level interactions between NPs and endothelial cell membranes. These simulations provide a molecular-level understanding of binding events, insertion mechanisms, and nanoparticle behavior within cellular environments [164-166].

NP-endothelial cell interaction is an important area of study in the field of nanotechnology and biomedical research. Understanding how NPs interact with endothelial cells is crucial for developing effective drug delivery systems and targeted therapies. Several studies have investigated the effects of flow rate on NP uptake into endothelial cells. It has been found that increased flow rates lead to diminished NP uptake by endothelial cells [154]. This is significant because in vivo, the transport of NPs would depend on the flow rate of the bloodstream, which may impact their overall biodistribution. Endothelial cells are mechanically responsive and their propensity to uptake NPs can be influenced by varying flow shear [154]. The influence of shear is particularly important in the context of tumor vasculature, which is disorganized and leads to varying flow rates [154]. Furthermore, the surface properties of NPs play a crucial role in their interaction with endothelial cells. Functionalizing NPs with endothelial-cell-binding ligands can enable them to resist the effects of flow and enhance their uptake by endothelial cells [154]. Additionally, interactions with serum proteins can promote the delivery of NPs to certain cell types, including endothelial cells [167]. The physiochemical interactions between nanomaterials and serum proteins can direct NPs to endothelial cells in vivo and in vitro models, such as the use of HUVECs, have been employed to study NP-endothelial cell interactions [154],[168]. HUVECs are a commonly used model endothelial cell line and have been developed into barriers to study

NP uptake kinetics. These models allow for the characterization of NP uptake in endothelial cells and the investigation of factors that influence this interaction. Overall, computational modeling and simulations can provide valuable insights into NP-endothelial cell interactions. These models can help predict NP behavior in vivo and guide the design of more effective drug delivery systems [168-169]. By understanding the factors that influence NP uptake into endothelial cells, researchers can develop strategies to enhance targeted delivery and improve the efficacy of NP-based therapies.

In summary, the integration of emerging technologies and techniques has revolutionized our ability to study NP-endothelial cell interactions in detail. These innovative approaches provide insights into the dynamic processes occurring at the cellular and molecular levels, aiding our perception of NP uptake mechanisms, intracellular trafficking, and cellular responses. As these techniques continue to advance, they hold promise for accelerating the development of NP-based therapies targeting endothelial cells for various vascular and cardiovascular applications.

13. Future directions in NP-endothelial cell interaction research

The realm of NP-endothelial cell interaction research is dynamic and evolving, propelled by rapid technological advancements and the growing need for innovative therapies targeting vascular and cardiovascular diseases. Future directions in this field are poised to drive breakthroughs that reshape our understanding of NP behavior, enhance therapeutic precision, and ultimately improve patient outcomes.

13.1   Integrative systems biology approaches

The integration of omics data (genomics, transcriptomics, proteomics, metabolomics) would facilitate a systems-level understanding of the intricate responses of endothelial cells to NP interactions. Comprehensive multi-omics analyses can uncover complex signaling networks, identify key regulatory nodes, and reveal novel therapeutic targets, fostering a more holistic perspective on NP effects [170].

13.2   Personalized nanomedicine strategies

Advancements in personalized medicine are extending to NP-based therapies. Tailoring NP properties to individual patients' genetic profiles, disease conditions, and immune responses will enhance treatment efficacy while minimizing side effects. This approach will optimize therapeutic outcomes through patient-specific NP formulations [171-173].

13.3   Multifunctional nanoplatforms

Future research will focus on engineering NPs with multifunctional capabilities. By combining targeting, imaging, and therapeutic functionalities within a single NP, researchers can achieve synergistic effects and maximize the therapeutic potential while minimizing off-target effects. Multifunctional nanoplatforms offer the potential for precision therapeutics in complex diseases.

13.4   Nanoparticle-microbiota interactions

Emerging evidence suggests that the gut microbiota plays a significant role in NP behavior and efficacy. Future research will delve into how NP interactions with endothelial cells are influenced by gut microbiota composition. This understanding would lead to strategies that modulate microbiota to enhance NP therapeutic responses [174-175].

13.5   3D bioprinting and organoids

The integration of 3D bioprinting and organoid technologies would enable the creation of more physiologically relevant models for studying nanoparticle-endothelial cell interactions. By replicating the complex cellular and structural architecture of tissues and organs, these models will offer insights into NP behavior in a more natural context [176-177].

13.6   Smart and responsive nanoparticles

The development of NPs that respond to specific stimuli within the cellular microenvironment would revolutionize therapeutic precision. Smart NPs can be designed to release therapeutic cargo or alter their properties upon encountering specific cues, such as pH changes or enzymatic activity, enhancing therapeutic efficacy while minimizing off-target effects [178-180].

13.7   Advanced imaging and sensing techniques

Continued advancements in imaging and sensing technologies will enable real-time and non-invasive monitoring of NP interactions with endothelial cells. Techniques such as single-molecule imaging, multi-modal imaging, and nanoscale sensors would provide unprecedented insights into dynamic processes at the cellular level [181].

13.8   Ethical and societal implications

As the field advances, addressing the ethical, legal, and societal implications of NP research and applications have become increasingly important. Researchers and policymakers must engage in dialogues with stakeholders to ensure that nanotechnology is deployed responsibly and equitably, and that its benefits are accessible to all segments of society [181-182].

The future of NP-endothelial cell interaction research holds immense promise, with cutting-edge technologies and novel approaches poised to drive transformative discoveries. Integrative systems biology, personalized nanomedicine, smart NPs, and advanced imaging techniques will revolutionize our ability to understand and harness the potential of NPs in vascular and cardiovascular applications. As the field evolves, ethical and societal considerations will play a crucial role in guiding responsible research, ensuring equitable access, and maximizing the positive impact of nanotechnology on human health.

Conclusion

The exploration of NP-endothelial cell interactions has far-reaching implications that spans from therapeutic interventions to fundamental understanding of cellular processes. These interactions have the potential to revolutionize the treatment of vascular and cardiovascular diseases while opening new avenues for scientific discovery. NPs offer unprecedented opportunities for precision therapeutics by enabling targeted drug delivery to endothelial cells. The ability to design NPs that selectively interact with specific endothelial cell receptors holds promise for treating a range of vascular disorders, such as atherosclerosis, angiogenesis-related cancers, and vascular inflammation. By minimizing off-target effects, NPs could enhance therapeutic efficacy and reduce adverse reactions. The multifunctional nature of NPs allows for the integration of multiple therapeutic modalities within a single platform. Combining therapies, such as chemotherapy and immunotherapy, within a single NP enables synergistic effects, where the individual treatments enhance each other's efficacy. This approach may lead to breakthroughs in treating complex diseases that involve intricate cellular processes

Furthermore, NPs equipped with imaging agents enable enhanced visualization of endothelial cells and vasculature. The precise targeting capabilities of NPs coupled with imaging modalities can provide real-time insights into disease progression, vascular changes, and treatment responses. This promises to revolutionize disease diagnostics, enabling early detection and personalized treatment monitoring.

NP-endothelial cell interactions shed light on fundamental cellular processes that extend beyond therapeutic applications. The intricate interplay between NPs and endothelial cells reveals insights into cellular uptake mechanisms, intracellular trafficking, and signaling pathways. This deepens our understanding of basic cellular biology and can lead to the discovery of novel cellular regulatory mechanisms. The study of NP-endothelial cell interactions is fostering the emergence of innovative therapeutic strategies. By understanding how NPs influence endothelial dysfunction, inflammation, and barrier integrity, researchers can design novel therapies that directly target the underlying mechanisms of vascular diseases. This approach holds promise for creating treatments that address the root causes of these disorders. By promoting angiogenesis and endothelial cell migration, NPs could contribute to the development of regenerative therapies. Furthermore, their potential to engineer vascularized tissues and support organ-on-a-chip systems opens possibilities for creating functional tissue models for drug testing and disease modeling.

References

[1]K. Lin, P.-P. Hsu, B.P. Chen, S. Yuan, S. Usami, J.Y.-J. Shyy, Y.-S. Li, S. Chien, Molecular mechanism of endothelial growth arrest by laminar shear stress, Proc. Natl. Acad. Sci. 97 (2000) 9385–9389. https://doi.org/10.1073/pnas.170282597.

[2]D. Gospodarowicz, J. Moran, D. Braun, C. Birdwell, Clonal growth of bovine vascular endothelial cells: fibroblast growth factor as a survival agent., Proc. Natl. Acad. Sci. 73 (1976) 4120–4124. https://doi.org/10.1073/pnas.73.11.4120.

[3]M.A. Gimbrone, G. García-Cardeña, Endothelial Cell Dysfunction and the Pathobiology of Atherosclerosis, Circ. Res. 118 (2016) 620–636. https://doi.org/10.1161/CIRCRESAHA.115.306301.

[4]B.T. Luk, L. Zhang, Cell membrane-camouflaged nanoparticles for drug delivery, J. Controlled Release 220 (2015) 600–607. https://doi.org/10.1016/j.jconrel.2015.07.019.

[5]J.-I. Koga, T. Matoba, K. Egashira, Anti-inflammatory Nanoparticle for Prevention of Atherosclerotic Vascular Diseases, J. Atheroscler. Thromb. 23 (2016) 757–765. https://doi.org/10.5551/jat.35113.

[6]A. Jayagopal, P.K. Russ, F.R. Haselton, Surface Engineering of Quantum Dots for In Vivo Vascular Imaging, Bioconjug. Chem. 18 (2007) 1424–1433. https://doi.org/10.1021/bc070020r.

[7] Y. Zhao, Y. Zhang, G. Qin, J. Cheng, W. Zeng, S. Liu, H. Kong, X. Wang, Q. Wang, H. Qu, In vivo biodistribution and behavior of CdTe/ZnS quantum dots, Int. J. Nanomedicine Volume 12 (2017) 1927–1939. https://doi.org/10.2147/IJN.S121075.

[8] Y. Yu, J. Gao, L. Jiang, J. Wang, Antidiabetic nephropathy effects of synthesized gold nanoparticles through mitigation of oxidative stress, Arab. J. Chem. 14 (2021) 103007. https://doi.org/10.1016/j.arabjc.2021.103007.

[9] Y. Pan, S. Neuss, A. Leifert, M. Fischler, F. Wen, U. Simon, G. Schmid, W. Brandau, W. Jahnen-Dechent, Size-Dependent Cytotoxicity of Gold Nanoparticles, Small 3 (2007) 1941–1949. https://doi.org/10.1002/smll.200700378.

[10] M.H. Sarfraz, M. Zubair, B. Aslam, A. Ashraf, M.H. Siddique, S. Hayat, J.N. Cruz, S. Muzammil, M. Khurshid, M.F. Sarfraz, A. Hashem, T.M. Dawoud, G.D. Avila-Quezada, E.F. Abd_Allah, Comparative analysis of phyto-fabricated chitosan, copper oxide, and chitosan-based CuO nanoparticles: antibacterial potential against Acinetobacter baumannii isolates and anticancer activity against HepG2 cell lines, Front. Microbiol. 14 (2023) 1188743. https://doi.org/10.3389/fmicb.2023.1188743.

[11] P. Gupta, E. Garcia, A. Sarkar, S. Kapoor, K. Rafiq, H.S. Chand, R.D. Jayant, Nanoparticle Based Treatment for Cardiovascular Diseases, Cardiovasc. Hematol. Disord.-Drug Targets 19 (2019) 33–44. https://doi.org/10.2174/1871529X18666180508113253.

[12] F. Li, H. Shao, G. Zhou, B. Wang, Y. Xu, W. Liang, L. Chen, The recent applications of nanotechnology in the diagnosis and treatment of common cardiovascular diseases, Vascul. Pharmacol. 152 (2023) 107200. https://doi.org/10.1016/j.vph.2023.107200.

[13] P. Mancuso, A. Calleri, G. Gregato, V. Labanca, J. Quarna, P. Antoniotti, L. Cuppini, G. Finocchiaro, M. Eoli, V. Rosti, F. Bertolini, A Subpopulation of Circulating Endothelial Cells Express CD109 and is Enriched in the Blood of Cancer Patients, PLoS ONE 9 (2014) e114713. https://doi.org/10.1371/journal.pone.0114713.

[14] Y.-J. Chiu, K. Kusano, T.N. Thomas, K. Fujiwara, Endothelial Cell-Cell Adhesion and Mechanosignal Transduction, Endothelium 11 (2004) 59–73. https://doi.org/10.1080/10623320490432489.

[15] C. Cerutti, A.J. Ridley, Endothelial cell-cell adhesion and signaling, Exp. Cell Res. 358 (2017) 31–38. https://doi.org/10.1016/j.yexcr.2017.06.003.

[16] D. Mehta, A.B. Malik, Signaling Mechanisms Regulating Endothelial Permeability, Physiol. Rev. 86 (2006) 279–367. https://doi.org/10.1152/physrev.00012.2005.

[17] F.H. Epstein, S. Moncada, A. Higgs, The L-Arginine-Nitric Oxide Pathway, N. Engl. J. Med. 329 (1993) 2002–2012. https://doi.org/10.1056/NEJM199312303292706.

[18] K. Ley, C. Laudanna, M.I. Cybulsky, S. Nourshargh, Getting to the site of inflammation: the leukocyte adhesion cascade updated, Nat. Rev. Immunol. 7 (2007) 678–689. https://doi.org/10.1038/nri2156.

[19] P. Carmeliet, R.K. Jain, Molecular mechanisms and clinical applications of angiogenesis, Nature 473 (2011) 298–307. https://doi.org/10.1038/nature10144.

[20]   N. Ferrara, K. Alitalo, Clinical applications of angiogenic growth factors and their inhibitors, Nat. Med. 5 (1999) 1359–1364. https://doi.org/10.1038/70928.

[21]   T.M. Kiio, S. Park, Physical properties of nanoparticles do matter, J. Pharm. Investig. 51 (2021) 35–51. https://doi.org/10.1007/s40005-020-00504-w.

[22]   X. Duan, Y. Li, Physicochemical Characteristics of Nanoparticles Affect Circulation, Biodistribution, Cellular Internalization, and Trafficking, Small 9 (2013) 1521–1532. https://doi.org/10.1002/smll.201201390.

[23]   Y. Zhao, Y. Wang, F. Ran, Y. Cui, C. Liu, Q. Zhao, Y. Gao, D. Wang, S. Wang, A comparison between sphere and rod nanoparticles regarding their in vivo biological behavior and pharmacokinetics, Sci. Rep. 7 (2017) 4131. https://doi.org/10.1038/s41598-017-03834-2.

[24]   P. Foroozandeh, A.A. Aziz, Insight into Cellular Uptake and Intracellular Trafficking of Nanoparticles, Nanoscale Res. Lett. 13 (2018) 339. https://doi.org/10.1186/s11671-018-2728-6.

[25]   S. Behzadi, V. Serpooshan, W. Tao, M.A. Hamaly, M.Y. Alkawareek, E.C. Dreaden, D. Brown, A.M. Alkilany, O.C. Farokhzad, M. Mahmoudi, Cellular uptake of nanoparticles: journey inside the cell, Chem. Soc. Rev. 46 (2017) 4218–4244. https://doi.org/10.1039/C6CS00636A.

[26]   E. Blanco, H. Shen, M. Ferrari, Principles of nanoparticle design for overcoming biological barriers to drug delivery, Nat. Biotechnol. 33 (2015) 941–951. https://doi.org/10.1038/nbt.3330.

[27]   A. Amruta, D. Iannotta, S.W. Cheetham, T. Lammers, J. Wolfram, Vasculature organotropism in drug delivery, Adv. Drug Deliv. Rev. 201 (2023) 115054. https://doi.org/10.1016/j.addr.2023.115054.

[28]   M.D. Howard, E.D. Hood, C.F. Greineder, I.S. Alferiev, M. Chorny, V. Muzykantov, Targeting to Endothelial Cells Augments the Protective Effect of Novel Dual Bioactive Antioxidant/Anti-Inflammatory Nanoparticles, Mol. Pharm. 11 (2014) 2262–2270. https://doi.org/10.1021/mp400677y.

[29]   S.H. Anastasiadis, K. Chrissopoulou, E. Stratakis, P. Kavatzikidou, G. Kaklamani, A. Ranella, How the Physicochemical Properties of Manufactured Nanomaterials Affect Their Performance in Dispersion and Their Applications in Biomedicine: A Review, Nanomaterials 12 (2022) 552. https://doi.org/10.3390/nano12030552.

[30]   S. Gelperina, O. Maksimenko, A. Khalansky, L. Vanchugova, E. Shipulo, K. Abbasova, R. Berdiev, S. Wohlfart, N. Chepurnova, J. Kreuter, Drug delivery to the brain using surfactant-coated poly(lactide-co-glycolide) nanoparticles: Influence of the formulation parameters, Eur. J. Pharm. Biopharm. 74 (2010) 157–163. https://doi.org/10.1016/j.ejpb.2009.09.003.

[31]   M.P. Monopoli, D. Walczyk, A. Campbell, G. Elia, I. Lynch, F. Baldelli Bombelli, K.A. Dawson, Physical–Chemical Aspects of Protein Corona: Relevance to in Vitro and in Vivo Biological Impacts of Nanoparticles, J. Am. Chem. Soc. 133 (2011) 2525–2534. https://doi.org/10.1021/ja107583h.

[32]   M. Pacurari, Y. Qian, W. Fu, D. Schwegler-Berry, M. Ding, V. Castranova, N.L. Guo, Cell Permeability, Migration, and Reactive Oxygen Species Induced by Multiwalled Carbon Nanotubes in Human Microvascular Endothelial Cells, J. Toxicol. Environ. Health A 75 (2012) 112–128. https://doi.org/10.1080/15287394.2011.615110.

[33]   B. Dehouck, M.P. Dehouck, J.C. Fruchart, R. Cecchelli, Upregulation of the low density lipoprotein receptor at the blood-brain barrier: intercommunications between brain capillary endothelial cells and astrocytes., J. Cell Biol. 126 (1994) 465–473. https://doi.org/10.1083/jcb.126.2.465.

[34]   R.E. Serda, J. Gu, J.K. Burks, K. Ferrari, C. Ferrari, M. Ferrari, Quantitative mechanics of endothelial phagocytosis of silicon microparticles, Cytometry A 75A (2009) 752–760. https://doi.org/10.1002/cyto.a.20769.

[35]   S. Muro, ed., Drug Delivery Systems that Fuse with Plasmalemma, in: Drug Deliv. Physiol. Barriers, 0 ed., Jenny Stanford Publishing, 2016: pp. 309–330. https://doi.org/10.1201/b19907-16.

[36]   S. Xu, B.Z. Olenyuk, C.T. Okamoto, S.F. Hamm-Alvarez, Targeting receptor-mediated endocytotic pathways with nanoparticles: Rationale and advances, Adv. Drug Deliv. Rev. 65 (2013) 121–138. https://doi.org/10.1016/j.addr.2012.09.041.

[37]   Z. Wang, C. Tiruppathi, R.D. Minshall, A.B. Malik, Size and Dynamics of Caveolae Studied Using Nanoparticles in Living Endothelial Cells, ACS Nano 3 (2009) 4110–4116. https://doi.org/10.1021/nn9012274.

[38]   S. Mayor, R.E. Pagano, Pathways of clathrin-independent endocytosis, Nat. Rev. Mol. Cell Biol. 8 (2007) 603–612. https://doi.org/10.1038/nrm2216.

[39]   J. Yoo, C. Park, G. Yi, D. Lee, H. Koo, Active Targeting Strategies Using Biological Ligands for Nanoparticle Drug Delivery Systems, Cancers 11 (2019) 640. https://doi.org/10.3390/cancers11050640.

[40]   A.K. Pearce, R.K. O'Reilly, Insights into Active Targeting of Nanoparticles in Drug Delivery: Advances in Clinical Studies and Design Considerations for Cancer Nanomedicine, Bioconjug. Chem. 30 (2019) 2300–2311. https://doi.org/10.1021/acs.bioconjchem.9b00456.

[41]   T. Dos Santos, J. Varela, I. Lynch, A. Salvati, K.A. Dawson, Quantitative Assessment of the Comparative Nanoparticle-Uptake Efficiency of a Range of Cell Lines, Small 7 (2011) 3341–3349. https://doi.org/10.1002/smll.201101076.

[42]   C. Guo, M. Yang, L. Jing, J. Wang, Y. Yu, Y. Li, J. Duan, X. Zhou, Y. Li, Z. Zwsun@Ccmu.Edu.Cn, Amorphous silica nanoparticles trigger vascular endothelial cell injury through apoptosis and autophagy via reactive oxygen species-mediated MAPK/Bcl-2 and PI3K/Akt/mTOR signaling, Int. J. Nanomedicine Volume 11 (2016) 5257–5276. https://doi.org/10.2147/IJN.S112030.

[43]   P. Laux, C. Riebeling, A.M. Booth, J.D. Brain, J. Brunner, C. Cerrillo, O. Creutzenberg, I. Estrela-Lopis, T. Gebel, G. Johanson, H. Jungnickel, H. Kock, J. Tentschert, A. Tlili, A. Schäffer, A.J.A.M. Sips, R.A. Yokel, A. Luch, Biokinetics of nanomaterials: The role of biopersistence, NanoImpact 6 (2017) 69–80. https://doi.org/10.1016/j.impact.2017.03.003.

[44]   B. Bahrami, M. Hojjat-Farsangi, H. Mohammadi, E. Anvari, G. Ghalamfarsa, M. Yousefi, F. Jadidi-Niaragh, Nanoparticles and targeted drug delivery in cancer therapy, Immunol. Lett. 190 (2017) 64–83. https://doi.org/10.1016/j.imlet.2017.07.015.

[45]   S. Raj, S. Khurana, R. Choudhari, K.K. Kesari, M.A. Kamal, N. Garg, J. Ruokolainen, B.C. Das, D. Kumar, Specific targeting cancer cells with nanoparticles and drug delivery in cancer therapy, Semin. Cancer Biol. 69 (2021) 166–177. https://doi.org/10.1016/j.semcancer.2019.11.002.

[46]   L. Bareford, P. Swaan, Endocytic mechanisms for targeted drug delivery☆, Adv. Drug Deliv. Rev. 59 (2007) 748–758. https://doi.org/10.1016/j.addr.2007.06.008.

[47]   R.V. Stan, Endothelial stomatal and fenestral diaphragms in normal vessels and angiogenesis, J. Cell. Mol. Med. 11 (2007) 621–643. https://doi.org/10.1111/j.1582-4934.2007.00075.x.

[48]   R. Bawa, Nanoparticle-based therapeutics in humans: A survey, Nanotechnol. Law Bus. 5 (2008) 135–155.

[49]   N.D. Donahue, H. Acar, S. Wilhelm, Concepts of nanoparticle cellular uptake, intracellular trafficking, and kinetics in nanomedicine, Adv. Drug Deliv. Rev. 143 (2019) 68–96. https://doi.org/10.1016/j.addr.2019.04.008.

[50]   S.A. Smith, L.I. Selby, A.P.R. Johnston, G.K. Such, The Endosomal Escape of Nanoparticles: Toward More Efficient Cellular Delivery, Bioconjug. Chem. 30 (2019) 263–272. https://doi.org/10.1021/acs.bioconjchem.8b00732.

[51]   G. Sahay, W. Querbes, C. Alabi, A. Eltoukhy, S. Sarkar, C. Zurenko, E. Karagiannis, K. Love, D. Chen, R. Zoncu, Y. Buganim, A. Schroeder, R. Langer, D.G. Anderson, Efficiency of siRNA delivery by lipid nanoparticles is limited by endocytic recycling, Nat. Biotechnol. 31 (2013) 653–658. https://doi.org/10.1038/nbt.2614.

[52]   P. Vader, E.A. Mol, G. Pasterkamp, R.M. Schiffelers, Extracellular vesicles for drug delivery, Adv. Drug Deliv. Rev. 106 (2016) 148–156. https://doi.org/10.1016/j.addr.2016.02.006.

[53]   O.M. Elsharkasy, J.Z. Nordin, D.W. Hagey, O.G. De Jong, R.M. Schiffelers, S.E. Andaloussi, P. Vader, Extracellular vesicles as drug delivery systems: Why and how?, Adv. Drug Deliv. Rev. 159 (2020) 332–343. https://doi.org/10.1016/j.addr.2020.04.004.

[54]   J.A. Champion, S. Mitragotri, Role of target geometry in phagocytosis, Proc. Natl. Acad. Sci. 103 (2006) 4930–4934. https://doi.org/10.1073/pnas.0600997103.

[55]   S.G. Han, B. Newsome, B. Hennig, Titanium dioxide nanoparticles increase inflammatory responses in vascular endothelial cells, Toxicology 306 (2013) 1–8. https://doi.org/10.1016/j.tox.2013.01.014.

[56]   T.-C. Tsou, S.-C. Yeh, F.-Y. Tsai, H.-J. Lin, T.-J. Cheng, H.-R. Chao, L.-A. Tai, Zinc oxide particles induce inflammatory responses in vascular endothelial cells via NF-κB signaling, J. Hazard. Mater. 183 (2010) 182–188. https://doi.org/10.1016/j.jhazmat.2010.07.010.

[57]   K. Peters, R.E. Unger, A.M. Gatti, E. Sabbioni, R. Tsaryk, C.J. Kirkpatrick, Metallic Nanoparticles Exhibit Paradoxical Effects on Oxidative Stress and Pro-Inflammatory Response in Endothelial Cells in Vitro, Int. J. Immunopathol. Pharmacol. 20 (2007) 685–695. https://doi.org/10.1177/039463200702000404.

[58]   M.D. Mauricio, S. Guerra-Ojeda, P. Marchio, S.L. Valles, M. Aldasoro, I. Escribano-Lopez, J.R. Herance, M. Rocha, J.M. Vila, V.M. Victor, Nanoparticles in Medicine: A Focus on Vascular Oxidative Stress, Oxid. Med. Cell. Longev. 2018 (2018) 1–20. https://doi.org/10.1155/2018/6231482.

[59]   G. Taneja, A. Sud, N. Pendse, B. Panigrahi, A. Kumar, A.K. Sharma, Nano-medicine and Vascular Endothelial Dysfunction: Options and Delivery Strategies, Cardiovasc. Toxicol. 19 (2019) 1–12. https://doi.org/10.1007/s12012-018-9491-x.

[60]   M.L. Carmo Bastos, J.V. Silva-Silva, J. Neves Cruz, A.R. Palheta da Silva, A.A. Bentaberry-Rosa, G. da Costa Ramos, J.E. de Sousa Siqueira, M.R. Coelho-Ferreira, S. Percário, P. Santana Barbosa Marinho, A.M. do R. Marinho, M. de Oliveira Bahia, M.F. Dolabela, Alkaloid from Geissospermum sericeum Benth. & Hook.f. ex Miers (Apocynaceae) Induce Apoptosis by Caspase Pathway in Human Gastric Cancer Cells, Pharmaceuticals 16 (2023) 765. https://doi.org/10.3390/ph16050765.

[61]   Y. Cao, J. Long, L. Liu, T. He, L. Jiang, C. Zhao, Z. Li, A review of endoplasmic reticulum (ER) stress and nanoparticle (NP) exposure, Life Sci. 186 (2017) 33–42. https://doi.org/10.1016/j.lfs.2017.08.003.

[62]   Q. Li, Y. Feng, R. Wang, R. Liu, Y. Ba, H. Huang, Recent insights into autophagy and metals/nanoparticles exposure, Toxicol. Res. 39 (2023) 355–372. https://doi.org/10.1007/s43188-023-00184-2.

[63]   L. Jia, S.-L. Hao, W.-X. Yang, Nanoparticles induce autophagy via mTOR pathway inhibition and reactive oxygen species generation, Nanomed. 15 (2020) 1419–1435. https://doi.org/10.2217/nnm-2019-0387.

[64]   W. Osburn, T. Kensler, Nrf2 signaling: An adaptive response pathway for protection against environmental toxic insults, Mutat. Res. Mutat. Res. 659 (2008) 31–39. https://doi.org/10.1016/j.mrrev.2007.11.006.

[65]   E. Dejana, Endothelial cell–cell junctions: happy together, Nat. Rev. Mol. Cell Biol. 5 (2004) 261–270. https://doi.org/10.1038/nrm1357.

[66]   C.-H. Li, M.-K. Shyu, C. Jhan, Y.-W. Cheng, C.-H. Tsai, C.-W. Liu, C.-C. Lee, R.-M. Chen, J.-J. Kang, Gold Nanoparticles Increase Endothelial Paracellular Permeability by Altering Components of Endothelial Tight Junctions, and Increase Blood-Brain Barrier Permeability in Mice, Toxicol. Sci. 148 (2015) 192–203. https://doi.org/10.1093/toxsci/kfv176.

[67]   S. Tenzer, D. Docter, J. Kuharev, A. Musyanovych, V. Fetz, R. Hecht, F. Schlenk, D. Fischer, K. Kiouptsi, C. Reinhardt, K. Landfester, H. Schild, M. Maskos, S.K. Knauer, R.H. Stauber, Rapid formation of plasma protein corona critically affects nanoparticle pathophysiology, Nat. Nanotechnol. 8 (2013) 772–781. https://doi.org/10.1038/nnano.2013.181.

[68]   A. Aliyandi, C. Reker-Smit, R. Bron, I.S. Zuhorn, A. Salvati, Correlating Corona Composition and Cell Uptake to Identify Proteins Affecting Nanoparticle Entry into Endothelial Cells, ACS Biomater. Sci. Eng. 7 (2021) 5573–5584. https://doi.org/10.1021/acsbiomaterials.1c00804.

[69]   A. Nel, T. Xia, L. Mädler, N. Li, Toxic Potential of Materials at the Nanolevel, Science 311 (2006) 622–627. https://doi.org/10.1126/science.1114397.

[70]   U. Förstermann, Nitric oxide and oxidative stress in vascular disease, Pflüg. Arch. - Eur. J. Physiol. 459 (2010) 923–939. https://doi.org/10.1007/s00424-010-0808-2.

[71]   Y. Cao, The Toxicity of Nanoparticles to Human Endothelial Cells, in: Q. Saquib, M. Faisal, A.A. Al-Khedhairy, A.A. Alatar (Eds.), Cell. Mol. Toxicol. Nanoparticles, Springer International Publishing, Cham, 2018: pp. 59–69. https://doi.org/10.1007/978-3-319-72041-8_4.

[72]   M. Shilo, A. Sharon, K. Baranes, M. Motiei, J.-P.M. Lellouche, R. Popovtzer, The effect of nanoparticle size on the probability to cross the blood-brain barrier: an in-vitro endothelial cell model, J. Nanobiotechnology 13 (2015) 19. https://doi.org/10.1186/s12951-015-0075-7.

[73]   A. Aliyandi, S. Satchell, R.E. Unger, B. Bartosch, R. Parent, I.S. Zuhorn, A. Salvati, Effect of endothelial cell heterogeneity on nanoparticle uptake, Int. J. Pharm. 587 (2020) 119699. https://doi.org/10.1016/j.ijpharm.2020.119699.

[74]   I. Rahman, S.K. Biswas, P.A. Kirkham, Regulation of inflammation and redox signaling by dietary polyphenols, Biochem. Pharmacol. 72 (2006) 1439–1452. https://doi.org/10.1016/j.bcp.2006.07.004.

[75]   B. Li, M. Tang, Research progress of nanoparticle toxicity signaling pathway, Life Sci. 263 (2020) 118542. https://doi.org/10.1016/j.lfs.2020.118542.

[76]   M. Valko, D. Leibfritz, J. Moncol, M.T.D. Cronin, M. Mazur, J. Telser, Free radicals and antioxidants in normal physiological functions and human disease, Int. J. Biochem. Cell Biol. 39 (2007) 44–84. https://doi.org/10.1016/j.biocel.2006.07.001.

[77]   T.A. Springer, Adhesion receptors of the immune system, Nature 346 (1990) 425–434. https://doi.org/10.1038/346425a0.

[78]   I.F. Charo, R.M. Ransohoff, The Many Roles of Chemokines and Chemokine Receptors in Inflammation, N. Engl. J. Med. 354 (2006) 610–621. https://doi.org/10.1056/NEJMra052723.

[79]   A. Sukhanova, S. Bozrova, P. Sokolov, M. Berestovoy, A. Karaulov, I. Nabiev, Dependence of Nanoparticle Toxicity on Their Physical and Chemical Properties, Nanoscale Res. Lett. 13 (2018) 44. https://doi.org/10.1186/s11671-018-2457-x.

[80]   V. Lenders, X. Koutsoumpou, A. Sargsian, B.B. Manshian, Biomedical nanomaterials for immunological applications: ongoing research and clinical trials, Nanoscale Adv. 2 (2020) 5046–5089. https://doi.org/10.1039/D0NA00478B.

[81]   M.A. Dobrovolskaia, S.E. McNeil, Immunological properties of engineered nanomaterials, Nat. Nanotechnol. 2 (2007) 469–478. https://doi.org/10.1038/nnano.2007.223.

[82]   T. Heitzer, T. Schlinzig, K. Krohn, T. Meinertz, T. Münzel, Endothelial Dysfunction, Oxidative Stress, and Risk of Cardiovascular Events in Patients With Coronary Artery Disease, Circulation 104 (2001) 2673–2678. https://doi.org/10.1161/hc4601.099485.

[83]   A. Abdal Dayem, M. Hossain, S. Lee, K. Kim, S. Saha, G.-M. Yang, H. Choi, S.-G. Cho, The Role of Reactive Oxygen Species (ROS) in the Biological Activities of Metallic Nanoparticles, Int. J. Mol. Sci. 18 (2017) 120. https://doi.org/10.3390/ijms18010120.

[84]   Z. Wang, M. Tang, Research progress on toxicity, function, and mechanism of metal oxide nanoparticles on vascular endothelial cells, J. Appl. Toxicol. 41 (2021) 683–700. https://doi.org/10.1002/jat.4121.

[85]   J.N. Cruz, S.G. Silva, D.S. Pereira, A.P. da S. Souza Filho, M.S. de Oliveira, R.R. Lima, E.H. de A. Andrade, In Silico Evaluation of the Antimicrobial Activity of Thymol—Major Compounds in the Essential Oil of Lippia thymoides Mart. & Schauer (Verbenaceae), Molecules 27 (2022). https://doi.org/10.3390/molecules27154768.

[86]   J. Duan, Y. Yu, Y. Li, Y. Yu, Y. Li, X. Zhou, P. Huang, Z. Sun, Toxic Effect of Silica Nanoparticles on Endothelial Cells through DNA Damage Response via Chk1-Dependent G2/M Checkpoint, PLoS ONE 8 (2013) e62087. https://doi.org/10.1371/journal.pone.0062087.

[87]   M. Zhu, L. Du, R. Zhao, H.Y. Wang, Y. Zhao, G. Nie, R.-F. Wang, Cell-Penetrating Nanoparticles Activate the Inflammasome to Enhance Antibody Production by Targeting Microtubule-Associated Protein 1-Light Chain 3 for Degradation, ACS Nano 14 (2020) 3703–3717. https://doi.org/10.1021/acsnano.0c00962.

[88]   J. Wu, Z. Zhu, W. Liu, Y. Zhang, Y. Kang, J. Liu, C. Hu, R. Wang, M. Zhang, L. Chen, L. Shao, How Nanoparticles Open the Paracellular Route of Biological Barriers: Mechanisms, Applications, and Prospects, ACS Nano 16 (2022) 15627–15652. https://doi.org/10.1021/acsnano.2c05317.

[89]   J.K. Tee, L.X. Yip, E.S. Tan, S. Santitewagun, A. Prasath, P.C. Ke, H.K. Ho, D.T. Leong, Nanoparticles' interactions with vasculature in diseases, Chem. Soc. Rev. 48 (2019) 5381–5407. https://doi.org/10.1039/C9CS00309F.

[90]   Y. Liu, E. Yoo, C. Han, G.J. Mahler, A.L. Doiron, Endothelial barrier dysfunction induced by nanoparticle exposure through actin remodeling via caveolae/raft-regulated calcium signalling, NanoImpact 11 (2018) 82–91. https://doi.org/10.1016/j.impact.2018.02.007.

[91]   Duan, Y. Yu, Y. Yu, Y. Li, J. Wang, W. Geng, L. Jiang, Q. Li, X. Zhou, Z. Sun, Silica nanoparticles induce autophagy and endothelial dysfunction via the PI3K/Akt/mTOR signaling pathway, Int. J. Nanomedicine (2014) 5131. https://doi.org/10.2147/IJN.S71074.

[92]   Y. Cao, M. Roursgaard, P.H. Danielsen, P. Møller, S. Loft, Carbon Black Nanoparticles Promote Endothelial Activation and Lipid Accumulation in Macrophages Independently of Intracellular ROS Production, PLoS ONE 9 (2014) e106711. https://doi.org/10.1371/journal.pone.0106711.

[93]   H. Ota, M. Eto, M.R. Kano, T. Kahyo, M. Setou, S. Ogawa, K. Iijima, M. Akishita, Y. Ouchi, Induction of Endothelial Nitric Oxide Synthase, SIRT1, and Catalase by Statins

Inhibits Endothelial Senescence Through the Akt Pathway, Arterioscler. Thromb. Vasc. Biol. 30 (2010) 2205–2211. https://doi.org/10.1161/ATVBAHA.110.210500.

[94]   M. Lasak, K. Ciepluch, Overview of mechanism and consequences of endothelial leakiness caused by metal and polymeric nanoparticles, Beilstein J. Nanotechnol. 14 (2023) 329–338. https://doi.org/10.3762/bjnano.14.28.

[95]   F. Zhao, Y. Zhao, Y. Liu, X. Chang, C. Chen, Y. Zhao, Cellular Uptake, Intracellular Trafficking, and Cytotoxicity of Nanomaterials, Small 7 (2011) 1322–1337. https://doi.org/10.1002/smll.201100001.

[96]   C.M. Beddoes, C.P. Case, W.H. Briscoe, Understanding nanoparticle cellular entry: A physicochemical perspective, Adv. Colloid Interface Sci. 218 (2015) 48–68. https://doi.org/10.1016/j.cis.2015.01.007.

[97]   G. Wardani, J. Nugraha, R. Kurnijasanti, M.R. Mustafa, S.A. Sudjarwo, Molecular Mechanism of Fucoidan Nanoparticles as Protector on Endothelial Cell Dysfunction in Diabetic Rats' Aortas, Nutrients 15 (2023) 568. https://doi.org/10.3390/nu15030568.

[98]   S. Muzammil, J. Neves Cruz, R. Mumtaz, I. Rasul, S. Hayat, M.A. Khan, A.M. Khan, M.U. Ijaz, R.R. Lima, M. Zubair, Effects of Drying Temperature and Solvents on In Vitro Diabetic Wound Healing Potential of Moringa oleifera Leaf Extracts, Molecules 28 (2023). https://doi.org/10.3390/molecules28020710.

[99]   J.N. Cruz, S. Muzammil, A. Ashraf, M.U. Ijaz, M.H. Siddique, R. Abbas, M. Sadia, Saba, S. Hayat, R.R. Lima, A review on mycogenic metallic nanoparticles and their potential role as antioxidant, antibiofilm and quorum quenching agents, Heliyon 10 (2024). https://doi.org/10.1016/j.heliyon.2024.e29500.

[100]   S. Allen, Y.-G. Liu, E. Scott, Engineering Nanomaterials to Address Cell-Mediated Inflammation in Atherosclerosis, Regen. Eng. Transl. Med. 2 (2016) 37–50. https://doi.org/10.1007/s40883-016-0012-9.

[101]   Q. Hu, Z. Fang, J. Ge, H. Li, Nanotechnology for cardiovascular diseases, Innov. Camb. Mass 3 (2022) 100214. https://doi.org/10.1016/j.xinn.2022.100214.

[102]   E.S. Ali, S.Md. Sharker, M.T. Islam, I.N. Khan, S. Shaw, Md.A. Rahman, S.J. Uddin, M.C. Shill, S. Rehman, N. Das, S. Ahmad, J.A. Shilpi, S. Tripathi, S.K. Mishra, M.S. Mubarak, Targeting cancer cells with nanotherapeutics and nanodiagnostics: Current status and future perspectives, Semin. Cancer Biol. 69 (2021) 52–68. https://doi.org/10.1016/j.semcancer.2020.01.011.

[103]   R.K. Jain, P. Au, J. Tam, D.G. Duda, D. Fukumura, Engineering vascularized tissue, Nat. Biotechnol. 23 (2005) 821–823. https://doi.org/10.1038/nbt0705-821.

[104]   S.S. Katta, V. Nagati, A.S.V. Paturi, S.P. Murakonda, A.B. Murakonda, M.K. Pandey, S.C. Gupta, A.K. Pasupulati, K.B. Challagundla, Neuroblastoma: Emerging trends in pathogenesis, diagnosis, and therapeutic targets, J. Controlled Release 357 (2023) 444–459. https://doi.org/10.1016/j.jconrel.2023.04.001.

[105]   J. Kim, H.S. Kim, N. Lee, T. Kim, H. Kim, T. Yu, I.C. Song, W.K. Moon, T. Hyeon, Multifunctional Uniform Nanoparticles Composed of a Magnetite Nanocrystal Core and a Mesoporous Silica Shell for Magnetic Resonance and Fluorescence Imaging and for Drug

Delivery, Angew. Chem. Int. Ed. 47 (2008) 8438–8441.
https://doi.org/10.1002/anie.200802469.

[106] A.M. Nyström, B. Fadeel, Safety assessment of nanomaterials: Implications for nanomedicine, J. Controlled Release 161 (2012) 403–408.
https://doi.org/10.1016/j.jconrel.2012.01.027.

[107] M.R.C. Marques, Q. Choo, M. Ashtikar, T.C. Rocha, S. Bremer-Hoffmann, M.G. Wacker, Nanomedicines - Tiny particles and big challenges, Adv. Drug Deliv. Rev. 151–152 (2019) 23–43. https://doi.org/10.1016/j.addr.2019.06.003.

[108] A.E. Nel, L. Mädler, D. Velegol, T. Xia, E.M.V. Hoek, P. Somasundaran, F. Klaessig, V. Castranova, M. Thompson, Understanding biophysicochemical interactions at the nano–bio interface, Nat. Mater. 8 (2009) 543–557. https://doi.org/10.1038/nmat2442.

[109] Y. Liu, J. Hardie, X. Zhang, V.M. Rotello, Effects of engineered nanoparticles on the innate immune system, Semin. Immunol. 34 (2017) 25–32.
https://doi.org/10.1016/j.smim.2017.09.011.

[110] Q. Muhammad, Y. Jang, S.H. Kang, J. Moon, W.J. Kim, H. Park, Modulation of immune responses with nanoparticles and reduction of their immunotoxicity, Biomater. Sci. 8 (2020) 1490–1501. https://doi.org/10.1039/C9BM01643K.

[111] J. Wolfram, M. Zhu, Y. Yang, J. Shen, E. Gentile, D. Paolino, M. Fresta, G. Nie, C. Chen, H. Shen, M. Ferrari, Y. Zhao, Safety of Nanoparticles in Medicine, Curr. Drug Targets 16 (2015) 1671–1681. https://doi.org/10.2174/1389450115666140804124808.

[112] S. Schöttler, G. Becker, S. Winzen, T. Steinbach, K. Mohr, K. Landfester, V. Mailänder, F.R. Wurm, Protein adsorption is required for stealth effect of poly(ethylene glycol)- and poly(phosphoester)-coated nanocarriers, Nat. Nanotechnol. 11 (2016) 372–377.
https://doi.org/10.1038/nnano.2015.330.

[113] Z. Liu, C. Davis, W. Cai, L. He, X. Chen, H. Dai, Circulation and long-term fate of functionalized, biocompatible single-walled carbon nanotubes in mice probed by Raman spectroscopy, Proc. Natl. Acad. Sci. 105 (2008) 1410–1415.
https://doi.org/10.1073/pnas.0707654105.

[114] S. Bremer-Hoffmann, B. Halamoda-Kenzaoui, S.E. Borgos, Identification of regulatory needs for nanomedicines, J. Interdiscip. Nanomedicine 3 (2018) 4–15.
https://doi.org/10.1002/jin2.34.

[115] D. Van Der Zwaag, N. Vanparijs, S. Wijnands, R. De Rycke, B.G. De Geest, L. Albertazzi, Super Resolution Imaging of Nanoparticles Cellular Uptake and Trafficking, ACS Appl. Mater. Interfaces 8 (2016) 6391–6399. https://doi.org/10.1021/acsami.6b00811.

[116] X. Yao, X. Fan, N. Yan, Cryo-EM analysis of a membrane protein embedded in the liposome, Proc. Natl. Acad. Sci. 117 (2020) 18497–18503.
https://doi.org/10.1073/pnas.2009385117.

[117] M. Piffoux, A. Nicolás-Boluda, V. Mulens-Arias, S. Richard, G. Rahmi, F. Gazeau, C. Wilhelm, A.K.A. Silva, Extracellular vesicles for personalized medicine: The input of physically triggered production, loading and theranostic properties, Adv. Drug Deliv. Rev. 138 (2019) 247–258. https://doi.org/10.1016/j.addr.2018.12.009.

[118]   W.J.M. Mulder, G.J. Strijkers, G.A.F. Van Tilborg, D.P. Cormode, Z.A. Fayad, K. Nicolay, Nanoparticulate Assemblies of Amphiphiles and Diagnostically Active Materials for Multimodality Imaging, Acc. Chem. Res. 42 (2009) 904–914. https://doi.org/10.1021/ar800223c.

[119]   J. Kim, Y. Piao, T. Hyeon, Multifunctional nanostructured materials for multimodal imaging, and simultaneous imaging and therapy, Chem Soc Rev 38 (2009) 372–390. https://doi.org/10.1039/B709883A.

Advances in Healthcare and Nanoparticle Toxicology
Materials Research Foundations 171 (2024) 251-278

Materials Research Forum LLC
https://doi.org/10.21741/9781644903339-9

Chapter 9

Nanoparticles in Focus: Understanding Genotoxicity and Carcinogenicity

Harishkumar Madhyastha[1,a*], Remya Varadarajan[2,b], Pallavi Baliga[2,c]

[1]Department of Cardio-Vascular Physiology, Faculty of Medicine, University of Miyazaki, Miyazaki 8891692, Japan.

[2]Department of Biochemistry, St Aloysius (Deemed to be University), Mangaluru, Karnataka, India

[a]hkumar@med.miyazaki-u.ac.jp, [b]remyavaradarajan18@gmail.com, [c]pallavibaliga1@gmail.com

Abstract

Genotoxicity is the damage caused by substances to genetic material, leading to gene mutations, chromosomal rearrangements, and aberrations. Nanoparticles can have primary or secondary genotoxic effects depending on their interaction with genetic material. Primary genotoxicity occurs when nanoparticles interact directly with the genetic material and proteins without invoking an inflammatory response. Indirect primary genotoxicity which results from the generation of reactive oxygen species (ROS), can cause structural modifications and inactivation of proteins. NPs can also interfere with the proper functioning of protein kinases involved in cell cycle regulation, leading to aneuploidy and multinucleation. Secondary genotoxicity results from DNA damage induced by free radicals generated by activated inflammatory cells. NPs can induce epigenetic changes in the DNA by altering methylation status, histone modifications, and activation of regulatory miRNAs. They can also impact DNA repair by downregulating repair enzymes and sequestering DNA repair proteins in a "corona" in the nucleoplasm. Metal oxide nanoparticles cause an increase in intracellular reactive oxygen species (ROS) levels, potentially resulting in toxicity or immunological reactions. DNA damage is a crucial stage in carcinogenesis, with ROS regulating cell proliferation through cell-signaling networks.

Keywords

Metal Oxide Nanoparticles, Genotoxicity, Carcinogenecis, ROS, Oxidative Stress, Inflammation, Cell-Signaling

List of Abbreviations:

ADME: Absorption, distribution, metabolism, and excretion
AP: Apurinic sites
AP-1: Activator protein -1
ATM: Ataxia Telangiectasia Mutated
ATR: Rad3-related
BER: Base excision repair
CeO_2: Cerium oxide

CHO: Chinese hamster ovary
CuONPs: Copper oxide nanoparticles
DNA: Deoxyribonucleic acid
DNA: PKcs-DNA-dependent protein kinase catalytic subunit
DSBs: Double-stranded breaks
GLP: Good Laboratory Practices
HGPRT: Hypoxanthine-guanine phosphoribosyltransferase
hMSCs: Human mesenchymal stem cells
hOGG1: Human 8-oxoguanine DNA glycosylase
HTS: High-throughput screening
IONP: Iron oxide nanoparticles
MAPK: Mitogen Activated Protein Kinase
MN assay: Micronucleus assay
MWCNTs: Multi-walled carbon nanotubes
NER: Nucleotide excision repair
NF-KB: Nuclear Factor Kappa B
NP: Nanoparticle
Nrf-2: Nuclear factor erythroid 2-related factor 2
OECD: Organization for Economic Co-operation and Development
OGG1: 8-Oxoguanine DNA glycosylase 1
PBL: Peripheral blood leukocytes
PIKK: Phosphatidylinositol 3-kinase-like-kinase
PPE: Personal Protective Equipment
ROS: Reactive oxygen species
SiNP: Silica Nanoparticle
SMART: Somatic mutation and recombination test
SSB: Single-strand breaks
SWCNTs: Single-wall carbon nanotubes
TFT: Trifluorothymidine
TG: 6-thioguanine
TiO_2: Titanium dioxide
TK: Thymidine kinase
ZnO : Zinc oxide
4-HNE: 4-hydroxy-2-nonenal
8-oxoG: 8-oxo-7,8-dihydro-2-deoxyguanine

Contents

1. Introduction

1.1 Background

1.1.1 Nanoparticles: Definition

The term nanoparticle (NP) refers to particles whose size ranges from 1 to 100 nanometers [1]. Nanoparticles (NPs) exhibit properties that are distinct from their bulk or macroscale counterparts [2]. The high surface area-to-volume ratio contributes to an exponential increase in reactivity, making it suitable for myriad scientific applications [1].

1.1.2 Increasing use of NPs in various industries

Nanotechnology has gradually but profoundly taken over several industries, such as healthcare, agriculture, food, textiles, energy, electronics, computing, and others. It has enabled groundbreaking advancements and progress in several industrial sectors across the globe [2]. In the healthcare sector, NPs have enabled the development of newer devices, the effective delivery of drugs, and the faster diagnosis of several diseases [3]. Nanotechnology based research in agriculture focuses mainly on sensors to obtain up-to-date information on soil conditions, plant growth, and the application of fertilizers and pesticides based on the requirements. Furthermore, NPs of Au, Ag, Cu, CuO, SiO_2, Fe_2O_3 and ZnO, fullerenes, and carbon nanotubes are being investigated to be incorporated into fertilizers and pesticides due to their growth-promoting and antibacterial properties [4]. Zinc oxide-based nanoplastic wrappers are inherently antibacterial, temperature and UV-resistant, and fire-proof. As a result, food makers greatly value these packaging materials that help enhance food nutrition value, retain product odor, and flavor, and extend shelf life [4]. NPs are crucial in the textile industry for improving textile properties like fabric's softness, resilience, breathability, water resistance, fire resistance, and antibacterial nature [5]. Due to the high efficiency and cost-effectiveness of nanomaterials, nanotechnology will play an important role in the energy sector. It will be used for improving solar cells, batteries, and fuel cells [3]. Nanotechnology has enabled the design of lightweight, energy-efficient, and faster computers and electronic devices [2]. NPs can be used to improve the efficiency of manufacturing processes and create new products. As the science of nanotechnology advances, we can expect to see even more innovative applications in the years to come. As a result of this technological advancement, humans are more likely to be exposed to NPs. This has raised concerns about the consequences of such exposures for human health [1].

1.1.3 Routes of exposure

NPs can enter host systems through cutaneous, olfactory, respiratory, and gastrointestinal routes [9]. Their ingestion may be deliberate or accidental. Their presence in the body can bring about a range of harmful effects [9].

1.1.4 Properties that influence toxicity

Scientific evidence suggests that NPs can exert their toxic effects in various ways. Factors that influence the toxicity of NPs include composition, size, shape, surface charge, solubility, and agglomeration status [4]. Composition is the main factor governing the genotoxicity of NPs [5]. It influences the distribution, dissolution rate, and nature of the biological interaction of the NPs [5] (Fig. 1).

From a toxicological perspective, particle size and surface area are important material properties because interactions between NPs and biological organisms mainly take place at the surface of the NP. As size decreases, the surface area to volume ratio increases, enhancing the reactivity towards cellular components such as nucleic acids, proteins, fatty acids, and carbohydrates [6]. Typically, NPs are shaped as spheres, cylinders, cubes, sheets, or rods. The shape of the NP has a considerable impact on its cellular absorption. Spherical NPs are more readily absorbed by cells than other shapes [7]. However, needle-shaped NPs are more dangerous than spherical ones because they have better multiple endocytic processes and internalization rates and are more efficient at adhering to the surface of the target cells [7]. Star-shaped NPs are most lethal to human cells [7]. Likewise, the surface charge of particles also influences cytotoxicity [1]. Positively charged NPs will interact more strongly with negatively charged glycosaminoglycans on mammalian cell surfaces, leading to greater internalization rates [1]. Furthermore, they can also interact with the negatively charged DNA, leading to damage [1]. Often, NPs become agglomerated due to the action of weak forces like Van der Waals forces or electrostatic forces. This increases their visibility to macrophages, consequently leading to the production of free radicals [8].

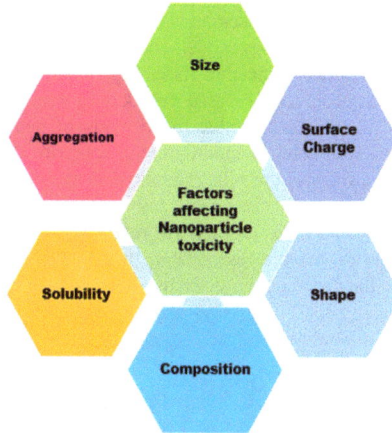

Figure 1: Factors influencing genotoxicity of nanoparticles

1.2 Purpose

Nanotechnology is currently the most trending technology across several sectors and has become indispensable in several applications that benefit society. However, it has become evident that there is also a flipside to this marvelous technology that requires thorough investigation and introspection. The rise in the number of paper publications on the toxicity aspects of NPs in recent years highlights the importance of this topic (Fig. 2).

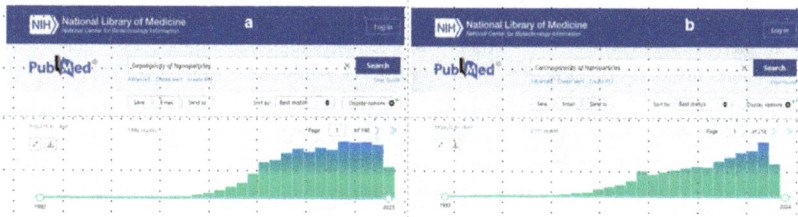

Figure. 2: The trends in publications on the topic of genotoxicity and carcinogenicity of nanoparticles in the PubMed literature database.

Due to their extremely small size, they can easily overcome several biological barriers, including the blood-brain barrier. Interaction of NPs with subcellular components could be damaging and can often lead to DNA damage, mutation, and the initiation of cancer, posing serious risks to human health. Recent studies have suggested that NPs tend to accumulate in the human body and

Advances in Healthcare and Nanoparticle Toxicology   Materials Research Forum LLC
Materials Research Foundations 171 (2024) 251-278   https://doi.org/10.21741/9781644903339-9

environment due to their small size and unique features. It is therefore critical to analyze the dangers of chronic NP exposure. A clear understanding of the associated risks would be beneficial to researchers and manufacturers to mitigate risks, design safer products, and chart out effective guidelines for long-term exposure. Risk assessment of nanotechnology-based products also helps in the development of predictive toxicology models and computational tools. The use of genotoxic and carcinogenic NPs raises ethical concerns. Assessing genotoxicity and carcinogenicity is a prerequisite for the approval of novel NP-based products by regulatory agencies. Therefore, a thorough risk-benefit analysis will ensure that the benefits of nanotechnology can be tapped without jeopardizing human well-being or environmental integrity. The objective of this chapter is to expand our knowledge of how the NPs interact with the cellular machinery, to provide mechanistic insights into the toxic outcomes, and to assess associated risks, regulatory implications, and future directions. This chapter will reveal the potential risks and challenges of current approaches. It aims to provide researchers, policymakers, and stakeholders with the insights needed to navigate the nanotechnology landscape responsibly and prudently.

2.   Genotoxicity of NPs

2.1.1 Genotoxicity mechanisms

Genotoxicity refers to the ability of a substance to cause damage to the genetic material, leading to gene mutations, chromosomal rearrangements, and aberrations [4]. Based on the kind of interaction, the genotoxic effects of NPs can be classified as primary or secondary (Fig. 3). Primary genotoxicity occurs when the NPs interact directly with the genetic material and associated proteins without invoking an inflammatory response [1]. It can be further classified as direct or indirect. Primary direct genotoxicity results when the NPs closely interact with chromosomes or DNA molecules, interfering with the normal replication and transcription processes [4]. Additionally, the interaction between NP and mitotic chromosomes can result in chromosomal breakage (clastogenic effect) or loss (aneugenic effect). Indirect primary genotoxicity can be attributed to the generation of reactive oxygen species (ROS), which result in structural modifications and inactivation of proteins associated with the centrioles or mitotic spindle apparatus. This can lead to an aneugenic effect, resulting in the loss or gain of chromosomes in newly formed cells following cell division [9]. NPs can also interfere with the proper functioning of the protein kinases involved in cell cycle regulation. Dysregulation of the cytokinesis process can lead to aneuploidy and multinucleation in cells [9].

In addition, NPs may interfere with the cell's ability to remove endogenous ROS, leading to oxidative damage to the deoxyribose backbone as well as to the bases. This leads to the formation of oxidized DNA bases, single-strand breaks (SSBs), and apurinic (AP) sites [10]. Double-stranded breaks (DSBs), formed directly or through replication of SSBs, as well as blocking base lesions, are the most serious forms of damage, as they can lead to chromosomal aberrations if left without repair [10]. DSBs activate a series of damage signals [11]. Following a DSB, numerous histone H2AX molecules are phosphorylated at serine 139, and form foci named gamma-H2AX (γ-H2AX) in the vicinity of the DNA damage site [11]. The members of the phosphatidylinositol 3-kinase-like-kinase (PIKK) family, including Ataxia Telangiectasia Mutated (ATM), Rad3-related (ATR), and DNA-dependent protein kinase catalytic subunit (DNA-PKcs), are recruited to the site and mediate the phosphorylation event [11]. H2AX deficiency or suppression of the H2AX phosphorylation reaction impairs the DNA repair pathway, leading to genomic instability [11].

Advances in Healthcare and Nanoparticle Toxicology   Materials Research Forum LLC
Materials Research Foundations 171 (2024) 251-278   https://doi.org/10.21741/9781644903339-9

The low redox potential of guanine makes it vulnerable to oxidative damage, converting it to 8-oxo-7,8-dihydro-2′-deoxyguanosine (8-oxo-dGuo). Due to its ability to pair with adenine (A) or cytosine (C), it can lead to G: C to thymine (T): A transversion [10]. In addition, DNA damage can also be induced by NP-generated lipid peroxidation products like ketonic and aldehydic derivatives [malondialdehyde, acrolein, 4-hydroxy-2-nonenal (4-HNE) or its epoxidized derivatives] [10]. Similarly, secondary genotoxicity results from the DNA damage induced by the free radicals generated by activated inflammatory cells such as macrophages and neutrophils [1]. NPs can also induce epigenetic changes in the DNA by altering the methylation status, histone modifications, and activation of regulatory miRNAs [12]. Gene expression in DNA repair and apoptosis pathways is suppressed when promoter regions (CpG islands) are hypermethylated, whereas hypomethylation occurs when a CpG dinucleotide in the global genome is hypomethylated [13]. DNA methylation can be affected by nanoparticles in two basic ways: by reducing the availability of methyl donors and by altering the activity of enzymes involved in DNA methyl transfer [14]. The covalent modification of histones (e.g., acetylation, methylation, phosphorylation, ubiquitylation, and sumoylation) can regulate gene expression by defining the conformation of chromatin or indirectly recruiting transcription factors. Exposure to nanomaterials and nanoparticles disrupts histone modification patterns substantially [15]. miRNAs are involved in the regulation of several pathways associated with DNA damage repair [16]. Engagement of miRNAs by NPs can have adverse outcomes for the DNA damage repair process [16].

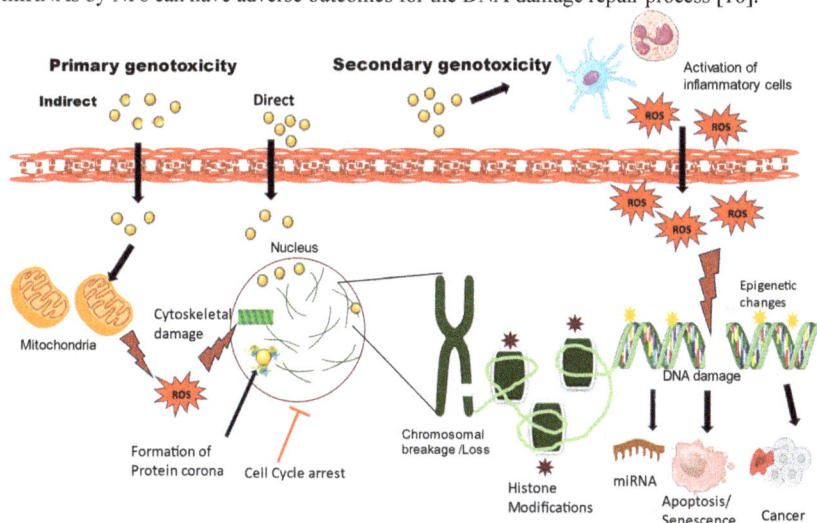

Figure 3: Mechanism of nanoparticle induced genotoxicity

2.1.2 Influence of nanoparticles on DNA repair

The DNA repair process can be impeded by the unavailability of repair enzymes. NPs might influence the downregulation of repair enzymes [10]. 8-Oxoguanine DNA glycosylase 1 (OGG1) is a DNA repair enzyme that excises 8-oxo-7,8-dihydro-2-deoxyguanine (8-oxoG) residues. Downregulation of OGG1 by AgNPs results in genotoxicity [17]. In the nucleoplasm, the NPs sequester DNA repair proteins in a "corona," so the cell cannot repair DNA damage [1]. There is evidence that metals such as arsenic, cobalt, and cadmium have adverse effects on DNA repair by interfering with the base excision repair (BER) and nucleotide excision repair (NER) processes [1]. In the cytosol, NPs release metal ions when exposed to an acidic pH. Metal cofactors are required for the activation of a few repair proteins in the vicinity of the cytosol. Metallic NPs interact with these proteins and remodel them [1]. All essential processes in the cell can also be altered by NPs by bringing about a change in the status of "metal homeostasis" in the cell [1]. Metal ions released from ZnNPs, for instance, can replace zinc ions in zinc finger motifs, which are pivotal molecules in DNA protein interactions, causing protein distortions and functional inactivation [1]. The surface reactions of active NPs produce rapid changes in their oxidation states, leading to the modification and inactivation of DNA repair proteins [1].

2.1.3 Notable examples of genotoxic NPs

Single-wall carbon nanotubes (SWCNTs), multi-walled carbon nanotubes (MWCNTs), and carbon fullerenes (C60 fullerenes) are the major carbon-based NPs that find a wide range of applications in the biomedical industry [18]. A study involving V79 Chinese hamster fibroblast cells indicated that the functionalization of SWCNTs renders them genotoxic. This was confirmed by the occurrence of mutations in the HPRT locus of these cells [18]. The Comet assay indicated the genotoxic nature of MWCNTs. A single pharyngeal aspiration of MWCNTs (1-200 μg/mouse) resulted in DNA breaks in C57Bl/6 mice [19]. Exposing mice to C60 fullerene via the intratracheal route resulted in DNA damage due to a 2-5-fold increase in the formation of DNA adducts [20].

Silver NPs (AgNPs) were found to exert genotoxic effects on human glioblastoma cells (U251) by disrupting the mitochondrial respiratory chain and generating ROS. It also caused cell cycle arrest in the G2/M phase [21]. AgNPs induced significant DNA damage in the liver cells of Swiss albino mice in a concentration-dependent manner [22]. AgNPs induce DNA damage in human mesenchymal stem cells (hMSCs) at concentrations as low as 0.1 μg/ml [23]. Various *in vivo* studies involving rodent exposure to titanium dioxide (TiO$_2$) NPs have shown that they bind with DNA phosphate residues and insert themselves into base pairs [24]. Zinc oxide (ZnO) NPs have been reported to induce both primary and secondary genotoxicity on human neuronal cells in a time and dose-dependent manner. In addition, ZnO exposure can result in micronuclei formation and H2AX phosphorylation [25]. In the bone marrow and peripheral blood leukocytes (PBL) of Wistar rats, cerium oxide (CeO$_2$) NPs induced significant DNA damage and micronucleus formation [26,27]. Iron oxide nanoparticles (IONP) of size <50 nm displayed genotoxicity when fed to *D. melanogaster* at concentrations between 1 and 10 mM [28]. *P. reticulata* exposed to IONPs for long periods displayed high rates of clastogenic and aneugenic effects, such as DNA damage, mutation, and an increased frequency of nuclear abnormalities in the erythrocytes [29]. Significant DNA fragmentation was seen in rats exposed to SiO$_2$ NPs for 24 hours [30]. Several factors contribute to the genotoxicity of SiO$_2$ NPs, including modifications of chromatin, ROS induction, and the release of DNase from lysozomes [31]. Exposure of mice to cobalt nanoparticles (Nano-Co) resulted in elevated levels of 8-oxodG residues in the DNA of lung tissues [32]. Furthermore, it increased the rate of G:C to T:A transversion mutations and levels of DNA damage-specific markers such as γ-H2AX [32]. Extensive inflammation and lung injuries were also reported [32]. Significant reduction in glutathione, catalase, and superoxide dismutase was observed in human skin epidermal (HaCaT) cells exposed to copper oxide nanoparticles (CuONPs) [33].

2.2 Testing and evaluation

2.2.1 *In vitro* assays for genotoxicity assessment

The Organization for Economic Co-operation and Development (OECD) has formulated several test guidelines for genotoxicity tests. These standard testing methods can be used by regulatory agencies, industries, and laboratories to determine the safety of NPs or materials derived from them. These test guidelines provide specifications for the registration, notification, and regulatory safety assessment of new compounds. These guidelines are subject to modification with advancements in science and changes in national policies [34]. A few important testing methods recommended by the OECD for screening NPs for genotoxicity are listed below.

2.2.2 Ames test

This is a popular and well-established bioassay that can be employed to determine if a given substance is mutagenic. *Salmonella typhimurium* or *Escherichia coli* that contain various types of mutations for amino acid synthesis (His for *S. typhimurium* and Trp for *E. coli*) are used as test strains in this assay [35]. These test strains fail to grow in media deficient in histidine and tryptophan unless the mutation is reversed by a mutagen or a spontaneous event. Therefore, substances that produce noticeably more reversed mutants or colonies than naturally occurring reverse mutants of the negative control are considered mutagenic [35].

2.2.3 Comet assay

In this assay, fragmented DNA can be visualized as "comets" following lysis of individual cells, DNA denaturation, and electrophoresis [36]. There have been several modifications to the alkaline comet assay to enhance its sensitivity, reduce variability, and make it more robust and suitable for use in different applications [37]. Several modified assays are available that utilize human 8-oxoguanine DNA glycosylase (hOGG1) or formamidopyrimidine DNA glycosylase, which introduce breaks at sites of oxidative damage in the DNA, making it easier to detect them [38].

2.2.4 Micronucleus assay (MN assay)

This assay can be used to detect both clastogenic and aneugenic effects of NPs [39]. Chromosomes or chromosomal fragments that do not segregate during anaphase form micronuclei, which can be visualized by staining [9]. The sensitivity of this assay can be increased by the incorporation of cytochalasin B, which blocks cell division but not mitosis, resulting in the accumulation of binucleated cells. Only scoring binucleated cells will reduce the chances of scoring MN that existed before treatment. MN assays can be performed within a short timeframe and are easier to automate [9].

2.2.5 Mammalian mutation assay

In the HPRT (Hypoxanthine-guanine phosphoribosyltransferase) test, mutant V79 cell lines of Chinese hamster pulmonary fibroblasts or Chinese hamster ovary (CHO) cell lines lacking the HGPRT enzyme are resistant to the cytostatic effects of the purine analogue 6-thioguanine (TG). Cells with active HGPRT are sensitive to TG, which inhibits cellular metabolism and prevents further cell division. As a result, mutant cells can multiply in the presence of TG, whereas normal cells, which possess the HGPRT enzyme, cannot [40,41].

2.2.6 The mouse lymphoma essay

Mutant cells lacking thymidine kinase (TK⁻) activity are resistant to the cytotoxic effects of the pyrimidine analogue trifluorothymidine (TFT). TFT sensitizes TK^+ by inhibiting cellular metabolism and halting cell division. The mutant cells formed by exposure to a genotoxic compound can multiply and form visible colonies even in the presence of TFT, but the non-mutant cells cannot [42].

2.2.7 *In vivo* models and their relevance

Genotoxicity testing can be performed with *in vivo* models whenever the results obtained from *in vitro* testing are not adequate. Such assays are particularly relevant when the response to a test compound is influenced by the ADME (absorption, distribution, metabolism, and excretion) phenomena and also DNA repair [43].

When it can be reasonably expected that the target tissue will be adequately exposed, an *in vivo* comet assay can be performed on it [43]. In this essay, the animals are exposed to the test compound via the desired route. At the predetermined sampling time, the target tissue is collected, processed into individual cells or nuclear suspensions, and embedded in soft agar for subsequent immobilization [43]. The cells are lysed and exposed to strong alkali to allow DNA unwinding for gel electrophoresis. The non-fragmented DNA does not migrate in the gel, whereas the fragmented DNA migrates towards the anode [43]. The gel can be visualized with fluorescent stains, as in the *in vitro* comet assay [43]. Similarly, the mammalian *in vivo* micronucleus assay is used to determine the damage caused by the test compound to the chromosomes or mitotic apparatus of the erythroblasts. This test measures the rate of appearance of micronuclei in the erythrocytes sampled from either the bone marrow or peripheral blood cells of the test animals [44].

Due to the high homology between human and fly genes, *Drosophila melanogaster* is an ideal organism to test the genotoxicity of NPs [45]. The somatic mutation and recombination test (SMART) is based on the loss of heterozygosity in two genetic markers, multiple wing hair (mwh) and flare³ (flr³), located on the third chromosome of *D. melanogaster* and involved in the formation of wing hairs [45]. Any mutation occurring in these markers while in the imaginal disc stage reflects on the wing surface of the adult fly. Genotoxicity of the test compound is indicated by an increase in mutations in the treatment group compared to the control group [45].

2.2.8 High-throughput method

The purpose of high-throughput screening (HTS) is to determine biological activity at the organism, cellular, molecular, or pathway level rapidly and efficiently by using automated equipment [46]. DNA damage and repair can be evaluated at high throughput with CometChip technology [47]. Based on the classical comet assay, this assay uses microfabrication technology to build an array of agarose microwells that can be adjusted to fit a single cell, thus allowing the testing of the genotoxicity of multiple compounds simultaneously [48]. Similarly, the micronucleus test can also be run on the HTS platform to achieve high sensitivity and specificity for detecting genotoxic compounds. This is achieved by culturing cells in 96-well plates preloaded with a dye that stains the cytosol. The test compound is added to the wells, and after the incubation, the micronuclei can be visualized under a fluorescent microscope by staining the nuclei with a DNA-specific stain [49].

2.2.9 *In silico* methods

By studying the physicochemical attributes that contribute to genotoxicity, certain *in silico* predictive models have been developed recently [50,51]. The genotoxicity of new, untested compounds could be predicted using such models.

2.3 Challenges in assessing genotoxicity

Assessment of the toxicity and biological impact of NPs is extremely challenging. During the *in vivo* assessment of genotoxicity, there is an interplay of several critical factors, such as hormonal variations and cell-cell and cell-matrix interactions. These cannot be replicated in the *in vitro* tests, and hence the data may not be a true representation of the dynamics occurring in the biological system [52,53]. Furthermore, *in vitro* systems are not suited to studies on the long-term chronic effects of NP exposure. Some *in vitro* methods do not yield reliable data or are not accurate. Several studies have revealed that the Ames test is not a good assay for assessing nanoparticle genotoxicity since NPs that tested negative for genotoxicity in the Ames test tested positive for genotoxicity in *in vitro* mammalian test systems [14]. This could be because the bacterial cell wall might limit the entry of NPs into the bacterial cell [15] and also due to the absence of the endocytotic mechanism [48]. Comet assays have limited their acceptance due to their low throughput, poor reproducibility, and labor-intensive data processing and analysis [47]. Similarly, the MN test could also yield false-positive results due to reagent interference by cytochalasin B [48]. The TK assay is not reliable for detecting aneugens [42]. In addition, the cell lines used for *in vitro* testing of genotoxicity are usually cancer cells, which express altered metabolic pathways in comparison to their normal counterparts and therefore may not be a true representative of the biological system [54]. Lack of standardized protocols, differing sensitivities of various cell lines, inconsistencies in media components, and a lack of positive controls are some of the challenges faced with *in vitro* testing systems [54]. Determining an appropriate dose for *in vivo* challenge studies is a daunting task due to the minuscule nature and unpredictable behavior of NPs in the biological system [52]. NPs tend to aggregate, and this can lead to unequal distribution and undesirable outcomes in biological systems [52]. Furthermore, interaction with proteins in the body can lead to the formation of protein corona and alter biodistribution [52].

2.4 Nanocarcinogenecity: An interplay of oxidative stress, mitochondrial dysfunction cell signaling & transcription factors

2.4.1 Nanoparticle-induced oxidative stress

Increased oxidative stress is one of the vital events in nanoparticle-induced toxicity. The definition of oxidative stress is "a disturbance in the prooxidant/antioxidant balance favoring the former." [55]. Superoxide anion (O_2^-), hydrogen peroxide (H_2O_2), and hydroxyl radical (HO) are only a few of the chemical species that are created when oxygen is incompletely reduced and are collectively referred to as ROS [56,57]. Endoplasmic reticulum (ER), peroxisomes, mitochondria, and cytoplasm all have the ability to create ROS. Due to their higher chemical reactivity as compared to oxygen, ROS are hypothesized to mediate the toxicity of oxygen. ROS are crucial for controlling cellular signaling and physiological processes in vivo [58]. Oxidative stress is frequently brought

Advances in Healthcare and Nanoparticle Toxicology   Materials Research Forum LLC
Materials Research Foundations 171 (2024) 251-278   https://doi.org/10.21741/9781644903339-9

on by the excessive buildup of ROS, which damages the cells' oxidative-antioxidant system. Enzymes like glutathione peroxidase, catalase, and superoxide dismutase are important antioxidant defence mechanisms in cells [59]. These enzymes aid in scavenging ROS and stopping their ability to do harm.

According to the underlying processes, there are primarily two forms of NP-induced oxidative stress [60]. Direct oxidative stress, also known as primary oxidative stress, is one form. It has been demonstrated that NPs with bigger surface areas have more reactive sites on their surfaces, increasing chemical reactivity and boosting ROS production [61]. The generation of ROS can also be considerably accelerated by the reactive surface with oxidants and free radicals. Secondary oxidative stress, also known as indirect oxidative stress, is another form. In this form, oxidative stress is not directly brought on by NPs. The processes underlying NP-induced oxidative stress will be covered in more detail in the next sections.

Metal oxide nanoparticles cause an increase in intracellular reactive oxygen species (ROS) levels as a result of following factors:

- Cellular uptake of the nanoparticles
- Intracellular metal ion release

When NPs get to a cell's outer membrane, they are able to interact with the plasma membrane or extracellular matrix and then mostly enter the cell by endocytosis [62,63]. Following their budding and pinching off to create endocytic vesicles, which are then transported to specialised intracellular sorting/trafficking compartments, NPs are first engulfed in membrane invaginations during endocytosis. Endocytosis may be divided into several categories depending on the type of cell as well as the proteins, lipids, and other molecules involved in the process [62]. Phagocytosis, clathrin-mediated endocytosis, caveolin-mediated endocytosis, clathrin/caveolae-independent endocytosis, and micropinocytosis are the five primary endocytosis [64,65]. Nanomaterials can interact with various biological elements, such as organelles or certain proteins, once they have entered cells. These interactions may have an impact on cellular processes, signaling pathways, and may result in toxicity or immunological reactions.

Metallic NPs cause Fenton-type reactions that result in toxicity caused by free radicals [66]. Iron NPs and hydrogen peroxide interact to create hydroxyl ions and redox-active iron, which undergoes the Fenton reaction to produce hydroxyl radicals. These free radicals have been connected to a number of conditions and have the potential to harm biological macromolecules and cell organelles. Oxidative stress is produced by the Fenton-type processes, which produce free radicals and interact with cellular macromolecules. A transition metal ion usually undergoes a Fenton reaction with H_2O_2 to produce $\cdot OH$ and an oxidized metal ion, which is particularly reactive and harmful to biological molecules. The Haber-Weiss reaction, on the other hand, includes a reaction between an oxidized metal ion and H_2O_2 to produce $\cdot OH$. Oxidative stress is specifically brought on by the Fenton-type processes, which produce free radicals and interact with cellular macromolecules [67].

FENTON REACTION

$$Fe^{3+} + O_2^- \longrightarrow Fe^{2+} + O_2$$

$$Fe^{2+} + H_2O_2 \longrightarrow Fe^{3+} + O\dot{H} + HO^-$$

HABER-WEISS REACTION

$$O_2 + H_2O_2 \longleftrightarrow Fe^{3+} + O\dot{H} + HO^-$$

Haber-Weiss reactions indicate the interactions between oxidised metal ions and H_2O_2 to form HO•, whereas metallic NPs, such as Cu and Fe, impact oxidative stress (O2•- and HO•) via Fenton reactions [68,69]. Fenton and Haber-Weiss reactions may both be catalyzed by nanoparticles (NPs) made of Cr, Co, and Va. Metal-based NP cause oxidative damage to biological macromolecules such proteins, lipids, and DNA via Fenton-type and Haber Weiss-type processes because of their chemical reactivity [70].

2.4.2 Nanoparticles and mitochondrial dysfunction

Internalised metal-based NPs may release intracellular metal ions that can depolarize the mitochondrial membrane, disrupt the electron transport chain, and ultimately cause mitochondrial malfunction [71]. In the physiological process, a series of linked proton and electron transport events in mitochondria use molecular oxygen to synthesize adenosine triphosphate. However, exposure to metal-based NP would alter the mitochondrial electron-transport chain (METC), which would raise the overall amount of intracellular ROS [72]. The oxidative imbalance, also known as ROS-induced ROS release, may be exacerbated by the greater amount of intracellular ROS because it may harm the mitochondrial membrane and the electron-transport chain [73][74]. Various metal oxide NP including Zn, Cu, Ti, and Si elicit ROS-mediated cell death via mitochondrial dysfunction [75][76][77].

2.4.3 Cell signaling & carcinogenesis

DNA damage is a crucial stage in carcinogenesis. ROS can regulate cell proliferation through cell-signaling networks, playing a crucial role in cancer development[78]. According to investigations, excessive ROS production and metal ions released by internalized metal-based NPs, can be affected by the activation or inhibition of various signaling pathways [79] has a direct effect on physiological processes, such as cell proliferation, differentiation, survival, and immunological control. Most significant pathways are related to Janus kinase/signal transducers and activators of transcription (JAK-STAT), nuclear factor-kappa-light chain enhancer of activated B cells (NF-KB), phosphoinositide 3-kinase (PI3K)/protein serine threonidase (Akt), mitogen-activated protein kinase (MAPK), and nuclear factor erythroid 2-related factor 2 (Nrf2) [80–82]. The MAPK/ERK pathway is a chain of proteins in cells that communicates a signal from a cell's surface to the DNA in the cell nucleus. It starts when a signaling molecule binds to the epidermal growth factor receptor (EGFR), which is activated by extracellular ligands like epidermal growth factors. This leads to the formation of mRNA, which is encoded into various proteins [83]. The MAPK signaling pathway is divided into three components: ERK1/2, JNK, and p38 MAPK. Serine-threonine protein kinases (MAPK) regulate various physiological responses, including cell division, proliferation, gene expression, differentiation, and death [70].

Free radicals can cause oxidative alteration of MAPK signaling proteins, activating MAPK once ROS generation surpasses the capacity of antioxidant proteins. ROS can also activate MAPK pathways through the inhibition or destruction of MAPK phosphatases (MKP). Through the inhibition and/or destruction of MAPK phosphatases (MKP), ROS can activate MAPK pathways [84] [85]. In this situation, ROS and occasionally direct NPs interact with these receptors, and because to alterations in the DNA coding, aberrant proteins may or may not be produced. One of the pathway's proteins can become trapped in the "on" or "off" state when it mutates, which is a crucial stage in the development of many malignancies. The body's production of 8-hydroxy-2-deoxyguanosine (8-OHdG) is greatly influenced by the introduction of NPs. 8-OHdG can result in G-to-T transversion mutations in the important genes known to be involved in the growth of cancer [86]. These mutations give birth to tumors. Thus, the development of carcinogenicity that impairs immunity is directly correlated with an elevated level of 8-OHdG. The levels of 8-OHdG were measured in the urine and white blood cells of 130 indium tin oxide employees who were often exposed to metal oxide nanoparticles [87]. p53 inactivation and caveolin-1 overexpression are two other mechanisms. When NPs lower the phosphate levels in cells, which are necessary to activate proteins, p53 is inactivated, which increases the likelihood that tumors will grow [88].

The body's tumor suppressor genes are assigned to generating proteins that stop unchecked cell development that results in cancers, reducing the likelihood of genomic mutability. Due to the genotoxicity brought on by NPs, mutated tumor suppressor genes are unable to carry out their intended role, ultimately resulting in the loss of growth control [89]. Alveolar papillary neoplasia and carcinoma are caused by ceria NPs' capacity to produce ROS and damage DNA. Particle size, bio-persistence, and a person's co-morbidity are other factors affecting cancer. Other in vivo investigations support the iron oxide NPs' role as a significant contributor to lung hyperplasia and fibrosis [70]. Ag-NP-hydrogel exposure caused toxicity in HeLa cells by upregulating the JAK-STAT cascade-related gene and activating the JAK-STAT signaling pathway [90]. It has been shown that SiO_2 NPs caused lung damage by inducing apoptosis in lung alveolar epithelial cells via ROS-regulated PI3K/AKT-mediated mitochondrial and ER stress-dependent signaling pathways [91].

2.4.4 Cell signaling & transcription factors

According to Dröge and Holmstrom et al., ROS can activate many signaling cascades, such as the nuclear factor-KB (NF-κB), the transcription factor activator protein-1 (AP-1), and the mitogen-activated protein kinase (MAPK) cascades, as well as contribute to mammalian growth, proliferation, and differentiation [92][93]. Inflammation and consequent diseases are brought on by the production of proinflammatory mediators, induced by ROS through a variety of signaling pathways including transcriptional regulators (Fig: 4)

NF-κB

Nuclear factor κB (NF-κB) is an important nuclear transcription factor family in cells, including five members: NF-κB1 (P50), NF-κB2 (p52), RelA(p65), RelB and c-Rel [94] [79].In cells, NF-

Advances in Healthcare and Nanoparticle Toxicology   Materials Research Forum LLC
Materials Research Foundations 171 (2024) 251-278   https://doi.org/10.21741/9781644903339-9

KB can be found as dimers. The cytoplasm of healthy cells contains NF-KB dimers and members of the inhibitor of NF-KB (IκB) inhibitor protein family. NF-KB activity in cells is inhibited by the IB protein. The inhibitor of nuclear factor kappa-B kinase (IKK) complex is activated when cells receive external stimulation. IKK is released by ROS, activating NF-KB and causing it to translocate into the nucleus [95]. Once within the nucleus, NF-KB triggers the transcription of proinflammatory mediators, causing oxidative stress and inflammation. The production of proinflammatory TNF-, IL-8, IL-2, and IL-6 from macrophages and lung epithelial cells is modulated during NP-mediated lung damage by ROS activating NF-KB [96]. Numerous metal oxide nanoparticles (NP), including Zn, Cd, Si, and Fe, activate NF-KB in a ROS-dependent manner to cause toxicity [97][98]. Nickel oxide (Nio) NPs triggered the NF-KB pathway after intratracheal instillation in male Wistar rats by increasing NF-KB, an inhibitor of B kinase and nuclear factor-inducing kinase, and partially damaging the lungs [99].

Activator Protein -1 (AP-1)

AP-1, a dimeric complex composed of Jun and Fos protein families, can be activated by extracellular stimuli. It is triggered by oxidants, cytokines, growth factors, and bacterial and viral infections [100]. This activation triggers downstream (MAPK members) signal cascades [101], including Jun and Fos target genes, which regulate cell responses. It act as an intermediate transcriptional regulator during signal transduction, leading to cell proliferation and malignant transformation. There is growing evidence that AP-1 up-regulation plays a role for the development of tumors [102][103]. The primary processes behind the induction of inflammation and/or carcinogenesis by TiO₂ NPs showed stimulation of intracellular ROS production and subsequent induction of AP-1 activation through p38 MAPK and ERK pathways [104]. In JB6 cells, titanium dioxide nanoparticles (TiO₂ NPs) had a dose-dependent impact on the production of hydroxyl radicals (OH), which in turn increased AP-1 activity[104]. This study highlighted the importance of understanding the role of TiO₂ nanoparticles in promoting inflammation and carcinogenesis. Protooncogene c-jun is thought to be phosphorylated in order to activate AP-1 under oxidative circumstances [105].

Nrf2

Nuclear factor (erythroid-derived 2)-like 2 (Nrf2), transcription factor is regarded as a master regulator of redox homeostasis and is essential for the defense against inflammation and free radicals. It has been widely documented that Nrf2 is involved in the oxidative stress-induced toxicity caused by nanoparticles [106]. Under conditions of oxidative stress, Nrf2 is stabilized and relocated to the nucleus, where it can transactivate a broad array of genes controlled by antioxidant response elements, such as transporters, phase II detoxifying enzymes, endogenous antioxidants, such as SOD, CAT, GSH-Px, and Trx reductase [TrxR], and antioxidant response element-regulated genes [107]. An adaptive intracellular response to defend against NP-induced oxidative stress and associated oxidative damage is the stimulation of Nrf2 and the production of its downstream genes. The induction of HO-1 (heme oxygenase-1—an antioxidant enzyme) via the Nrf-2-ERK MAP kinase signaling pathway along with an increase in oxidative stress as a result of silica nanoparticle exposure were both clearly demonstrated in an in vitro study on BEAS-2B cell lines [67].

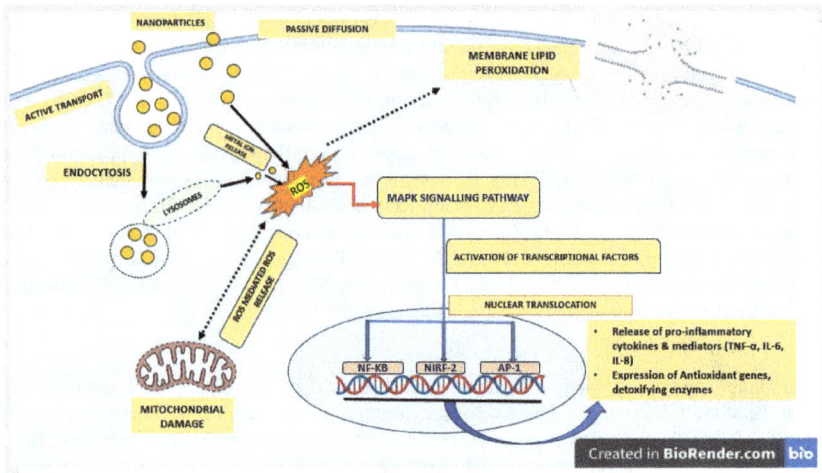

Figure 4: ROS produced by metal nanoparticle uptake, mitochondria triggers MAPK pathways that result in translocation of NF-KB, Nrf-2 & AP-1 and induces gene transcription of pro-inflammatory mediators and antioxidant genes. Biorender (https://biorender.com/) approved the usage of the figure elements.

But as of yet, no one mechanism or signaling route has been identified that completely explains the NP toxicity. The precise mechanism of NP toxicity is further complicated by the interaction of many signaling pathways.

Implications for risk assessment.

- Dose-response relationships: Establishing an appropriate dose-response relationship is challenging due to particle size, surface chemistry, and diverse routes of exposure. The tendency to agglomerate can affect biological uptake, cellular barrier penetration, or sedimentation, altering the effective concentration and hence overall action [108].

- Extrapolation to human risk: The metabolism, distribution, and response to DNA damage activation and detoxification of NPs are species-specific [109]. Therefore, translating genotoxicity data from animal to human models requires careful consideration.

- Cumulative effects: NPs often coexist with environmental stressors such as radiation or pollutants. A clear understanding of how these two elements interact to enhance or mitigate the genotoxic and carcinogenic effects is important for accurate risk assessment [110].

- Regulatory framework: The lack of a unified formal regulation on the testing of NP toxicity is a global problem. To ensure the safety of NPs worldwide, recent data on NP genotoxicity needs to be included in existing safety regulations. Developing standardized testing protocols and thresholds is imperative for safeguarding public health [111].

2.5   Safety measures

Given the risks of NP exposure to human health and the environment, it is imperative to develop and implement safe handling and disposal procedures for NP-containing materials.

The following precautionary measures must be practiced while handling NPs:

- Engineering controls: Make use of local exhaust ventilation systems and containment enclosures (e.g., glove bags, glove boxes, fume hoods, biological safety cabinets) fitted with high-efficiency particulate air filters during production, handling, and processing to minimize exposure risk [112]. This reduces the possibility of inhalation and skin contact.

- Personal Protective Equipment (PPE): Ensure that personnel working with NP-containing materials are equipped with protective gear such as respirators, lab coats, gloves, face shields, goggles, etc. The choice of PPE should be based on the hazards associated with NP in use [112].

- Training and Education: Comprehensive training programs need to be organised to educate personnel on the potential hazards of NPs, safe handling techniques, and emergency procedures [112]. Regular updates on the latest research findings are crucial to maintaining a safe working environment.

- Good Laboratory Practices (GLP): Adhere to GLP guidelines for NP-containing materials, including labelling, storage, and record-keeping.

Proper disposal of NP-containing materials is an important consideration for anyone working with these materials. To minimize the risk associated with NP exposure, the following guidelines need to be followed:

- Characterization of Waste: The NP-containing materials should be sufficiently characterized to devise acceptable disposal procedures and reduce environmental impact.

- Segregation: NP-containing waste must be isolated from general waste streams and disposed of in specialized containers labelled with the nature of NPs and associated dangers.

- Specialized disposal: The disposal method must meet the specifications of local environmental agencies. Specialized facilities may be required for the disposal of certain waste streams. Where appropriate, chemical treatments may be used to neutralize or stabilize the NP waste and mitigate the risk associated with disposal. Recycling is an excellent option to minimize waste. Incineration or high-temperature treatment can convert NP-containing waste into inert materials safe for disposal [113,114]. When other treatment options have failed, landfill disposal might be the only viable option, provided it complies with regulatory requirements [115]

2.6   Encouraging responsible innovation in nanotechnology

Nanotechnologists often face the enigma of striking a healthy balance between innovation and safety. Here are some guidelines for pursuing responsible innovation:

- Transparency and Open Communication: It is mandatory to maintain transparency and open communication at every stage of development. Scientific communities and regulatory bodies must be informed of the findings, methodologies, and risk assessments [116].

- Ethical Considerations: The creation of novel nanotechnological applications raises ethical questions. During the inception of newer products, researchers must think about the wider ethical implications of their work, such as concerns about safety and equitable access [117].

- Risk Assessment and Management: Identifying hazards and taking steps to mitigate them is of paramount importance. Researchers should continuously monitor the risk associated with products derived from NPs and develop strategies for containment and management [118].

- Regulatory Compliance: Responsible innovation requires abiding by current regulatory guidelines and creating new ones when required. Researchers should interact with regulatory bodies to ensure that emerging nanotechnologies are subject to appropriate oversight.

- Sustainability: Responsible innovation and the sustainability of nanotechnology are interlinked. To find environmentally sustainable solutions, researchers must take the environmental effects of NP manufacture, disposal, and recycling into account [119].

- Public Engagement: Building trust and promoting acceptance of nanotechnology require active public engagement. Discussions concerning the advantages, dangers, and ethical concerns of nanotechnology should actively involve the public.

Conclusion

Nanoparticles have transformed various sectors but also raise concerns about genotoxicity and carcinogenicity. The study of nanoparticle genotoxicity faces challenges like lack of standardized testing methods, diverse nanoparticle characteristics, and in vivo data scarcity. Existing tests were designed for larger substances, making them unsuitable for nanoparticles. To ensure reliable findings, established techniques for nanoparticles are needed. The size, shape, surface charge, and coatings of nanoparticles significantly influence their interaction with biological systems. More research is needed to determine the most important nanoparticle properties for genotoxicity. Understanding nanoparticle carcinogenicity requires more in vivo data and long-term investigations. To improve knowledge and create safety laws, interdisciplinary research, regulatory cooperation, and standardized testing protocols are needed.

References

[1] R.K. Shukla, A. Badiye, K. Vajpayee, N. Kapoor, Genotoxic Potential of Nanoparticles: Structural and Functional Modifications in DNA, Front. Genet. 12 (2021) 1–16. https://doi.org/10.3389/fgene.2021.728250.

[2] T. Adhikari, Nanotechnology in Environmental Soil Science, in: R.K. et al. Rattan (Ed.), Soil Sci. Fundam. to Recent Adv., 2021: pp. 297–310. https://doi.org/10.1007/978-981-16-0917-6_14.

[3] S.K. Kulkarni, Nanotechnology: Principles and Practices, Third Edition, 2014. https://doi.org/10.1007/978-3-319-09171-6.

[4] N. Zhang, G. Xiong, Z. Liu, Toxicity of metal-based nanoparticles: Challenges in the nano era, Front. Bioeng. Biotechnol. 10 (2022) 1–16. https://doi.org/10.3389/fbioe.2022.1001572.

[5]H. Barabadi, M. Najafi, H. Samadian, A. Azarnezhad, H. Vahidi, M.A. Mahjoub, M. Koohiyan, A. Ahmadi, A systematic review of the genotoxicity and antigenotoxicity of biologically synthesized metallic nanomaterials: Are green nanoparticles safe enough for clinical marketing?, Med. 55 (2019). https://doi.org/10.3390/medicina55080439.

[6]Y.W. Huang, M. Cambre, H.J. Lee, The Toxicity of Nanoparticles Depends on Multiple Molecular and Physicochemical Mechanisms, Int. J. Mol. Sci. 18 (2017). https://doi.org/10.3390/ijms18122702.

[7]C. Egbuna, V.K. Parmar, J. Jeevanandam, S.M. Ezzat, K.C. Patrick-Iwuanyanwu, C.O. Adetunji, J. Khan, E.N. Onyeike, C.Z. Uche, M. Akram, M.S. Ibrahim, N.M. El Mahdy, C.G. Awuchi, K. Saravanan, H. Tijjani, U.E. Odoh, M. Messaoudi, J.C. Ifemeje, M.C. Olisah, N.J. Ezeofor, C.J. Chikwendu, C.G. Ibeabuchi, Toxicity of Nanoparticles in Biomedical Application: Nanotoxicology, J. Toxicol. 2021 (2021). https://doi.org/10.1155/2021/9954443.

[8]S. Sharifi, S. Behzadi, S. Laurent, M.L. Forrest, P. Stroeve, M. Mahmoudi, Toxicity of nanomaterials, Chem. Soc. Rev. 41 (2012) 2323–2343. https://doi.org/10.1039/c1cs15188f.

[9]Z. Magdolenova, A. Collins, A. Kumar, A. Dhawan, V. Stone, M. Dusinska, Mechanisms of genotoxicity. A review of in vitro and in vivo studies with engineered nanoparticles, Nanotoxicology. 8 (2014) 233–278. https://doi.org/10.3109/17435390.2013.773464.

[10] M. Carriere, S. Sauvaigo, T. Douki, J.L. Ravanat, Impact of nanoparticles on DNA repair processes: Current knowledge and working hypotheses, Mutagenesis. 32 (2017) 203–213. https://doi.org/10.1093/mutage/gew052.

[11] R. Wan, Y. Mo, R. Tong, M. Gao, Q. Zhang, Determination of phosphorylated histone H2AX in nanoparticle-induced genotoxic studies, in: Q. Zhang (Ed.), Methods Mol. Biol., Springer New York, New York, NY, 2019: pp. 145–159. https://doi.org/10.1007/978-1-4939-8916-4_9.

[12] J. Bi, C. Mo, S. Li, M. Huang, Y. Lin, P. Yuan, Z. Liu, B. Jia, S. Xu, Immunotoxicity of metal and metal oxide nanoparticles: from toxic mechanisms to metabolism and outcomes, Biomater. Sci. 11 (2023) 4151–4183. https://doi.org/10.1039/d3bm00271c.

[13] M.R. Gedda, P.K. Babele, K. Zahra, P. Madhukar, Epigenetic aspects of engineered nanomaterials: Is the collateral damage inevitable?, Front. Bioeng. Biotechnol. 7 (2019). https://doi.org/10.3389/fbioe.2019.00228.

[14] K. Haliloğlu, A. Türkoğlu, Ö. Balpınar, H. Nadaroğlu, A. Alaylı, P. Poczai, Effects of Zinc, Copper and Iron Oxide Nanoparticles on Induced DNA Methylation, Genomic Instability and LTR Retrotransposon Polymorphism in Wheat (Triticum aestivum L.), Plants. 11 (2022). https://doi.org/10.3390/plants11172193.

[15] M. Pogribna, G. Hammons, Epigenetic Effects of Nanomaterials and Nanoparticles, J. Nanobiotechnology. 19 (2021) 2. https://doi.org/10.1186/s12951-020-00740-0.

[16] M. Hu, D. Palić, Role of MicroRNAs in regulation of DNA damage in monocytes exposed to polystyrene and TiO2 nanoparticles, Toxicol. Reports. 7 (2020) 743–751. https://doi.org/10.1016/j.toxrep.2020.05.007.

[17] S. Nallanthighal, C. Chan, T.M. Murray, A.P. Mosier, N.C. Cady, R. Reliene, Differential effects of silver nanoparticles on DNA damage and DNA repair gene expression in Ogg1-

deficient and wild type mice, Nanotoxicology. 11 (2017) 996–1011.
https://doi.org/10.1080/17435390.2017.1388863.

[18]   M. Mrakovcic, C. Meindl, G. Leitinger, E. Roblegg, E. Fröhlich, Carboxylated short
single-walled carbon nanotubes but not plain and multi-walled short carbon nanotubes show
in vitro genotoxicity, Toxicol. Sci. 144 (2015) 114–127.
https://doi.org/10.1093/toxsci/kfu260.

[19]   J. Catalán, K.M. Siivola, P. Nymark, H. Lindberg, S. Suhonen, H. Järventaus, A.J.
Koivisto, C. Moreno, E. Vanhala, H. Wolff, K.I. Kling, K.A. Jensen, K. Savolainen, H.
Norppa, In vitro and in vivo genotoxic effects of straight versus tangled multi-walled carbon
nanotubes, Nanotoxicology. 10 (2016) 794–806.
https://doi.org/10.3109/17435390.2015.1132345.

[20]   Y. Totsuka, T. Kato, S.I. Masuda, K. Ishino, Y. Matsumoto, S. Goto, M. Kawanishi, T.
Yagi, K. Wakabayashi, In vitro and in vivo genotoxicity induced by Fullerene (C 60) and
Kaolin, Genes Environ. 33 (2011) 14–20. https://doi.org/10.3123/jemsge.33.14.

[21]   P. V. AshaRani, G.L.K. Mun, M.P. Hande, S. Valiyaveettil, Cytotoxicity and
genotoxicity of silver nanoparticles in human cells, ACS Nano. 3 (2009) 279–290.
https://doi.org/10.1021/nn800596w.

[22]   K.K. Awasthi, R. Verma, A. Awasthi, K. Awasthi, I. Soni, P.J. John, In vivo genotoxic
assessment of silver nanoparticles in liver cells of Swiss albino mice using comet assay, Adv.
Mater. Lett. 6 (2015) 187–193. https://doi.org/10.5185/amlett.2015.5640.

[23]   S. Hackenberg, A. Scherzed, M. Kessler, S. Hummel, A. Technau, K. Froelich, C.
Ginzkey, C. Koehler, R. Hagen, N. Kleinsasser, Silver nanoparticles: Evaluation of DNA
damage, toxicity and functional impairment in human mesenchymal stem cells, Toxicol. Lett.
201 (2011) 27–33. https://doi.org/10.1016/j.toxlet.2010.12.001.

[24]   M.R. Wani, G.G.H.A. Shadab, Titanium dioxide nanoparticle genotoxicity: A review of
recent in vivo and in vitro studies, Toxicol. Ind. Health. 36 (2020) 514–530.
https://doi.org/10.1177/0748233720936835.

[25]   V. Valdiglesias, C. Costa, G. Kiliç, S. Costa, E. Pásaro, B. Laffon, J.P. Teixeira,
Neuronal cytotoxicity and genotoxicity induced by zinc oxide nanoparticles, Environ. Int. 55
(2013) 92–100. https://doi.org/10.1016/j.envint.2013.02.013.

[26]   M. Kumari, S.I. Kumari, P. Grover, Genotoxicity analysis of cerium oxide micro and
nanoparticles in Wistar rats after 28 days of repeated oral administration, Mutagenesis. 29
(2014) 467–479. https://doi.org/10.1093/mutage/geu038.

[27]   S. Nithya, R.R. Krishnan, N.R. Rao, K. Naik, N. Praveen, V.L. Vasantha, Microwave-
assisted extraction of phytochemicals, in: J.N. Cruz (Ed.), Drug Discov. Des. Using Nat.
Prod., Springer Nature Switzerland, Cham, 2023: pp. 209–238. https://doi.org/10.1007/978-3-
031-35205-8_8.

[28]   Ş.Y. Kaygisiz, I.H. Ciğerci, Genotoxic evaluation of different sizes of iron oxide
nanoparticles and ionic form by SMART, Allium and comet assay, Toxicol. Ind. Health. 33
(2017) 802–809. https://doi.org/10.1177/0748233717722907.

[29]   G. Qualhato, T.L. Rocha, E.C. de Oliveira Lima, D.M. e Silva, J.R. Cardoso, C. Koppe Grisolia, S.M.T. de Sabóia-Morais, Genotoxic and mutagenic assessment of iron oxide (maghemite-Γ-Fe2O3) nanoparticle in the guppy Poecilia reticulata, Chemosphere. 183 (2017) 305–314. https://doi.org/10.1016/j.chemosphere.2017.05.061.

[30]   J. Jiménez-Villarreal, D.I. Rivas-Armendáriz, R.D. Arellano Pérez-Vertti, E. Olivas Calderón, R. García-Garza, N.D. Betancourt-Martínez, L.B. Serrano-Gallardo, J. Morán-Martínez, Relationship between lymphocyte DNA fragmentation and dose of iron oxide (Fe2O3) and silicon oxide (SiO2) nanoparticles, Genet. Mol. Res. 16 (2017). https://doi.org/10.4238/gmr16019206.

[31]   S. Rajiv, J. Jerobin, V. Saranya, M. Nainawat, A. Sharma, P. Makwana, C. Gayathri, L. Bharath, M. Singh, M. Kumar, A. Mukherjee, N. Chandrasekaran, Comparative cytotoxicity and genotoxicity of cobalt (II, III) oxide, iron (III) oxide, silicon dioxide, and aluminum oxide nanoparticles on human lymphocytes in vitro, Hum. Exp. Toxicol. 35 (2016) 170–183. https://doi.org/10.1177/0960327115579208.

[32]   R. Wan, Y. Mo, Z. Zhang, M. Jiang, S. Tang, Q. Zhang, Cobalt nanoparticles induce lung injury, DNA damage and mutations in mice, Part. Fibre Toxicol. 14 (2017) 38. https://doi.org/10.1186/s12989-017-0219-z.

[33]   S. Alarifi, D. Ali, A. Verma, S. Alakhtani, B.A. Ali, Cytotoxicity and genotoxicity of copper oxide nanoparticles in human skin keratinocytes cells, Int. J. Toxicol. 32 (2013) 296–307. https://doi.org/10.1177/1091581813487563.

[34]   Y.H. Chung, M. Gulumian, R.C. Pleus, I.J. Yu, Animal Welfare Considerations When Conducting OECD Test Guideline Inhalation and Toxicokinetic Studies for Nanomaterials, Animals. 12 (2022). https://doi.org/10.3390/ani12233305.

[35]   X. Pan, Mutagenicity Evaluation of Nanoparticles by the Ames Assay, in: X. Pan, B. Zhang (Eds.), Methods Mol. Biol., Springer US, New York, NY, 2021: pp. 275–285. https://doi.org/10.1007/978-1-0716-1514-0_20.

[36]   A. Ávalos, A.I. Haza, D. Mateo, P. Morales, In vitro and in vivo genotoxicity assessment of gold nanoparticles of different sizes comet and SMART assays, Food Chem. Toxicol. 120 (2018) 81–88. https://doi.org/10.1016/j.fct.2018.06.061.

[37]   'N. El Yamani, E. Rundén-Pran, A.R. Collins, E.M. Longhin, E. Elje, P. Hoet, I. Vinković Vrček, S.H. Doak, V. Fessard, M. Dusinska, The miniaturized enzyme-modified comet assay for genotoxicity testing of nanomaterials, Front. Toxicol. 4 (2022). https://doi.org/10.3389/ftox.2022.986318.

[38]   S. Pfuhler, T.R. Downs, A.J. Allemang, Y. Shan, M.E. Crosby, Weak silica nanomaterial-induced genotoxicity can be explained by indirect DNA damage as shown by the OGG1-modified comet assay and genomic analysis, Mutagenesis. 32 (2017) 5–12. https://doi.org/10.1093/MUTAGE/GEW064.

[39]   N.V.S. Vallabani, H.L. Karlsson, Primary and Secondary Genotoxicity of Nanoparticles: Establishing a Co-Culture Protocol for Assessing Micronucleus Using Flow Cytometry, Front. Toxicol. 4 (2022) 845987. https://doi.org/10.3389/ftox.2022.845987.

Materials Research Forum LLC
https://doi.org/10.21741/9781644903339-9

[40]   Test No. 476: In Vitro Mammalian Cell Gene Mutation Tests using the Hprt and xprt genes, OECD, 2015. https://doi.org/10.1787/9789264243088-en.

[41]   Y. Yang, W. Li, E. Kroner, E. Arzt, B. Bhushan, L. Benameur, L. Wei, A. Botta, Y. Lu, J. Lou, D. Jena, M. Nosonovsky, B. Bhushan, T. Søndergaard, P.K. Sekhar, S. Bhansali, A.A. Trusov, Genotoxicity of Nanoparticles, in: B. Bhushan (Ed.), Encycl. Nanotechnol., Springer Netherlands, Dordrecht, 2012: pp. 952–962. https://doi.org/10.1007/978-90-481-9751-4_335.

[42]   Test No. 490: In Vitro Mammalian Cell Gene Mutation Tests Using the Thymidine Kinase Gene, 2015. https://doi.org/10.1787/9789264242241-en.

[43]   OECD, Test No. 489: In Vivo Mammalian Alkaline Comet Assay, 2014. https://doi.org/10.1787/9789264224179-en.

[44]   OCDE, Guideline 474: Mammalian Erythrocyte Micronucleus Test, 2016. https://doi.org/https://doi.org/https://doi.org/10.1787/9789264264762-en.

[45]   A. Ávalos, A.I. Haza, E. Drosopoulou, P. Mavragani-Tsipidou, P. Morales, In vivo genotoxicity assesment of silver nanoparticles of different sizes by the Somatic Mutation and Recombination Test (SMART) on Drosophila, Food Chem. Toxicol. 85 (2015) 114–119. https://doi.org/10.1016/j.fct.2015.06.024.

[46]   M.S. Attene-Ramos, C.P. Austin, M. Xia, High Throughput Screening, in: P.B.T.-E. of T. (Third E. Wexler (Ed.), Encycl. Toxicol. Third Ed., Academic Press, Oxford, 2014: pp. 916–917. https://doi.org/10.1016/B978-0-12-386454-3.00209-8.

[47]   J. Ge, D.K. Wood, D.M. Weingeist, S.N. Bhatia, B.P. Engelward, CometChip: Single-Cell Microarray for High-Throughput Detection of DNA Damage, Elsevier, 2012. https://doi.org/10.1016/B978-0-12-405914-6.00013-5.

[48]   C. Watson, J. Ge, J. Cohen, G. Pyrgiotakis, B.P. Engelward, P. Demokritou, High-throughput screening platform for engineered nanoparticle-mediated genotoxicity using cometchip technology, ACS Nano. 8 (2014) 2118–2133. https://doi.org/10.1021/nn404871p.

[49]   A.R. Collins, B. Annangi, L. Rubio, R. Marcos, M. Dorn, C. Merker, I. Estrela-Lopis, M.R. Cimpan, M. Ibrahim, E. Cimpan, M. Ostermann, A. Sauter, N. El Yamani, S. Shaposhnikov, S. Chevillard, V. Paget, R. Grall, J. Delic, F.G. de-Cerio, B. Suarez-Merino, V. Fessard, K.N. Hogeveen, L.M. Fjellsbø, E.R. Pran, T. Brzicova, J. Topinka, M.J. Silva, P.E. Leite, A.R. Ribeiro, J.M. Granjeiro, R. Grafström, A. Prina-Mello, M. Dusinska, High throughput toxicity screening and intracellular detection of nanomaterials, Wiley Interdiscip. Rev. Nanomedicine Nanobiotechnology. 9 (2017). https://doi.org/10.1002/wnan.1413.

[50]   N.A. Subramanian, A. Palaniappan, NanoTox: Development of a Parsimonious in Silico Model for Toxicity Assessment of Metal-Oxide Nanoparticles Using Physicochemical Features, ACS Omega. 6 (2021) 11729–11739. https://doi.org/10.1021/acsomega.1c01076.

[51]   T. Adhikary, P. Basak, Software for drug discovery and protein engineering: A comparison between the alternatives and recent advancements in computational biology, in: J.N. Cruz (Ed.), Drug Discov. Des. Using Nat. Prod., Springer Nature Switzerland, Cham, 2023: pp. 241–269. https://doi.org/10.1007/978-3-031-35205-8_9.

Advances in Healthcare and Nanoparticle Toxicology   Materials Research Forum LLC
Materials Research Foundations 171 (2024) 251-278   https://doi.org/10.21741/9781644903339-9

[52]   A. Gupta, S. Kumar, V. Kumar, Challenges for Assessing Toxicity of Nanomaterials, in: M. Ince, O.K. Ince, G. Ondrasek (Eds.), Biochem. Toxicol. - Heavy Met. Nanomater., IntechOpen, Rijeka, 2020: p. Ch. 4. https://doi.org/10.5772/intechopen.89601.

[53]   F.S. Alves, J.N. Cruz, I.N. de Farias Ramos, D.L. do Nascimento Brandão, R.N. Queiroz, G.V. da Silva, G.V. da Silva, M.F. Dolabela, M.L. da Costa, A.S. Khayat, J. de Arimatéia Rodrigues do Rego, D. do Socorro Barros Brasil, Evaluation of Antimicrobial Activity and Cytotoxicity Effects of Extracts of Piper nigrum L. and Piperine, Separations. 10 (2023). https://doi.org/10.3390/separations10010021.

[54]   V. Forest, Experimental and Computational Nanotoxicology— Complementary Approaches for Nanomaterial Hazard Assessment, Nanomaterials. 12 (2022). https://doi.org/10.3390/nano12081346.

[55]   S. H, Physiological Society Symposium : Impaired Endothelial and Smooth Muscle Cell Function in Oxidative Stress Oxidative Stress : Oxidants and Antioxidants, Exp. Physiol. 82 (1996) 291–295.

[56]   B. D'Autréaux, M.B. Toledano, ROS as signaling molecules: Mechanisms that generate specificity in ROS homeostasis, Nat. Rev. Mol. Cell Biol. 8 (2007) 813–824. https://doi.org/10.1038/nrm2256.

[57]   J.N. Cruz, S. Muzammil, A. Ashraf, M.U. Ijaz, M.H. Siddique, R. Abbas, M. Sadia, Saba, S. Hayat, R.R. Lima, A review on mycogenic metallic nanoparticles and their potential role as antioxidant, antibiofilm and quorum quenching agents, Heliyon. 10 (2024). https://doi.org/10.1016/j.heliyon.2024.e29500.

[58]   X. lihui, G. Jinming, G. Yalin, W. Hemeng, W. Hao, C. Ying, Albicanol inhibits the toxicity of profenofos to grass carp hepatocytes cells through the ROS/PTEN/PI3K/AKT axis, Fish Shellfish Immunol. 120 (2022) 325–336. https://doi.org/10.1016/j.fsi.2021.11.014.

[59]   J. Suski, M. Lebiedzinska, M. Bonora, P. Pinton, J. Duszynski, M.R. Wieckowski, Relation between mitochondrial membrane potential and ROS formation, Methods Mol. Biol. 1782 (2018) 357–381. https://doi.org/10.1007/978-1-4939-7831-1_22.

[60]   M. Horie, Y. Tabei, Role of oxidative stress in nanoparticles toxicity, Free Radic. Res. 55 (2021) 331–342. https://doi.org/10.1080/10715762.2020.1859108.

[61]   P.P. Fu, Q. Xia, H.M. Hwang, P.C. Ray, H. Yu, Mechanisms of nanotoxicity: Generation of reactive oxygen species, J. Food Drug Anal. 22 (2014) 64–75. https://doi.org/10.1016/j.jfda.2014.01.005.

[62]   D. Manzanares, V. Ceña, Endocytosis: The nanoparticle and submicron nanocompounds gateway into the cell, Pharmaceutics. 12 (2020) 139–148. https://doi.org/10.3390/pharmaceutics12040371.

[63]   K. Vidwathpriya, S. Sriranjani, P.K. Niharika, N.V.A. Kumar, Supercritical fluid for extraction and isolation of natural compounds, in: J.N. Cruz (Ed.), Drug Discov. Des. Using Nat. Prod., Springer Nature Switzerland, Cham, 2023: pp. 177–208. https://doi.org/10.1007/978-3-031-35205-8_7.

[64]   J. Zhu, L. Liao, L. Zhu, P. Zhang, K. Guo, J. Kong, C. Ji, B. Liu, Size-dependent cellular uptake efficiency, mechanism, and cytotoxicity of silica nanoparticles toward HeLa cells, Talanta. 107 (2013) 408–415. https://doi.org/10.1016/j.talanta.2013.01.037.

[65]   I.N. de F. Ramos, M.F. da Silva, J.M.S. Lopes, J.N. Cruz, F.S. Alves, J. de A.R. do Rego, M.L. da Costa, P.P. de Assumpção, D. do S. Barros Brasil, A.S. Khayat, Extraction, Characterization, and Evaluation of the Cytotoxic Activity of Piperine in Its Isolated form and in Combination with Chemotherapeutics against Gastric Cancer, Molecules. 28 (2023). https://doi.org/10.3390/molecules28145587.

[66]   A. Martin, A. Sarkar, Overview on biological implications of metal oxide nanoparticle exposure to human alveolar A549 cell line, Nanotoxicology. 11 (2017) 713–724. https://doi.org/10.1080/17435390.2017.1366574.

[67]   Y.W. Huang, C.H. Wu, R.S. Aronstam, Toxicity of transition metal oxide nanoparticles: Recent insights from in vitro studies, Materials (Basel). 3 (2010) 4842–4859. https://doi.org/10.3390/ma3104842.

[68]   M. Valko, C.J. Rhodes, J. Moncol, M. Izakovic, M. Mazur, Free radicals, metals and antioxidants in oxidative stress-induced cancer, Chem. Biol. Interact. 160 (2006) 1–40. https://doi.org/10.1016/j.cbi.2005.12.009.

[69]   M.H. Sarfraz, M. Zubair, B. Aslam, A. Ashraf, M.H. Siddique, S. Hayat, J.N. Cruz, S. Muzammil, M. Khurshid, M.F. Sarfraz, A. Hashem, T.M. Dawoud, G.D. Avila-Quezada, E.F. Abd_Allah, Comparative analysis of phyto-fabricated chitosan, copper oxide, and chitosan-based CuO nanoparticles: antibacterial potential against Acinetobacter baumannii isolates and anticancer activity against HepG2 cell lines, Front. Microbiol. 14 (2023). https://doi.org/10.3389/fmicb.2023.1188743.

[70]   A. Manke, L. Wang, Y. Rojanasakul, Mechanisms of nanoparticle-induced oxidative stress and toxicity, Biomed Res. Int. 2013 (2013). https://doi.org/10.1155/2013/942916.

[71]   M.J. Hosseini, F. Shaki, M. Ghazi-Khansari, J. Pourahmad, Toxicity of Copper on Isolated Liver Mitochondria: Impairment at Complexes I, II, and IV Leads to Increased ROS Production, Cell Biochem. Biophys. 70 (2014) 367–381. https://doi.org/10.1007/s12013-014-9922-7.

[72]   A. Ashrafi Hafez, P. Naserzadeh, A.M. Mortazavian, B. Mehravi, K. Ashtari, E. Seydi, A. Salimi, Comparison of the effects of MnO 2 -NPs and MnO 2 -MPs on mitochondrial complexes in different organs, Toxicol. Mech. Methods. 29 (2019) 86–94. https://doi.org/10.1080/15376516.2018.1512693.

[73]   M. Mishra, M. Panda, Reactive oxygen species: the root cause of nanoparticle-induced toxicity in Drosophila melanogaster, Free Radic. Res. 55 (2021) 671–687. https://doi.org/10.1080/10715762.2021.1914335.

[74]   D.B. Zorov, M. Juhaszova, S.J. Sollott, Mitochondrial reactive oxygen species (ROS) and ROS-induced ROS release, Physiol. Rev. 94 (2014) 909–950. https://doi.org/10.1152/physrev.00026.2013.

[75] Y. Shi, F. Wang, J. He, S. Yadav, H. Wang, Titanium dioxide nanoparticles cause apoptosis in BEAS-2B cells through the caspase 8/t-Bid-independent mitochondrial pathway, Toxicol. Lett. 196 (2010) 21–27. https://doi.org/10.1016/j.toxlet.2010.03.014.

[76] P. Manna, M. Ghosh, J. Ghosh, J. Das, P.C. Sil, Contribution of nano-copper particles to in vivo liver dysfunction and cellular damage: Role of IκBα/NF-κB, MAPKs and mitochondrial signal, Nanotoxicology. 6 (2012) 1–21. https://doi.org/10.3109/17435390.2011.552124.

[77] X.Q. Zhang, L.H. Yin, M. Tang, Y.P. Pu, ZnO, TiO 2, SiO 2, and Al 2O 3 nanoparticles-induced toxic effects on human fetal lung fibroblasts, Biomed. Environ. Sci. 24 (2011) 661–669. https://doi.org/10.3967/0895-3988.2011.06.011.

[78] P. Nymark, H.L. Karlsson, S. Halappanavar, U. Vogel, Adverse Outcome Pathway Development for Assessment of Lung Carcinogenicity by Nanoparticles, Front. Toxicol. 3 (2021) 1–40. https://doi.org/10.3389/ftox.2021.653386.

[79] B. Li, M. Tang, Research progress of nanoparticle toxicity signaling pathway, Life Sci. 263 (2020). https://doi.org/10.1016/j.lfs.2020.118542.

[80] J.W. Ko, J.W. Park, N.R. Shin, J.H. Kim, Y.K. Cho, D.H. Shin, J.C. Kim, I.C. Lee, S.R. Oh, K.S. Ahn, I.S. Shin, Copper oxide nanoparticle induces inflammatory response and mucus production via MAPK signaling in human bronchial epithelial cells, Environ. Toxicol. Pharmacol. 43 (2016) 21–26. https://doi.org/10.1016/j.etap.2016.02.008.

[81] S. Muzammil, J. Neves Cruz, R. Mumtaz, I. Rasul, S. Hayat, M.A. Khan, A.M. Khan, M.U. Ijaz, R.R. Lima, M. Zubair, Effects of Drying Temperature and Solvents on In Vitro Diabetic Wound Healing Potential of Moringa oleifera Leaf Extracts, Molecules. 28 (2023). https://doi.org/10.3390/molecules28020710.

[82] X. Chang, A. Zhu, F. Liu, L. Zou, L. Su, S. Li, Y. Sun, Role of NF-κB activation and Th1/Th2 imbalance in pulmonary toxicity induced by nano NiO, Environ. Toxicol. 32 (2017) 1354–1362. https://doi.org/10.1002/tox.22329.

[83] S. Sonwani, S. Madaan, J. Arora, S. Suryanarayan, D. Rangra, N. Mongia, T. Vats, P. Saxena, Inhalation Exposure to Atmospheric Nanoparticles and Its Associated Impacts on Human Health: A Review, Front. Sustain. Cities. 3 (2021) 1–20. https://doi.org/10.3389/frsc.2021.690444.

[84] K.Z. Guyton, Y. Liu, M. Gorospe, Q. Xu, N.J. Holbrook, Activation of mitogen-activated protein kinase by H2O2: Role in cell survival following oxidant injury, J. Biol. Chem. 271 (1996) 4138–4142. https://doi.org/10.1074/jbc.271.8.4138.

[85] C. Tournier, G. Thomas, J. Pierre, C. Jacquemin, M. Pierre, B. Saunier, Mediation by arachidonic acid metabolites of the H2O2-induced stimulation of mitogen-activated protein kinases (extracellular-signal-regulated kinase and c-Jun NH2-terminal kinase), Eur. J. Biochem. 244 (1997) 587–595. https://doi.org/10.1111/j.1432-1033.1997.00587.x.

[86] C. Guo, J. Wang, M. Yang, Y. Li, S. Cui, X. Zhou, Y. Li, Z. Sun, Amorphous silica nanoparticles induce malignant transformation and tumorigenesis of human lung epithelial cells via P53 signaling, Nanotoxicology. 11 (2017) 1176–1194. https://doi.org/10.1080/17435390.2017.1403658.

[87]   L.M. Falcone, A. Erdely, R. Salmen, M. Keane, L. Battelli, V. Kodali, L. Bowers, A.B. Stefaniak, M.L. Kashon, J.M. Antonini, P.C. Zeidler-Erdely, Pulmonary toxicity and lung tumorigenic potential of surrogate metal oxides in gas metal arc welding–stainless steel fume: Iron as a primary mediator versus chromium and nickel, PLoS One. 13 (2018). https://doi.org/10.1371/journal.pone.0209413.

[88]   T.A. Stueckle, D.C. Davidson, R. Derk, T.G. Kornberg, D. Schwegler-Berry, S. V. Pirela, G. Deloid, P. Demokritou, S. Luanpitpong, Y. Rojanasakul, L. Wang, Evaluation of tumorigenic potential of CeO_2 and Fe_2O_3 engineered nanoparticles by a human cell in vitro screening model, NanoImpact. 6 (2017) 39–54. https://doi.org/10.1016/j.impact.2016.11.001.

[89]   S. Luanpitpong, S.J. Talbott, Y. Rojanasakul, U. Nimmannit, V. Pongrakhananon, L. Wang, P. Chanvorachote, Regulation of lung cancer cell migration and invasion by reactive oxygen species and caveolin-1, J. Biol. Chem. 285 (2010) 38832–38840. https://doi.org/10.1074/jbc.M110.124958.

[90]   L. Xu, X. Li, T. Takemura, N. Hanagata, G. Wu, L.L. Chou, Genotoxicity and molecular response of silver nanoparticle (NP)-based hydrogel, J. Nanobiotechnology. 10 (2012). https://doi.org/10.1186/1477-3155-10-16.

[91]   G.H. Lee, Y.S. Kim, E. Kwon, J.W. Yun, B.C. Kang, Toxicologic evaluation for amorphous silica nanoparticles: Genotoxic and non-genotoxic tumor-promoting potential, Pharmaceutics. 12 (2020) 1–17. https://doi.org/10.3390/pharmaceutics12090826.

[92]   W. Dröge, Free radicals in the physiological control of cell function, Physiol. Rev. 82 (2002) 47–95. https://doi.org/10.1152/physrev.00018.2001.

[93]   K.M. Holmström, T. Finkel, Cellular mechanisms and physiological consequences of redox-dependent signaling, Nat. Rev. Mol. Cell Biol. 15 (2014) 411–421. https://doi.org/10.1038/nrm3801.

[94]   M.H. Park, J.T. Hong, Roles of NF-κB in cancer and inflammatory diseases and their therapeutic approaches, Cells. 5 (2016). https://doi.org/10.3390/cells5020015.

[95]   Chromium(VI)-induced nuclear factor-kappa B activation in intact cells via free radical reactions - PubMed, (n.d.).

[96]   J.D. Byrne, J.A. Baugh, The significance of nanoparticles in particle-induced pulmonary fibrosis, McGill J. Med. 11 (2008) 43–50. https://doi.org/10.26443/mjm.v11i1.455.

[97]   I. Pujalté, I. Passagne, B. Brouillaud, M. Tréguer, E. Durand, C. Ohayon-Courtès, B. L'Azou, Cytotoxicity and oxidative stress induced by different metallic nanoparticles on human kidney cells, Part. Fibre Toxicol. 8 (2011) 1–16. https://doi.org/10.1186/1743-8977-8-10.

[98]   Activation of NF-kappaB-dependent gene expression by silica in lungs of luciferase reporter mice - PubMed, (n.d.).

[99]   L. Capasso, M. Camatini, M. Gualtieri, Nickel oxide nanoparticles induce inflammation and genotoxic effect in lung epithelial cells, Toxicol. Lett. 226 (2014) 28–34. https://doi.org/10.1016/j.toxlet.2014.01.040.

[100] M. Ding, X. Shi, Y.J. Lu, C. Huang, S. Leonard, J. Roberts, J. Antonini, V. Castranova, V. Vallyathan, Induction of Activator Protein-1 through Reactive Oxygen Species by Crystalline Silica in JB6 Cells, J. Biol. Chem. 276 (2001) 9108–9114. https://doi.org/10.1074/jbc.M007666200.

[101] M. Ding, X. Shi, Z. Dong, F. Chen, Y. Lu, V. Castranova, V. Vallyathan, Freshly fractured crystalline silica induces activator protein-1 activation through ERKs and p38 MAPK, J. Biol. Chem. 274 (1999) 30611–30616. https://doi.org/10.1074/jbc.274.43.30611.

[102] X.H. Jiang, B.C.Y. Wong, M.C.M. Lin, G.H. Zhu, H.F. Kung, S.H. Jiang, D. Yang, S.K. Lam, Functional p53 is required for triptolide-induced apoptosis and AP-1 and nuclear factor-κB activation in gastric cancer cells, Oncogene. 20 (2001) 8009–8018. https://doi.org/10.1038/sj.onc.1204981.

[103] Y. Gu, Y. Wang, Q. Zhou, L. Bowman, G. Mao, B. Zou, J. Xu, Y. Liu, K. Liu, J. Zhao, M. Ding, Correction: Inhibition of nickel nanoparticles-induced toxicity by epigallocatechin-3-gallate in JB6 cells may be through down-regulation of the MAPK signaling pathways, PLoS One. 11 (2016). https://doi.org/10.1371/journal.pone.0154978.

[104] L. Kong, T. Barber, J. Aldinger, L. Bowman, S. Leonard, J. Zhao, M. Ding, ROS generation is involved in titanium dioxide nanoparticle-induced AP-1 activation through p38 MAPK and ERK pathways in JB6 cells, Environ. Toxicol. 37 (2022) 237–244. https://doi.org/10.1002/tox.23393.

[105] H. Shi, L.G. Hudson, K.J. Liu, Oxidative stress and apoptosis in metal ion-induced carcinogenesis, Free Radic. Biol. Med. 37 (2004) 582–593. https://doi.org/10.1016/j.freeradbiomed.2004.03.012.

[106] F. Zheng, H. Li, Evaluation of Nrf2 with exposure to nanoparticles, Methods Mol. Biol. 1894 (2019) 229–246. https://doi.org/10.1007/978-1-4939-8916-4_13.

[107] C. Guo, Y. Xia, P. Niu, L. Jiang, J. Duan, Y. Yu, X. Zhou, Y. Li, Z. Sun, Silica nanoparticles induce oxidative stress, inflammation, and endothelial dysfunction in vitro via activation of the MAPK/Nrf2 pathway and nuclear factor-κB signaling, Int. J. Nanomedicine. 10 (2015) 1463–1477. https://doi.org/10.2147/IJN.S76114.

[108] A.G. Oomen, K.G. Steinhäuser, E.A.J. Bleeker, F. van Broekhuizen, A. Sips, S. Dekkers, S.W.P. Wijnhoven, P.G. Sayre, Risk assessment frameworks for nanomaterials: Scope, link to regulations, applicability, and outline for future directions in view of needed increase in efficiency, NanoImpact. 9 (2018) 1–13. https://doi.org/10.1016/j.impact.2017.09.001.

[109] A. Hartwig, M. Arand, B. Epe, S. Guth, G. Jahnke, A. Lampen, H.J. Martus, B. Monien, I.M.C.M. Rietjens, S. Schmitz-Spanke, G. Schriever-Schwemmer, P. Steinberg, G. Eisenbrand, Mode of action-based risk assessment of genotoxic carcinogens, Arch. Toxicol. 94 (2020) 1787–1877. https://doi.org/10.1007/s00204-020-02733-2.

[110] V. Forest, Combined effects of nanoparticles and other environmental contaminants on human health - an issue often overlooked, NanoImpact. 23 (2021) 100344. https://doi.org/10.1016/j.impact.2021.100344.

[111] T.I. Ramos, C.A. Villacis-Aguirre, K. V. López-Aguilar, L.S. Padilla, C. Altamirano, J.R. Toledo, N.S. Vispo, The Hitchhiker's Guide to Human Therapeutic Nanoparticle Development, Pharmaceutics. 14 (2022). https://doi.org/10.3390/pharmaceutics14020247.

[112] C.H. Plan, S. Manual, O. Safety, Guidelines for Safety during Nanomaterials Research, (2018) 1–8.

[113] Z. Zahra, Z. Habib, S. Hyun, M. Sajid, Nanowaste: Another Future Waste, Its Sources, Release Mechanism, and Removal Strategies in the Environment, Sustain. 14 (2022). https://doi.org/10.3390/su14042041.

[114] G.B. Pinto, A. dos Reis Corrêa, G.N.C. da Silva, J.S. da Costa, P.L.B. Figueiredo, Drug development from essential oils: New discoveries and perspectives, in: J.N. Cruz (Ed.), Drug Discov. Des. Using Nat. Prod., Springer Nature Switzerland, Cham, 2023: pp. 79–101. https://doi.org/10.1007/978-3-031-35205-8_4.

[115] R. Gupta, H. Xie, Nanoparticles in daily life: Applications, toxicity and regulations, J. Environ. Pathol. Toxicol. Oncol. 37 (2018) 209–230. https://doi.org/10.1615/JEnvironPatholToxicolOncol.2018026009.

[116] A.E. Kokotovich, J. Kuzma, C.L. Cummings, K. Grieger, Responsible Innovation Definitions, Practices, and Motivations from Nanotechnology Researchers in Food and Agriculture, Nanoethics. 15 (2021) 229–243. https://doi.org/10.1007/s11569-021-00404-9.

[117] P.A. Schulte, F. Salamanca-Buentello, Ethical and scientific issues of nanotechnology in the workplace, Environ. Health Perspect. 115 (2007) 5–12. https://doi.org/10.1289/ehp.9456.

[118] S. Bakand, A. Hayes, Toxicological considerations, toxicity assessment, and risk management of inhaled nanoparticles, Int. J. Mol. Sci. 17 (2016). https://doi.org/10.3390/ijms17060929.

[119] A. Ramanathan, Toxicity of nanoparticles_ challenges and opportunities, Appl. Microsc. 49 (2019) 2. https://doi.org/10.1007/s42649-019-0004-6.

Keyword Index

About the Editor

Jorddy Neves Cruz is a professor and researcher in the multidisciplinary areas of chemistry and molecular modeling. In 2021, 2022, 2023 and 2024 he was included in the list of Eminent Researchers in the area of Pharmaceutical Sciences at Federal University of Pará (AD Scientific Index). In 2023 and 2024 he was included among the Top 10,000 Scientists in Latin America (AD Scientific Index). He works as an Editor for the journals Frontiers in Chemistry, PLOS One, PeerJ, Molecules, Discover Toxicology, Current Medicinal Chemistry, Frontiers in Oral Health, Evidence-Based Complementary and Alternative Medicine, Combinatorial Chemistry High Throughput Screening, Journal Computational Biophysics and Chemistry and Journal Medicine, in addition to being a Reviewer for 61 International Scientific Journals.